Dimensional Psychopathology

Massimo Biondi • Massimo Pasquini
Angelo Picardi
Editors

Dimensional Psychopathology

Springer

Editors
Massimo Biondi
Department of Human Neurosciences
Policlinico Umberto I Hospital
Sapienza University of Rome
Rome
Italy

Massimo Pasquini
Department of Human Neurosciences
Sapienza University of Rome
Rome
Italy

Angelo Picardi
Centre for Behavioural Sciences and Mental Health
Italian National Institute of Health
Rome
Italy

ISBN 978-3-030-08640-4 ISBN 978-3-319-78202-7 (eBook)
https://doi.org/10.1007/978-3-319-78202-7

Printed on acid-free paper

This Springer imprint is published by the registered company Springer International Publishing AG part of Springer Nature
The registered company address is: Gewerbestrasse 11, 6330 Cham, Switzerland

Foreword

In 2010–2011, the World Psychiatric Association (WPA) and the World Health Organization (WHO) conducted a survey on a very large ($N = 4887$) and highly representative (randomly selected members of national psychiatric associations) sample of practicing psychiatrists from 44 countries in all WHO regions. The aim was to collect psychiatrists' views about the most significant issues in psychiatric diagnosis and classification, in order to inform the development of the 11th edition of the International Classification of Diseases and Related Health Problems (ICD-11). The results of the survey were published in 2011 in *World Psychiatry* [1].

One of the questions of the survey was: "Should a diagnostic system incorporate a dimensional component?" The response to the question was very clear: over 70% of the clinicians responded positively (with the most frequently mentioned motivations being that a dimensional component would be "a more accurate reflection of psychopathology" and would allow "a more detailed and personalized diagnosis"). Among the less than 30% clinicians who responded negatively, the most frequently mentioned motivations were that such a dimensional component would be "too complicated in clinical settings" and that there is at the moment "insufficient research on reliability."

The fact is, however, that dimensional assessment is already part of the diagnostic practice of many psychiatrists worldwide, especially when they have to make their choices concerning pharmacological treatment. Indeed, although the requirements of regulatory agencies have produced some generations of clinical trials aiming to demonstrate the "equivalence" between a new antidepressant or antipsychotic and a consolidated reference medication rather than any "differences" among them, clinicians are well aware that antidepressants and antipsychotics are not all equal, and that differences in their mechanisms of action lead to differences in the profile of their therapeutic efficacy that can be caught only by a dimensional approach.

Nevertheless, the situation concerning the application of a dimensional approach in ordinary clinical practice is not different from that existing in psychiatry in the 1970s concerning categorical diagnosis. In the absence of clear guidance by current diagnostic systems, the use of psychopathological dimensions in ordinary practice is very heterogeneous, and the interrater reliability of the dimensional assessment is likely to be somewhat low.

In the intention of the developers of the DSM-5, the dimensional component was going to achieve a much more significant prominence in that system. The section on personality disorders, in particular, was intended to be completely dimensional, and the section on psychotic disorders was expected to include a very visible dimensional characterization of patients with a categorical diagnosis of psychosis. These expectations, however, have not been fulfilled. The dimensional classification of personality disorders produced by the relevant work group has not been approved by the DSM-5 Task Force, and that section of the system has finally remained identical to the DSM-IV, while the dimensional characterization of psychoses has been relegated to an appendix of the manual.

The arguments which have led to these final decisions are emblematic of the resistances which still exist—at a level which is more academic than clinical—against the introduction of a dimensional component (complementary to the categorical one) in diagnostic systems. These arguments include difficulties in the identification and definition of the dimensions to be considered in each section of the system; doubts about the feasibility of the use of dimensions in ordinary clinical practice and about the likelihood that clinicians would be appropriately trained in that use; concerns about the interrater reliability in the application of the dimensions; and the idea that the identification of a cut-off for caseness along each dimension would in fact reproduce a categorical approach.

There are, however, a variety of clinical settings—not to mention research areas—in which the use of psychopathological dimensions appears today not postponable. Examples are the area of consultation-liaison psychiatry (in which categorical diagnoses are in the vast majority of cases not relevant), that of early detection of mental disorders (the evidence is growing that the early phases of development of virtually all disorders are marked by an aspecific symptomatology that cannot be described in categorical terms), and that of behavioral emergencies (which psychiatrists are increasingly called to deal with in both hospital and community settings, and for which a categorical approach appears in several cases reductive and clinically ineffective).

Furthermore, as previously noticed, the choice of treatment in psychiatry today cannot be guided just by the categorical diagnosis, but requires a further characterization of the individual patient with respect to several antecedent and concomitant variables, part of which should always be the assessment of all the relevant psychopathological dimensions (for instance, the positive, negative, disorganized, cognitive, manic, and depressive ones in a patient with a diagnosis of schizophrenia).

It is fair to state, however, that the need for a widely accepted instrument for a thorough dimensional characterization of individuals coming to the attention of psychiatrists remains unmet. This is the reason why this book represents a particularly useful addition to the literature.

The scale presented in this volume, the SVARAD (acronym for the Italian name "Scala per la VAlutazione Rapida Dimensionale", i.e., rapid dimensional assessment scale), was developed in the late 1990s by a group led by the late Paolo Pancheri and by Massimo Biondi, both Professors of Psychiatry at the University of Rome "La Sapienza," and has been then validated through a series of studies carried out in various settings, and subsequently applied to address a variety of issues of high clinical relevance, including the dimensional characterization of patients with major depression presenting different levels of anger/aggressiveness and impulsivity, the identification of the psychopathological dimensions influencing the decision to hospitalize and in particular to compulsorily admit psychiatric patients, and the dimensional approach to psychopharmacological treatment in somatizing patients.

I believe this book can be of great utility not only to researchers and scholars but also to the many clinicians who use a dimensional approach in their ordinary practice and perceive today the need to be guided in this approach so that it is as systematic and evidence based as possible.

Reference

1. Reed GM, Mendonca Correia J, Esparza P, Saxena S, Maj M. The WPA-WHO global survey of psychiatrists' attitudes towards mental disorders classification. World Psychiatry 2011;10:118–131.

Mario Maj
Department of Psychiatry
Campania University L. Vanvitelli
Caserta, Italy

World Psychiatric Association
Geneva, Switzerland

Preface

The introduction in the 1980s of the DSM and ICD systems of explicit diagnostic criteria to define diagnostic categories of mental disorders was a major step for psychiatry. These systems provided, for the first time, a common language for mental disorders for use in both clinical and research work. The systems were extremely helpful in that they served as a means to organize the intrinsic complexity of the topic through the convention of a choice of descriptive criteria based on a minimum number of symptoms (interchangeable from a predetermined list), decision trees with mutual exclusion rules, and the creation of diagnostic categories conceptualized as discrete entities. On the one hand, this approach has provided a solution to the problem of diagnostic variability and chaos. On the other hand, limitations and critical aspects of the categorical classifications have also emerged, such as the scarce descriptive versatility in individual patients, the high variability within categories, and, above all, their unsuitability for the design of tailored treatment for each individual patient.

Already in the 1980s, several clinicians began to show interest in the perspective of a diagnosis based on psychopathological "dimensions," rather than categories, as a way to better capture the true complexity of individuals suffering from mental illness. However, there were, and still are, many obstacles on this path. In its introduction, the DSM-IV mentioned interest in the dimensional perspective, but concluded that the time was still not ripe to introduce it. The DSM-5, too, as well as the upcoming ICD-11, maintains a categorical approach, although the possibility of adopting a dimensional approach for personality disorders has been discussed.

The DSM-5 and ICD-11 are useful resources, as they are the only internationally shared classification systems that we have available. However, alongside this practical consideration, it is also possible to attempt to develop dimensional systems. In the early 1990s, at the Psychiatric Clinic of the Sapienza University of Rome, Italy, some early efforts were made to outline a dimensional system characterized by simplicity, practicality, and, above all, speed of usage. Paolo Pancheri and Massimo Biondi (a co-editor in this book) outlined a first, rudimentary system, including ten dimensions selected on the basis of their clinical relevance and consistent identification in factor analytic studies of psychiatric symptoms. While more dimensions might have been included, their number was kept low for the sake of simplicity and ease of use. This first system, which was scored on a visual analogue scale, was subsequently refined and developed with the help of Paola Gaetano and two other co-editors of this book, Angelo Picardi and Massimo Pasquini. The refined system took the form of a validated rating scale that was named *Scala di Valutazione Rapida Dimensionale* (SVARAD), i.e., rapid dimensional assessment scale. An English version of the scale is known by its acronym, RADAS, but we will be using the original designation, SVARAD, throughout this book. Following its development and validation, this instrument began to be used for clinical evaluation in outpatient clinics. During the first years of its use, it was administered to many hundreds of psychiatric outpatients. The results were satisfactory and, in part, surprising. The rating scale, though somewhat raw, was simple, flexible, and well accepted by clinicians due to its ease and speed of use. Some findings from the study of this early use are reported in Chap. 2 of this book. In subsequent years, the instrument has continued to be routinely used for clinical evaluation in the outpatient and inpatient clinics, and it was also employed in several studies; these are described in Chap. 1.

This book presents a synthesis of the dimensional approach to psychopathology and of the research that has resulted from the use of the SVARAD. Certainly, it is not the only possible approach, nor is it perhaps the best one for all purposes. However, when used alongside the DSM-5 and ICD-11, it can provide a further, interesting perspective on psychopathology that can be useful in both clinical and research settings. In particular, in busy outpatient and inpatient services, it can be helpful in individualizing pharmacological and psychotherapeutic treatment, and in research settings it can provide a reliable and comprehensive, yet rapid, assessment of psychopathology.

This book is dedicated to the memory of Paolo Pancheri, M.D. (1938–2007), who was instrumental in introducing the dimensional perspective in our clinic and who played a key role in the development of the SVARAD and the subsequent clinical and research work.

We are grateful to Marina Duro, Psy.D., for her technical support, and to Nicoletta Gentili, M.D, and Neil Owens, M.Sc., for their valuable help in the development of the English version of the SVARAD. We also deeply thank Claudia S. Copeland, Ph.D., and Ena Konjolka, M.A., for their valuable help in editing the chapters. Finally, we express our gratitude to Professor Mario Maj, M.D., Past President of the World Psychiatric Association and Editor of *World Psychiatry*, for kindly writing an inspiring foreword to this book.

Rome, Italy

Massimo Biondi
Massimo Pasquini
Angelo Picardi

Contents

1 The SVARAD Scale for Rapid Dimensional Assessment: Development and Applications in Research 1
Massimo Biondi, Paola Gaetano, Massimo Pasquini, and Angelo Picardi

2 Dimensional Assessment with SVARAD in Clinical Practice 29
Massimo Biondi, Martina Valentini, Corinna Pancheri, Daria Piacentino, Massimo Pasquini, and Angelo Picardi

3 The Apprehension/Fear and Somatic Preoccupation/Somatisation Dimensions 83
Roberto Delle Chiaie and Amedeo Minichino

4 The Reality Distortion and Thought Disorganisation Dimensions . . . 127
Lorenzo Tarsitani and Annalisa Maraone

5 Psychopathological Dimensions in Emergency Psychiatry: Determinants of Admission, Compulsory Treatment, and Therapeutic Intervention 159
Federico Dazzi, Luigi Orso, Angelo Picardi, and Massimo Biondi

6 Sadness/Demoralisation and Apathy 175
Angelo Picardi, Paola Gaetano, and Elisa Fabi

7 The "Outer Dimensions": Impulsivity, Anger/Aggressiveness, Activation 211
Francesco Saverio Bersani and Massimo Pasquini

8 The Obsessive-Compulsive Dimension 233
Ines Taddei, Martina Valentini, and Massimo Pasquini

9 Psychopathological Dimensions and the Clinician's Subjective Experience 267
Mauro Pallagrosi, Angelo Picardi, Laura Fonzi, and Massimo Biondi

10 Phenomenology of Temporality and Dimensional Psychopathology 287
Thomas Fuchs and Mauro Pallagrosi

The SVARAD Scale for Rapid Dimensional Assessment: Development and Applications in Research

1

Massimo Biondi, Paola Gaetano, Massimo Pasquini, and Angelo Picardi

The SVARAD (acronym for the Italian name "Scala per la VAlutazione RApida Dimensionale") is an instrument for rapid dimensional assessment that was developed in the 1990s, during a period of progressive recognition in the psychiatric field of the limitations inherent in the traditional classification systems for mental disorders and the categorical approach to diagnosis. Psychiatric diagnosis is a complex and difficult issue and has been the subject of considerable discussion and debate over the past several decades. While a comprehensive treatment of this topic is beyond the scope of this chapter, some introductory remarks are appropriate.

1.1 Ontological and Epistemological Issues in Psychiatric Diagnosis

Ontological and epistemological questions permeate the literature on psychiatric nosology [1–3]. Questions of ontology deal with whether mental disorders really exist as abstract entities. Indeed, as noted by Pouncey [4], mental disorders generate ontological scepticism on a number of levels. First, they are abstract entities that

M. Biondi (✉)
Department of Human Neurosciences, Policlinico Umberto I Hospital, Sapienza University of Rome, Rome, Italy
e-mail: massimo.biondi@uniroma1.it

P. Gaetano
Italian Society of Cognitive and Behavioural Therapy, Rome, Italy

M. Pasquini
Department of Human Neurosciences, Sapienza University of Rome, Rome, Italy
e-mail: massimo.pasquini@uniroma1.it

A. Picardi
Centre for Behavioural Sciences and Mental Health, Italian National Institute of Health, Rome, Italy
e-mail: angelo.picardi@iss.it

M. Biondi et al. (eds.), *Dimensional Psychopathology*,
https://doi.org/10.1007/978-3-319-78202-7_1

cannot be directly appreciated with the human senses, even indirectly, as with, for instance, a microscope. Also, it is unclear if they should be considered as abstractions that exist in the world aside from the individuals who experience them and thus instantiate them. Moreover, they are not clearly natural processes whose detection is unaffected by human interpretation or value judgments.

It should be recognised that psychiatry is not alone in dealing with these issues. While most contemporary working scientists and philosophers subscribe to the opinion that there is an objective external reality, it is commonly acknowledged that there is a limit to our access to absolute reality. Given that humans' epistemic access to reality is limited, Pouncey has observed that all scientific constructs (be they phyla, subatomic particles, or diseases) are abstract entities, which can nevertheless be legitimate objects of scientific investigation. In this perspective, mental disorders can be viewed in a similar way to other medical diseases, as "a heterogeneous class of abstract entities that have uncertain ontic status aside from the persons who instantiate them" [4].

On the one hand, mental disorders could be "natural kinds", such as chemical elements, which reflect a deep structure in the universe that exists independently of any human action or will. On the other hand, they could be "social constructs" that do not actually exist in nature but, rather, are concepts that are created by humans. There are no strong arguments to support the position that any disease, let alone any mental disorder, is a natural kind. For instance, as noted by Greenberg [4], there is no difference, from nature's point of view, between the breaking of a tree branch and the breaking of a femur, as nature is indifferent. As nature does not intend hips to break in certain ways, things such as intracapsular, trochanteric, or subtrochanteric fractures do not exist in nature, any more than nature gives a branch different ways to snap off a tree. As observed by Phillips, while a broken bone may be a natural kind, declaring it a disease involves a human value judgment that is not inherent in this altered state of the bone [5]. Outside of psychiatry, the difference between fracture as an artificial vs. a natural category is negligible, aside from a philosophical perspective. As noted by Greenberg, designating a broken femur as a disease requires only assuming that it is in our nature to walk and to be out of pain, which are very broad and relatively uncontroversial assumptions about human nature. However, in psychiatry the issue becomes thorny. Much narrower and more controversial assumptions are needed to designate a state of fear as "generalised anxiety disorder" or sadness accompanied by sleep difficulties, lack of appetite, indecisiveness, and fatigue as "depression" [4]. Indeed, it is not easy to differentiate between mental disorders and homeostatic reaction to negative life events [6]. How much anxiety are humans supposed to feel, aware as they are of their inevitable death? How sad should we be about the human condition? How can such questions be answered? [4] It is equally difficult to sustain the position that mental disorders are merely social constructs with no basis in reality, as this would question the assumption that suffering is a real experience worthy of mitigation, or the existence of a mind that gives us the experience of suffering, or the usefulness of classifying mental suffering into categories in order to work towards alleviating it [4].

While each of these extreme positions hardly seems tenable, a more realistic middle ground, suggested by Zachar, is to consider mental disorders as "practical kinds," and embrace a pragmatic approach to developing diagnoses that best achieves the things psychiatrists need, both as scientists and clinical practitioners [7]. Such an approach may benefit from the adoption of a coherence theory of truth, by which disorders become more accepted as "true" when they grow increasingly valid over time, explain things about the world in a helpful way, and increasingly fit into our general knowledge about the world [3]. In this perspective, what might be considered the best classification would depend on the particular validator (e.g. genetic, outcome, treatment, neurobiology) that is emphasised. However, classification is more than a matter of preference or ideology; classifications can be invalid, and all classifications should be tested empirically [8]. Nevertheless, there is not a single right or wrong way to address the formidable problem of psychiatric classification. Different approaches have strengths and limitations.

Epistemological questions deal with how we can know anything about mental disorders and are particularly relevant in the field of psychiatric taxonomy. On the one hand, there are purely naturalistic definitions of mental disorders, which are exclusively based on objective, biological criteria, and do not refer to social or normative values. On the other hand, the normativist perspective emphasises the subjective and culturally driven nature of any definition of mental disorders. Indeed, definitions of disease often require value judgments, and even when the value judgment does have a physical explanation in terms of neurobiology, nothing physical can be the basis for deciding which judgment is correct. As noted by Cerullo, a look at areas of medicine outside psychiatry shows there is often a strong normativist element in how diseases are defined [4]. For instance, many diseases such as hypertension or hypercholesterolemia require making arbitrary decisions about cut-off points in laboratory values, based upon public health considerations and the risk/benefit ratio of any decision. In psychiatry, some conditions, such as mood or anxiety disorders, more easily lend themselves to a normativist definition, whereas others, like schizophrenia, seem to be better defined from the naturalist perspective, together with conditions such as Parkinson's disease [4].

Given that all definitions of disease have normativist and naturalist elements, hybrid approaches incorporating both a naturalist and a normative component have been advocated. The best-known of these is probably the "harmful dysfunction" approach proposed by Wakefield, which emphasises the disturbance of a healthy or satisfactory state of being as the basis of a disorder. This approach posits that the nature of the disturbance is simultaneously biological and social, and it situates disorders on the boundary between the given natural world and the constructed social world. A disorder is posited to exist when the failure of a person's internal mechanisms to perform their functions as optimised by nature has a harmful impact on the person's well-being as defined by social values and meanings [9].

While such hybrid approaches to the definition of mental disorder seem to identify a reasonable middle ground, they have also attracted criticism [10]. Indeed, any approach has counterexamples and can be alleged to define mental disorders either too broadly or too narrowly. As noted by Pierre [4], it should be acknowledged that developing an ironclad definition of mental disorder is an intimidating task. Inevitably, one has to face the subjective and relativistic nature of concepts such as "distress" and "suffering" and the value-ladenness of concepts such as "dysfunction".

All these considerations about the uncertain ontological status of psychiatric disorders and the difficulties inherent in coming up with an irreproachable definition of them should not be taken as philosophical evidence that mental disorders do not really exist or that any attempt at classifying them is flawed and unjustified. In fact, as observed by Frances [4], psychiatry is not alone in being "definitionally challenged", as there is really no indisputable operational definition in medicine for the concepts of "disease" or "illness" [4]. Rather, these considerations are useful to put the issue of psychiatric nosology into proper context in order to appreciate its subtleties and difficulties, as well as the fact that a nosological classification is necessary and can be useful despite being, by its very nature, flawed and limited in some ways.

1.2 The Traditional Categorical Approach to Psychiatric Classification

As noted by Berrios, modern psychiatric classification has a long history, stemming from the intense classificatory drive that appeared in the West during the seventeenth and eighteenth centuries. In the nineteenth century, developing a personal classification was part of professional growth and success for an alienist [11], and in subsequent times a myriad of classifications of mental disorders have been proposed, with varying degrees of acceptance and success.

In the last three decades, psychiatric nosology has undergone important developments. As observed by Jablensky, the introduction in the DSM and ICD systems of an internationally shared framework of concepts, a rule-based classification, and explicit diagnostic criteria has dramatically increased reliability and has played an essential role in linking psychiatry to science, keeping psychiatric diagnosis relevant, and furthering research. However imperfect they may be, these classification systems have provided clinicians with a common language for mental disorders, researchers with rigorous diagnostic standards, public health services and insurance companies with diagnostic codes, and judges and attorneys with reliable diagnoses of mental illness [12]. In both the DSM and the ICD systems, the diagnostic categories are defined in terms of syndromes, i.e. symptoms that cluster together and covary over time. Essentially, these systems build on Kraepelin's method of diagnosis, based on the careful examination of longitudinal history and current symptoms, which in turn was built on Kahlbaum's principles of classification of psychiatric disorders on the basis of symptoms, course, and outcome.

Although the introduction of internationally accepted operational diagnostic criteria has had many benefits for psychiatric practice and research, the current classification systems are the subject of much criticism and debate. Kendler and Zachar have noted that the use of the criteria has grown to the extent that they often tend to be reified, as if they represented all anyone would want to know about a given disorder, whereas the current diagnostic classifications are actually remarkably thin, descriptively. They have emphasised that the diagnostic criteria selected to detect a disorder with good reliability, sensitivity, and specificity should not be confused with the disorder itself [13]. Focusing exclusively on the symptoms and signs listed in the classification systems reflects the conceptual error of mistaking an index of something for the thing itself and may stifle conceptual innovation and thereby lead to a general impoverishment of psychopathology and the psychiatric culture [12, 14].

Criticism of the categorical approach includes claiming that the diagnostic categories often do not adequately reflect the heterogeneity of presentation in patients grouped under a particular category, that they are relatively unhelpful in distinguishing severity, that they do not accommodate subclinical cases usefully, and that they include highly heterogeneous "not otherwise specified" categories. Also, most diagnoses do not meet the validity standards set by Robins and Guze, who expected that each diagnostic category would ultimately be validated by its separation from other disorders, common clinical course, genetic aggregation in families, and differentiation by laboratory tests [15]. To these influential criteria for validating psychiatric diagnostic constructs, Kendler added differential response to treatment [16], which is also an unmet criterion as most pharmacological agents have been found to be effective for a variety of disorders, rather than matching up with specific diagnoses. Moreover, the current work in neuroscience, structural and functional neuroimaging, and genetics has not led to clear patterns that match up with the diagnostic categories [5, 17, 18]. Thus, as noted by Waterman, the assumption that psychopathology can be divided into discrete entities as defined in the classification systems, which is the basic assumption of the categorical approach to diagnosis, "is turning out to be inconsistent with the way genes and environments act and interact to produce brain function and dysfunction" [4].

Despite persistent doubts about the scientific legitimacy of psychiatric nosology [13], it should be recognised that psychiatry is not the only discipline that has worries about how to classify. In all scientific fields that rely on a taxonomy, no classificatory effort ever seems to do a perfect job of "carving nature at its joints". For instance, astronomers held a vote in 2006 to decide whether Pluto is really a planet, and they rewrote the definition of a planet [19]. Biology itself has been struggling with this problem since long before psychiatry came to be defined as a medical specialty. As observed by Zachar, we should not expect more clarity in a psychiatric nosology than we can achieve in a biological taxonomy. Failure to appreciate the complexity of biological taxonomies may lead to unrealistic standards for what counts as an adequate psychiatric nosology [8]. Even if there are important conceptual reasons why psychiatric classifications are not working well, it should not be inferred from this fact that classifying in psychiatry is a useless exercise. Instead, as noted by Berrios, when psychiatric classifications are not working optimally, this indicates that much more conceptual work is necessary to identify stable elements that anchor classifications to "nature" in order to develop classifications which do not only behave as "actuarial devices" [11].

Also, although most mental disorders cannot yet be described as valid disease categories, this does not mean that they are not valuable concepts. Kendell and Jablensky have suggested that a diagnostic rubric may be said to possess utility if it provides non-negligible information about prognosis and likely treatment outcome or testable propositions about biological and social correlates [18]. Many of the diagnostic concepts represented by the categories of disorder listed in the DSM and ICD nomenclatures are extremely useful to practicing clinicians [18] and may be viewed as possessing predictive validity [20]. However, given that utility may vary with the context in which these concepts are used, statements about utility must always be related to context, including who is using the diagnosis, in what circumstances, and for what purposes.

1.3 The Prototype-Matching Approach

An alternative approach to psychiatric diagnosis that does not rely on strict operational criteria is the prototype-matching approach or "prototype diagnosis", which has attracted considerable interest in recent years. In this context, the term "prototype" refers to the use of idealised models or archetypes of disease, and placement into a diagnostic category is determined by how much a given patient resembles the typical exemplars of the category in question. From a phenomenological perspective, Schwartz and Wiggins suggested that the clinician's experience is pervaded by "typifications" which help to structure the clinician's diagnosis meaningfully [21]. Husserl himself had indicated that perceptual meaning is itself based on such a typification process, as humans never perceive individual things or persons in isolation but instead perceive them in terms of the type that epitomises that individual entity [22]. Also, Westen has argued that research in cognitive science suggests that the prototype-matching approach is more congruent with the ways humans think and classify in general [23]. Indeed, it has been reported that clinicians tend to diagnose in their daily practice by pattern matching, rather than counting criteria for categorical diagnosis and applying cut-offs [24]. Schaffner has also noted that an approach that identifies the most robust categories as prototypes, related to other prototypes by similarity, is supported by the deep structure of biology [25].

In its operationalised form, prototype diagnosis involves assessing the extent to which the patient's clinical presentation matches paragraph-length descriptions of disorders "that weave together diagnostic criteria into a memorable *gestalt* designed to facilitate pattern recognition" [23]. The resemblance to the prototype is rated on a numerical scale, where the lowest score indicates no resemblance and the highest score indicates a resemblance so high that the patient exemplifies the disorder. High ratings (e.g. 4 or 5 on a 5-point scale) imply that the patient resembles the diagnosis enough to be described as having the disorder; middle ratings (e.g. 2 or 3 on a 5-point scale) mean that the patient has some or subthreshold features of the disorder; and low ratings (e.g. 1 on a 5-point scale) indicate that there is little or no match between the patient's clinical presentation and the prototype.

This approach has been the object of intense study in the field of personality disorders, where it was found to outperform diagnosis based on operational criteria in inter-rater reliability, validity, and ratings of clinical utility [26]. Studies on other classes of mental disorders, such as eating disorders or mood and anxiety disorders [27], corroborated the view that a diagnostic system based on refined prototypes may be as reliable as one based on operational criteria while being more user-friendly and having greater clinical utility. It may also reduce the portion of comorbidity that is an artifact of current diagnostic methods, as clinicians are required to make configural judgments, rather than judgments about isolated symptoms. In a sense, this system incorporates the advantages of both categorical and dimensional diagnosis, as patients can be described as having a given disorder and can also be rated for the extent to which they have the disorder in question.

However, there are also some potential disadvantages in the prototype-matching approach. As noted by Maj, some clinicians may be reluctant to change the templates of mental disorders they have built up in their mind over years of practice, and it cannot be taken for granted that they will not have difficulties memorising, recalling, and correctly applying the standardised prototypes proposed by a diagnostic system [28]. Also, prototype diagnosis may promote confirmatory biases and other heuristics that can lead clinicians to see what they expect to see, or to cling to hypotheses about a patient, despite disconfirming information. For instance, the expectation that a given patient will present the various features of a prototype may lead the clinician to form the erroneous opinion that certain clinical aspects are present in this patient, when they are actually absent. Finally, different clinicians may disagree in their conclusions; while a clinician may reason that a patient matches a given prototype because a number of components are present, another clinician may conclude that the same patient does not match that prototype because some other aspects are absent [28].

Although prototype diagnosis includes a dimensional element, it should be recognised that it is mostly a categorical approach to diagnosis. In fact, both the polythetic diagnostic criteria built into the DSM and, to an even greater extent, the clinical manual of the ICD-10 can be viewed as efforts to operationalise prototype matching. Indeed, although it lacks a way of operationalising clinical judgment to maximise reliability, the clinician version of the ICD-10 is close to a prototype-matching procedure, as clinicians are presented with what are usually paragraph-length descriptions of a disorder, frequently with an additional set of considerations, and they are instructed to diagnose the patient with whatever degree of certainty they feel comfortable [23]. Therefore, on the one hand, prototype diagnosis holds the promise of being clinically helpful and reliable and of allowing for clinically rich, empirically derived, and culturally relevant psychiatric classification. On the other hand, it mainly resides within the realm of the categorical approach to psychiatric diagnosis, the validity of which is itself under debate.

1.4 The Dimensional Approach

Although many of the diagnostic categories of psychiatric classificatory systems are quite useful for clinicians, it is a matter of fact that no ideal way of classifying even the common disorders has emerged. Further, some of the limitations of psychiatric classificatory systems are inherent in any taxonomy. As observed by Jablensky, the problem of drawing boundaries between psychiatric diagnostic entities has so far defeated all attempts at finding an optimal solution by various rearrangements of symptoms and signs [12]. Goldberg has stated that we appear to be drawing lines in the fog, rather than "carving nature at its joints" [29].

However, it is unclear if there are real "joints" between mental disorders, and dimensional approaches have been proposed in opposition to the categorical approach. The first way of using the concept of "dimension" in the context of psychiatric taxonomy is to contrast dimensions vs. categories in terms of which is the best way to conceptualise mental disorders. Categorical diagnostic systems, indeed, draw a sharp line between individuals meeting criteria for a disorder and those not meeting criteria, who may nonetheless have a form of illness. The question here is not whether psychiatric disorders are categorical or dimensional in nature, because, as noted by Kraemer and colleagues, every disorder is both [30]; each disorder is either present or not (categorical), but when it is present, patients may vary with respect to a variety of features of the disorder (dimensional). Indeed, every dimensional diagnosis can be transformed into a corresponding categorical one by judiciously applying a dichotomisation rule, while every categorical diagnosis can be transformed into a corresponding dimensional one by, for instance, requiring multiple assessments and using the percentage positive [30]. As observed by Zachar and Kendler, the really relevant question from a clinical and research perspective is whether psychiatric disorders are best understood as diseases with discrete boundaries or as the pathological ends of functional dimensions [31]. In considering this issue, it should be recognised that discrete disease entities and dimensions of continuous variation are not mutually exclusive ways of conceptualising mental disorders; both ways are consistent with a threshold model of disease and may account for different or even overlapping portions of psychiatric morbidity [18].

Dimensions can be used one at a time, for example, when the diagnosis of major depressive disorder is based on exceeding the cut-off score on a numerical scale of depression severity, rather than using rule-based diagnostic criteria. However, another way of making use of dimensions is to use many of them in order to construct a diagnostic system based on a numerically derived phenotypic classification. In psychiatry, systems of this kind are based on factorially derived structural models for representing the phenotypic variation found in the domain of mental disorders. Such systems work best in describing phenomena that are distributed continuously and that do not have clear boundaries, as is often the case with mental disorders. In fact, from a categorical perspective, various classes of disorders show relations of continuity rather than discontinuity. Kendell and Jablensky have noted that several attempts have been made to demonstrate natural boundaries between related syndromes, or between a common syndrome such as major depressive disorder and normality, either by identifying a zone of rarity between them or by demonstrating a nonlinear relationship between the symptom profiles and a validating variable such as outcome or heritability. Most such attempts have been unsuccessful [18]. As observed by Zachar, compared with the common classification systems, dimensional models offer a better solution to the problem of understanding the overlap that occurs between different groups of cases (i.e. diagnostic categories), although they cannot account for all the patterns that exist in any domain, and they do not eliminate classificatory conundrums [8].

Also, keeping in mind that categorical and dimensional models are not incompatible but complementary, the dimensions can be used not to construct an alternative taxonomy but rather to supplement the traditional categorical taxonomy in order to provide an enhanced characterisation of patients based on their most prominent symptom clusters. This approach aims at optimising decisions about treatment and providing opportunities for research activities that are not constrained by exclusive reliance on categorical diagnosis and the ensuing obligation to work within criterial boundaries.

Papers suggesting the use of a dimensional approach to psychiatric diagnosis began to appear with some frequency in the literature during the last decades of the twentieth century, following early seminal work in this direction [32]. For instance, Mundt suggested a transnosological psychopathology implying both biological functional entities and trans-symptomatological functional psychological entities [32], while van Praag and his colleagues proposed a functional psychopathology based on biological mechanisms [33, 34]. In the latter approach, psychiatric symptoms are viewed as the behavioural expression of a psychological dysfunction, putatively correlated with alterations in specific functional systems in the brain. The basic units of classification are these psychological dysfunctions, rather than syndromes or diagnostic categories. This approach is clearly dimensional in orientation, as it views each psychiatric disorder as a conglomerate of psychological dysfunctions, most of them nosologically non-specific and occurring in different severities and in different combinations in the various psychiatric syndromes. Conceptualised as complementary, rather than as an alternative to the categorical approach, this approach would allow for more refined treatment, from both a pharmacological and a psychotherapeutic perspective [35].

In recent years, the concepts of psychopathological dimensions and dimensional diagnosis have gained further interest. They are based on the observation that psychiatric disorders appear to occur along a range of dimensions, which cut across diagnostic boundaries [29]. It is the diverse combination of a number of symptom clusters, called psychopathological dimensions, that gives rise to the wide variety of clinical pictures that can be observed in patients receiving the same categorical diagnosis. A fertile ground for dimensional conceptualisations has been the field of personality disorders, where proposals have been made to provide dimensional profiles of the existing diagnostic categories, or to reorganise the existing sets of diagnostic criteria into more clinically useful and empirically valid dimensions of maladaptive personality functioning, or to integrate the classification of personality disorders with dimensional models of general personality structure [36].

Focusing our attention back on Axis I, the dimensional approach to diagnosis has received empirical support, which further stimulated interest in this approach. For instance, a large number of studies have investigated the symptom structure of psychotic disorders. Already decades ago, studies began to suggest that dimensional representations of psychopathological features were more useful than categorical representations as predictors of illness course and treatment decisions [37]. More recent studies came to similar conclusions in showing that symptom dimensions are superior to diagnostic categories in explaining illness-related characteristics, including risk factors, premorbid, clinical, and outcome variables [38]. Most of this literature agrees that either four or five dimensions can adequately describe the psychosis construct, with positive, negative, disorganisation, and affective symptom dimensions most frequently reported. Studies comparing the fit of dimensional and categorical models within the same data set have supported the value of dimensions. Also, studies comparing the predictive ability of empirically derived dimensions with existing diagnostic categories of psychotic disorders, using clinical or outcome measures as external validators, agreed that a complementary approach incorporating both dimensions and categories may provide the best system of classification, thus providing strong support for the utility of dimensions [39].

Further support for the dimensional approach comes from a recent study of 239 patients with schizophrenia. The patients had been admitted to a random sample of all Italian public and private acute inpatient units during an index period. Factor mixture analysis (FMA) with heteroscedastic components was used to explore unobserved population heterogeneity in this group of patients. The analysis indicated the presence of three heterogeneous groups and yielded a five-factor solution with Depression, Positive Symptoms, Disorganisation, Negative Symptoms, and Activation identified as the factors. As compared with traditional clinical subtypes, psychopathological dimensions displayed much greater discriminatory power between groups identified by FMA [40]. These findings are consistent with those of other studies using cluster analytic approaches that failed to identify the DSM-IV schizophrenia subtypes [41, 42] and form one of the pieces of evidence that led to the elimination of the subtypes from the DSM and the recommendation to use psychopathological dimensions in order to describe the heterogeneity of schizophrenia in a manner that is more valid and clinically useful [43].

It should be clear from the discussion above that there are many ways of conceptualising dimensions and using them in the context of psychiatric diagnosis. Apart from psychopathological dimensions, the term "dimension" is also used in the psychiatric literature to refer to basic dimensions of psychological functioning that have been the focus of neuroscience research over the past several decades. In this regard, it is worth mentioning the recent NIMH-sponsored Research Domain Criteria (RDoC) project, which focuses its pathophysiologic spotlight not so much on categorically defined disorders, but on endophenotypes and dimensions of symptoms, both within and across disorders. This project aims at shifting researchers towards a focus on dysregulated neurobiological systems, rather than categorical diagnoses, as the organising principle for selecting study populations. Therefore, the RDoC project is not intended to function as a diagnostic classification system, but rather as a research framework to assist researchers in relating the fundamental domains of behavioural functioning to their underlying neurobiological components, with the ultimate aim of linking dysfunctions in neurocircuitry with clinically relevant psychiatric conditions [44]. While this project traces new directions in aetiological research and holds hope for important advances in psychiatric diagnosis and in the understanding of psychopathology, at its current stage, it is still a long way from becoming or generating an alternative diagnostic system that may inform treatment decisions. Indeed, its distance from several issues relevant to clinical practice [45] is at the heart of the criticisms levelled against the RDoC approach, for example, the absence of consideration of environmental influences [46], and the lack of appreciation of clinically important concepts such as the difference between well and sick, and the importance of time in defining course or prognosis [47]. Possibly, as suggested by Jablensky, rather than clinical neuroscience replacing psychopathology in the diagnosis of mental disorders, clinical psychiatry will retain psychopathology as its core, and classification will evolve towards a dual system with an aetiological axis, using neurobiological and genetic organising concepts, and a behavioural-dimensional or syndromal axis, which would be isomorphic to clinical reality [12].

1.5 Development, Validation, and Use of the SVARAD

When, more than 20 years ago, we started to conceive the idea of developing a dimensional assessment system, the literature on the dimensional approach to psychiatric diagnosis was relatively scarce. Proceeding from the common-knowledge notion that clever clinicians commonly use symptomatic and severity dimensions to personalise treatment independent of diagnosis, we selected a limited number of symptom clusters, or "psychopathological dimensions", based on their clinical relevance and consistent identification in factor analytic studies of psychiatric symptoms, with the aim of developing a standardised assessment system that would enable clinicians to accurately characterise each patient for treatment purposes by the relative prominence of one or more psychopathological dimensions. For many of these dimensions, a putative underlying biological dysfunction had been hypothesised. However, we reasoned that a standardised dimensional assessment may be useful for individualised planning, not only of pharmacological treatment but also of psychotherapeutic treatment. It should be emphasised that our intention was not, and never has been, to replace categorical diagnosis with the dimensional assessment. Rather, we always viewed dimensional and categorical diagnoses as complementary, not antagonistic, in the firm belief that an optimal diagnostic process should make use of all available resources, be it dimensional or categorical.

We felt encouraged to undertake this work by the consideration that, in principle, clinicians view the dimensional approach to diagnosis favourably. Indeed, in the recent WPA-WHO global survey of attitudes towards mental disorders classification, involving nearly 5000 psychiatrists from over 40 countries, the majority of participants were favourable to the inclusion of a dimensional component in a diagnostic system, either because it would make the system more detailed and personalised or because it would be a more accurate reflection of the underlying psychopathology [48]. However, we were aware that a crucial issue in every proposal to incorporate dimensional measurements into a diagnostic process performed by a clinician is practicality. As noted by Whooley, between researchers and clinicians, there is, in fact, an epistemological tension that reflects the classic Aristotelian distinction between *episteme* and *phronesis*. While researchers understand psychiatric knowledge as aimed towards illuminating universal and general rules, clinicians understand it differently, adopting a more practical posture that aims not towards identifying a universal truth, but instead towards a particular one, namely, what will be the most effective intervention for a specific patient [49]. Therefore, adding a complex dimensional evaluation based on multiple scales would have likely been seen merely as a bureaucratic burden by clinicians, and would only have served to widen the divide between the *episteme* of researchers and the *phronesis* of clinicians, without any benefit to the patients. For clinicians to be interested in dimensions, they need to be measured in a practical way, and this is a key principle that has guided our work in developing the SVARAD.

The SVARAD is an observer-rated scale that consists of ten items, each scored on a 5-point scale, ranging from 0 ("not present") to 4 ("extremely severe"). For each item, a detailed description of the dimension being rated is included, along with defined anchor points for severity. To facilitate its use in clinical practice, scoring instructions were included directly into the scale, rather than being provided separately in a scoring manual [50]. The SVARAD, the English version of which (known by the acronym RADAS) is illustrated in Fig. 1.1, comprises the following items:

1. Apprehension/Fear: state of anxiety and worry; sense of constriction; perception of imminent threat; feelings of worry, fear, and anguish.
2. Sadness/Demoralisation: distrust in oneself and one's own abilities; decreased creativity and energy; pessimism; decreased interests and pleasure.
3. Anger/Aggressiveness: feelings of irritation, resentment, and anger; display of irritability, litigiousness, and hostility; verbal or physical violence.
4. Obsessiveness: doubtfulness, rigidity, meticulousness, and perfectionism; repetitive behaviours aimed at preventing, checking, and controlling; presence of obsessions and/or compulsions.
5. Apathy: indifference, detachment, affective flattening and blunting; decreased planning and initiative.
6. Impulsivity: tendency to suddenly act in ways that are improper or potentially harmful to oneself or others, without adequate reflection on the causes or the consequences of one's own actions.
7. Reality Distortion: difficulty distinguishing between reality and fantasy; tendency to attribute unusual and unshared meanings to events or experiences; presence of delusions or hallucinations.
8. Thought Disorganisation: disruption of connection between ideas and of principles governing the organisation of thought, which thereby becomes altered in its logical organisation and impaired in its communicative functions.
9. Somatic Preoccupation/Somatisation: preoccupation with one's own body; physical symptoms with no organic basis; excessive concern about one's own health; exaggerated and unjustified fear of being ill.
10. Activation: increased motor activity; racing thoughts; disinhibition; feelings of excessive energy and self-confidence; euphoria or irritability.

The validation study provided evidence of inter-rater reliability, content validity, and criterion validity for the SVARAD [51]. Content validity was formally

RA.D.A.S.
RAPID DIMENSIONAL ASSESSMENT SCALE
by Paolo Pancheri, Massimo Biondi, Paola Gaetano, Angelo Picardi

The **AIM** of this instrument is to quickly assess the degree of impairment of some basic psychological and behavioural functions ranging seamlessly from normalcy to pathology. It measures traits, signs and symptoms that describe psychopathological "trans-nosographic" dimensions. Each of these can have a different "relative weight" in the individual clinical presentation.
INSTRUCTIONS on how to carryout the assessment:

- The assessment must be based on what is reported by the patient and on the clinician's observation of the patient's behaviour;
- Completion of the instrument must not be influenced by the categorical diagnosis, as it measures the impairment of functions that are present in a variety of disorders or within specific stages of the same disorder.

Name... Surname... Age........... Date of completion..........................

APPREHENSION/FEAR

State of anxiety and worry; sense of constriction; perception of imminent threat; feelings of worry, fear and anguish.

0	**Absent**
1	**Mild**: present only occasionally or in response to specific stimuli, non-pervasive, with no impairment of the patient's social or occupational functioning.
2	**Moderate**:frequent, non-pervasive, appearing spontaneously or in response to unimportant stimuli, with no impairment of the patient's social or occupational functioning.
3	**Severe**: sub-continuous, pervasive, with a mild reduction of the social or occupational functioning.
4	**Profound**: continuous, pervasive, with a severe reduction of the social or occupational functioning.

SADNESS/DEMORALIZATION

Distrust in oneself and one's own abilities; decreased creativity and energy; pessimism; decreased interests and pleasure.

0	**Absent**
1	**Mild**: modifiable following pleasant stimuli, limited to some areas of experience, with no impairment of the patient's social or occupational functioning.
2	**Moderate**: poorly modifiable, extended to almost all areas of experience, mild reduction of the patient's social or occupational functioning.
3	**Severe**: non-modifiable, pervasive, with moderate reduction of the patient's social or occupational functioning.
4	**Profound**: non-modifiable, pervasive, with severe reduction of the patient's social or occupational functioning.

ANGER/AGGRESSIVENESS

Feelings of irritation, resentment and anger; display of irritability, litigiousness, hostility; verbal or physical violence.

0	**Absent**
1	**Mild**: only occasionally present, the patient can control his/her impulses.
2	**Moderate**: frequent, generally controlled.
3	**Severe**: pervasive, very frequent, little controlled, with problems in social relationships.
4	**Profound**: pervasive, continuous, poorly controlled, severe social consequences.

Fig. 1.1 The English version of the SVARAD

OBSESSIVENESS

Doubtfulness, rigidity, meticulousness, perfectionism; repetitive behaviors aimed at preventing, checking, controlling; presence of obsessionsand/orcompulsions.

- 0 **Absent**
- 1 **Mild**: present, with no clear structured obsessions or compulsions.
- 2 **Moderate**: obsessions or compulsions only occasionally present, non-invasive, partially controllable, and non-interfering with everyday activities.
- 3 **Severe**: frequent obsessions or compulsions, invasive, poorly controllable, interfering with the patient's everyday social and occupational activities without, however, compromising them.
- 4 **Profound**: invasive obsessions and compulsions, present for the vast majority of the day, non-controllable, with impairment of the social and occupational activities.

APATHY

Indifference, detachment, affective flattening and blunting; decreased planning and initiative.

- 0 **Absent**
- 1 **Mild**: slightly present, variable or modifiable, with a fair level of planning; social functioning is mildly altered.
- 2 **Moderate**: obvious, modifiable by specific stimuli, with reduced planning; reasonable social functioning.
- 3 **Severe**: dominant, hardly modified by even intense stimuli, with highly reduced planning; poor social functioning.
- 4 **Profound**: constant, non-modifiable, planning almost absent; severely impaired social functioning.

IMPULSIVITY

Tendency to suddenly act in ways that are improper or potentially harmful to oneself or others, without adequate reflection on the causes or the consequences of one's own actions.

- 0 **Absent**
- 1 **Mild**: generally controllable or changeable impulsive acts, they are rare, in response to significant stimuli.
- 2 **Moderate**: partially controllable or changeable impulsive acts, infrequent, in response even to mild stimuli, with moderate social interference.
- 3 **Severe**: poorly controllable or changeable impulsive acts, frequent, with serious social consequences.
- 4 **Profound**: lack of any impulse control, highly frequent impulsive acts, with severe social and legal consequences.

REALITY DISTORTION

Difficulty distinguishing between reality and fantasy; tendency to attribute unusual and unshared meanings to events or experiences; presence of delusions or hallucinations.

- 0 **Absent**
- 1 **Mild**: tendency to attribute out of the ordinary or uncommonly shared meaning to events, unusual perceptive experiences.
- 2 **Moderate**: delusions with partial criticism, fluctuating or not very congruous; or hallucinations experienced occasionally or in special conditions, with partial or fluctuating criticism.
- 3 **Severe**: clear but not pervasive delusions, with poor or no criticism; or hallucinations, frequent but not continuous, with poor or no criticism.
- 4 **Profound**: clear, continuous, pervasive delusions with no hint of criticism; or continuous and nagging or pervasive hallucinations that are not criticised.

Fig. 1.1 (continued)

THOUGHT DISORGANIZATION

Disruption of connection between ideas and of principles governing the organization of thought, which there by becomes altered in its logical organization and impaired in its communicative functions.

0 **Absent**

1 **Mild**: only occasionally present in spontaneous speech, or in response to specific stimuli.

2 **Moderate:** frequent in spontaneous speech, tends to diminish in a led conversation, a fairly effective communication is however possible.

3 **Severe**: constant in spontaneous speech, clear in the led discourse, communication is difficult.

4 **Profound**: continuous, pervasive, communication is impossible.

SOMATIC PREOCCUPATION/SOMATIZATION

Preoccupation with one's own body; physical symptoms with no organic basis; excessive concern about one's own health; exaggerated and unjustified fear of being ill.

0 **Absent**

1 **Mild**: rare, of mild intensity, sensitive to reassurances.

2 **Moderate**: frequent, clear, hardly sensitive to reassurances; little interference with the patient's social and occupational functioning.

3 **Severe**: sub-continuous, dominant, only temporarily sensitive to reassurances, significant interference with the patient's social and occupational functioning.

4 **Profound**: constant, pervasive, non-sensitive to any reassurance, disabling.

ACTIVATION

Increased motor activity; racing thoughts; disinhibition; feelings of excessive energy and self-confidence; euphoria or irritability.

0 **Absent**

1 **Mild**: mildly elated mood, irritability, disinhibition; psychomotor restlessness; judgement and critical thinking abilities are preserved.

2 **Moderate**: elated and irritable mood, obvious disinhibition, tendency toward a potentially risky or damaging hyperactivity; judgement and critical thinking abilities are fluctuating.

3 **Severe**: euphoric or highly irritable mood, marked disinhibition, hyperactivity that is poorly directed to any specific goal, exaggerated and potentially harmful; judgement and critical thinking abilities are reduced.

4 **Profound**: overexcited or severely irascible mood; activities clearly exaggerated, not directed to any specific goal, and severely interfering with social activities; judgement and critical thinking abilities severely impaired.

Name of the assessor ....

Fig. 1.1 (continued)

measured by asking 12 psychiatrists who had not been involved in the construction of the instrument to rate on a 5-point scale the adequacy of each item to measure the related construct and then by computing Aiken's V index [52]. Aiken's V index was statistically significant for all items, which supports content validity.

Inter-rater reliability was assessed in 68 psychiatric outpatients who were each independently rated by two psychiatrists. Criterion validity against selected items of the Positive and Negative Syndrome Scale (PANSS), the 21-item version of the Hamilton Depression Rating Scale (HDRS), and the Hamilton Anxiety Rating Scale was assessed in 70 psychiatric outpatients. Inter-rater reliability was found to be satisfactory, with values of the Cohen's kappa coefficient (measuring agreement after adjusting for chance) ranging from 0.48 to 0.68 and values of Spearman's rho coefficient (measuring correlation between assessments) ranging from 0.66 to 0.82 for the various items. Recent data collected on 22 psychiatrists, senior psychiatry residents, and clinical psychologists who independently rated videotaped clinical interviews of five patients as part of an ongoing study provided further support for the reliability of the SVARAD. For all items, the kappa coefficient was above 0.50, with very high values for Sadness/Demoralisation (0.94), Obsessiveness (0.93), Apathy (0.84), Reality Distortion (0.94), Somatic Preoccupation/Somatisation (0.73), and Activation (0.92).

In the validation study, the pattern of correlations between each SVARAD dimension and the relevant items of the PANSS and the Hamilton's scales provided evidence of criterion validity for all SVARAD items [51]. The criterion validity of the scale has subsequently been corroborated by unpublished data from a study that generated several publications [53–55] and from an ongoing study that is currently being performed at the Department of Human Neurosciences of the Sapienza University of Rome. In the first of these studies, 151 psychiatric inpatients were administered the SVARAD together with several other assessment instruments, among which the 24-item Brief Psychiatric Rating Scale (BPRS), the Bech-Rafaelsen Mania Scale, and the 21-item HDRS. In the second study, 105 psychiatric inpatients and outpatients were administered the SVARAD and a number of other assessment instruments, including the 24-item BPRS. In both these data sets, the patterns of correlation between the SVARAD items and the relevant items of the other rating scales were consistent with expectations and supported the criterion validity of the SVARAD. Table 1.1 summarises in detail the correlations between the SVARAD items and the criterion items in these three data sets.

Following its development and validation, the SVARAD began to be routinely used for clinical evaluation in the outpatient and inpatient clinics of the Department of Human Neurosciences of the Sapienza University of Rome, and it was also employed in several studies. Practical and research experience has suggested that, thanks to its brevity and ease of administration and scoring, the SVARAD can be used even in busy clinical settings where there is only a very limited amount of time devoted to standardised assessment or research. Using the SVARAD allows clinicians and researchers to broaden the scope of the assessment to encompass areas of psychopathology that rating scales with a narrower focus would neglect. The next chapter discusses in detail how the SVARAD enabled our group to collect

Table 1.1 Correlations between the SVARAD items and relevant items of the Positive and Negative Syndrome Scale (PANSS), 24-item Brief Psychiatric Rating Scale (BPRS), Bech-Rafaelsen Mania Scale (BRMS), 21-item Hamilton Depression Rating Scale (HDRS), Hamilton Anxiety Rating Scale (HARS)

	Apprehension/ Fear			Sadness/ Demoralisation			Anger/ Aggressiveness			Obsessiveness			Apathy			Impulsivity			Reality Distortion			Thought Disorganisation			Somatic Preoccupation/ Somatisation			Activation		
Source of data[a]	A	B	C	A	B	C	A	B	C	A	B	C	A	B	C	A	B	C	A	B	C	A	B	C	A	B	C	A	B	C
BPRS/PANSS anxiety	0.56	0.74	0.47																											
HDRS psychic anxiety	0.49	0.46																												
HARS anxiety	0.64																													
HARS tension	0.41																													
HARS fears	0.62																													
HARS depressed mood				0.80																										
HDRS depressed mood				0.83	0.70																									
HDRS guilt				0.52	0.31																									
HDRS motor retardation				0.51	0.41																									
HDRS work and interests				0.42	0.50								0.49	0.38																
BPRS/PANSS depression				0.83	0.71	0.79																								
BPRS suicidality					0.42	0.48																								

(continued)

Table 1.1 (continued)

	Apprehension/ Fear			Sadness/ Demoralisation			Anger/ Aggressiveness			Obsessiveness			Apathy			Impulsivity			Reality Distortion			Thought Disorganisation			Somatic Preoccupation/ Somatisation			Activation		
BPRS/PANSS guilt				0.42	0.42	0.35																								
BPRS/PANSS motor retardation				0.51	0.64	0.26							0.45	0.66	0.43															
BRMS agitation								0.34									0.31													
BRMS hostility/ destructiveness								0.37																						
BPRS/PANSS hostility							0.36	0.48	0.72								0.33	0.49												
BPRS/PANSS motor tension	0.50	0.42					0.31	0.27	0.47								0.25	0.51												
BPRS uncooperativeness								0.18	0.46									0.47												
HDRS obsessive-compulsive symptoms										0.62	0.70																			
BPRS self-neglect														0.51	0.38															
BPRS/PANSS blunted affect													0.45	0.59	0.51															
BPRS/PANSS emotional withdrawal													0.58	0.50	0.49															
PANSS apathetic social withdrawal													0.51																	

PANSS disturbance of volition												0.52																		
PANSS poor impulse control						0.56									0.83															
HDRS paranoid symptoms																		0.66	0.61											
PANSS delusions																		0.89												
BPRS/PANSS suspiciousness																		0.62	0.48	0.74										
BPRS/PANSS hallucinations																		0.73	0.51	0.51										
BPRS unusual thought content																		0.68	0.72	0.92										
BRMS flight of thoughts																						0.42								
BPRS/PANSS conceptual disorganization																					0.67	0.66	0.94							
PANSS difficulty in abstract thinking																					0.77									
PANSS stereotyped thinking																					0.63									
HDRS somatic anxiety																								0.43	0.59					
HDRS hypochondriasis																								0.80	0.69					
BPRS/PANSS somatic concern																								0.79	0.80	0.65				

(continued)

Table 1.1 (continued)

	Apprehension/ Fear			Sadness/ Demoralisation			Anger/ Aggressiveness			Obsessiveness			Apathy			Impulsivity			Reality Distortion			Thought Disorganisation			Somatic Preoccupation/ Somatisation			Activation		
BRMS motor activity																	0.28												0.70	
BRMS verbal activity																													0.56	
BRMS voice level																													0.51	
BRMS elevated mood																													0.53	
BPRS elevated mood																													0.57	0.67
BPRS/PANSS grandiosity																												0.62	0.32	0.75
BPRS/PANSS excitement																	0.28	0.46										0.71	0.63	0.73
BPRS motor hyperactivity																	0.26	0.51											0.45	0.70

[a]Data indicated as A come from the first validation study [51]; data indicated as B are unpublished data collected on 151 psychiatric inpatients as part of a study that generated several publications [53–55]; data indicated as C come from an ongoing study performed at Sapienza University of Rome on psychiatric inpatients and outpatients

Only statistically significant correlations are reported in the table standardised, quantitative data about psychopathological dimensions in a large sample of 1124 psychiatric outpatients [56] and 846 psychiatric inpatients.

Subsequently, in a series of studies, the SVARAD was used to investigate the symptom structure of unipolar depression. A first study [57] was carried out on 380 first-contact adult outpatients who had received a diagnosis of a DSM-IV unipolar depressive condition (major depressive disorder, dysthymic disorder, depressive disorder not otherwise specified, adjustment disorder with depressed mood, and adjustment disorder with mixed anxiety and depressed mood). The patients had no comorbid psychiatric diagnosis on DSM-IV Axis I or II, had not received treatment with antidepressant drugs in the preceding 2 months, and were free from severe medical illness. Exploratory factor analysis suggested that three main symptom domains underlay depressive symptomatology, namely, core depression (Sadness/Demoralisation, Apathy), anxiety (Apprehension/Fear, Somatic Preoccupation/Somatisation), and anger/irritability (Anger/Aggressiveness, Impulsivity, Activation). From a clinical perspective, the Anger/Aggressiveness dimension was particularly relevant, as 98 (26%) patients received a rating of 2 or more on the Anger/Aggressiveness item, as compared with 36 (9%) and 3 (1%) patients who were rated 2 or more on Impulsivity and on Activation, respectively.

Similar results were obtained in a subsequent study [58], which focused on major depressive disorder and involved 222 first-contact outpatients who had no comorbid psychiatric diagnosis on DSM-IV Axis I or II, had not been treated with antidepressants in the preceding 2 months, and were free from severe medical illness. In these patients, too, the anger/irritability domain appeared to be clinically relevant in a substantial proportion of patients, as 48 (22%) patients received a rating of 2 or more on the Anger/Aggressiveness item, 16 (7%) on the Impulsivity item, and 2 (1%) on the Activation item.

Interestingly, a related study [59] showed that the mean scores on the Anger/Aggressiveness item were significantly higher ($p < 0.01$) in these patients with major depressive disorder as compared with 258 patients with anxiety disorders and 26 patients with somatoform disorders. The difference remained significant ($p < 0.01$) after adjustment for age and gender. Also, about twice as many patients with major depression (22%) had a rating of 2 or more on Anger/Aggressiveness, compared with patients with anxiety (12%) or somatoform disorders (11%). The difference was significant ($p < 0.01$) in a multiple logistic regression model including age and gender.

Overall, these studies supported the notion that in depressive disorders there are psychopathological dimensions other than depressed mood that deserve greater clinical recognition and research. One of these is anxiety, which despite not being part of the diagnostic criteria for the major depressive episode, is nevertheless covered by the rating scales that are commonly used to assess depressed patients and thus, when present, is usually recognised. The other dimension is operationalised in the SVARAD Anger/Aggressiveness item and includes clinical features such as anger, irritability, aggressiveness, and hostility.

Neither concurrent antidepressant treatment nor misdiagnosis of bipolar II disorder was likely to explain our finding that a substantial proportion of depressed patients presented with clinically significant levels of anger, irritability, aggressiveness, and hostility. A link between depression and anger is indeed not surprising, as it was suggested by sources as diverse as psychoanalysts [60, 61], cognitive psychotherapists [62], neurobiologists [63], and attachment theorists [64]. However, the SVARAD was instrumental in providing quantitative evidence of the relevance of anger and aggressiveness in patients with unipolar depression, as most instruments that were available at that time for the assessment of depression did not assess these clinical features. Clearly, the proper recognition of significant levels of anger and related clinical phenomena is important, as it has substantial implications for treatment.

The SVARAD also allowed detection of treatment-related changes in Anger/Aggressiveness in a subsequent study of cancer patients who had been identified through a multistage screening process as suffering from a mood or anxiety disorder. Together with common measures such as the Hamilton Anxiety Rating Scale and the Beck Depression Inventory, the SVARAD enabled the detection of highly significant ($p < 0.001$) differences from baseline in patients treated with psychotropic drugs, not only in depressive and anxiety symptoms but also in the Anger/Aggressiveness dimension [65]. Apart from suggesting the usefulness of broad dimensional assessment via the SVARAD in psycho-oncology, this study provided preliminary evidence that the instrument is sensitive to clinical change.

Further evidence of responsiveness of the SVARAD was provided by a subsequent study on depressed patients with dysphoric mood [66]. A single-group, open-trial design was used to examine the effectiveness of a combination of a selective serotonin reuptake inhibitor (SSRI) and an anticonvulsant, mostly valproate, in unipolar depressed patients presenting with prominent symptoms of anger, irritability, and hostility. The participant group consisted of 35 consecutive outpatients with a unipolar depressive disorder and notable anger, aggressiveness, or hostility as attested by the SVARAD. The participants had neither comorbid cluster A personality disorder nor borderline personality disorder and were free from severe physical illness. At the 12-week follow-up visit, most patients (82%) were rated as "improved" or "very much improved" on the Global Improvement item of the Clinical Global Improvement (CGI) scale. Similarly, 80% of patients experienced a reduction in HDRS total score of at least 35%. There was a highly significant ($p < 0.001$) decrease in HDRS total score, HDRS and SVARAD items covering anxiety symptoms and core depression symptoms, and SVARAD anger/irritability symptoms. The average percentage of improvement in anger/irritability was 69%, while the average percentage of improvement in the depressive and anxiety domains was 56% and 36% on the HDRS and 69% and 35% on the SVARAD, respectively. Although limitations in the study design suggest caution in drawing inferences about the effectiveness of this drug combination, this study suggested that adding valproate and possibly other anticonvulsants to SSRI medication might be a profitable strategy when dealing with unipolar depressed patients presenting with prominent symptoms of anger, irritability, and hostility. With regard to the SVARAD, these findings provided not only further evidence of sensitivity to clinical change but also evidence of criterion validity, as changes in the HDRS core depression and anxiety factors closely paralleled changes in the SVARAD items covering related constructs.

Another study showed that the SVARAD can be useful for investigating subtle psychopathological issues. This study examined the association between psychopathological dimensions and specific obsession subtypes, such as aggressive, contamination, sexual, hoarding/saving, symmetry/exactness, religious, and somatic subtypes [67]. The study was carried out on 57 first-contact outpatients with severe obsessive-compulsive disorder (OCD) with a duration of at least 1 year. The patients were administered several assessment instruments, among which were the Yale-Brown Obsessive-Compulsive Scale and the SVARAD. Significant correlations were found between the Sadness/Demoralisation item and contamination and somatic obsessions; between the Apprehension/Fear item and contamination, religious, and somatic obsessions; and between the Somatic Preoccupation/Somatisation item and contamination and somatic obsessions. The most interesting findings concerned the Anger/Aggressiveness and Impulsivity items, which were correlated with aggressive, sexual, and, to a lesser degree, contamination obsessions. These findings are consistent with cognitive accounts of OCD, which emphasise that obsessive-compulsive phenomena are related to difficulties in identifying, understanding, expressing, and regulating anger [62] and that disgust and anger are important components of moral judgment and moral violation [68]. Freud himself [69] suggested that persistent unwanted aggressive, horrific, or sexual thoughts accompanied by ritualistic behaviours are the result of unsuccessful defence mechanisms against potential violations of moral standards.

Concerning obsessive-compulsive disorder, it is worth mentioning that more than a decade before its separation from anxiety disorders in DSM-5, we performed a study aimed at comparing its dimensional profile with that of other anxiety disorders [70]. The participants were consecutive adult outpatients with a DSM-IV anxiety disorder, free from psychiatric or medical comorbidity, of whom 33 received a diagnosis of OCD, 104 of panic disorder (PD), 18 of generalised anxiety disorder (GAD), and 67 of anxiety disorder not otherwise specified (ADNOS). All participants were rated on the SVARAD by a psychiatrist. On the one hand, the patients with OCD displayed higher scores on Sadness/Demoralisation and Apathy than those with PD and ADNOS. Also, they showed higher scores on Reality Distortion and Thought Disorganisation than patients with PD, GAD, and ADNOS. On the other hand, they displayed lower scores on Somatic Preoccupation/Somatisation than patients with other anxiety disorders, particularly PD. This study showed that there are several differences in psychopathology between OCD and the other anxiety disorders, thus questioning the appropriateness of the classification of OCD among anxiety disorders.

Finally, a recent study showed that the SVARAD can also be used in critical settings with limited time, information, and resources, such as emergency settings [71]. Indeed, a dimensional approach to acute psychopathology is particularly suitable to emergency settings, where clinicians are required to rapidly identify the psychopathological domains to be treated, independent of categorical diagnosis. The majority of the instruments allowing a comprehensive assessment of psychopathology require too much time to be routinely used in emergency settings, whereas the SVARAD can be completed quickly and covers more dimensions than disorder-specific rating scales. This study involved 312 consecutive patients undergoing psychiatric evaluation in the emergency room of the Policlinico Umberto I hospital in Rome over a 6-month period. A replication study was performed in another Rome hospital on a random sample of 118 patients. In both samples, the patients who were recommended for psychiatric hospitalisation displayed significantly higher levels of Anger/Aggressiveness, Apathy, Impulsivity, Reality Distortion, Thought Disorganisation, and Activation. Multivariate analysis pointed to Reality Distortion, Impulsivity, and Apathy as the most important psychopathological predictors. Also, other variables such as the almost self-fulfilling proposal for compulsory admission and, more importantly, the categorical diagnosis of psychotic or mood disorder were identified as independent predictors of hospitalisation. Hierarchical regression analysis revealed that the dimensional assessment was the strongest predictor of hospitalisation. This study suggests that, in emergency settings, a standardised dimensional assessment may usefully complement the categorical approach to psychopathology in the identification of the patients who need psychiatric hospitalisation and may also help select appropriate treatment more quickly and efficiently.

In addition to research on the validity and the clinical and research usefulness of the SVARAD, recent activities have included the development of foreign language versions of the instrument. Steps for the construction and validation of a Brazilian version of the SVARAD have recently been undertaken [72]. Also, an English version, named with the acronym RADAS (Rapid Dimensional Assessment Scale), has recently been developed according to established procedures for the cross-cultural adaptation of psychosocial measures [73], involving three independent translators fluent in both Italian and English, who followed an iterative process of reviewing and commenting aimed at converging on an optimal translation. We concentrated our efforts on producing a good translation while refraining from performing iterative back-translation. Iterative back-translation, which merely seeks to achieve linguistic and conceptual equivalence, has been criticised as a quality assurance measure by several authors for both theoretical and practical reasons [74], as it has been described as a suboptimal procedure with limited effectiveness in determining the accuracy of the target text in relation to the original source text [75]. It has also been accused of overlooking clarity and understandability and not taking into account context and milieu [76]. The previously presented description of the SVARAD items is based on this carefully developed English version, which is illustrated in Fig. 1.1.

1.6 Final Comments

In conclusion, more than 20 years of clinical and research experience with the SVARAD have corroborated its reliability, validity, and ease of use. Its dimensional nature may help in individualising treatment for a wide variety of clinical presentations, even for patients with clinical pictures that have fuzzy boundaries and are not well characterised in categorical terms, such as patients with somatic symptom disorders [77]. Its main limitation, which is inherent in all dimensional approaches to psychiatric diagnosis, lies in the cross-sectional nature of the assessment, which needs to be supplemented with longitudinal information in order to optimise evaluation and treatment. Also, the choice of using a single item to evaluate each dimension, while maximising ease and rapidity of use, involves some reduction in reliability and a restricted range of scores. Moreover, some areas of psychopathology, such as dissociative experiences, are not covered. With these limitations in mind, the instrument has proved to be suitable even for busy clinical practices where professionals have little time to devote to standardised assessment. While longer and more sophisticated rating scales might be preferable in specific settings and for other purposes, such as detailed evaluation of symptoms or outcome assessment in clinical trials, the SVARAD finds its sweet spot in clinical settings where a reliable, comprehensive, yet rapid assessment of psychopathology is needed. It is also a valuable resource in the training of residents in psychiatry and clinical psychologists, as it forces the rater to pay attention to all clinical aspects, rather than only to the diagnostic criteria relevant to each patient's specific disorder. It is our hope that the readers of this book will find something of interest in the following chapters, which provide a detailed presentation of the clinical, biological, and treatment aspects of all SVARAD dimensions.

References

1. Marková IS, Berrios GE. Epistemology of psychiatry. Psychopathology. 2012;45:220–7.
2. Parnas J, Sass LA, Zahavi D. Rediscovering psychopathology: the epistemology and phenomenology of the psychiatric object. Schizophr Bull. 2013;39:270–7.
3. Kendler KS. The nature of psychiatric disorders. World Psychiatry. 2016;15:5–12.
4. Phillips J, Frances A, Cerullo MA, Chardavoyne J, Decker HS, First MB, Ghaemi N, Greenberg G, Hinderliter AC, Kinghorn WA, LoBello SG, Martin EB, Mishara AL, Paris J, Pierre JM, Pies RW, Pincus HA, Porter D, Pouncey C, Schwartz MA, Szasz T, Wakefield JC, Waterman GS, Whooley O, Zachar P. The six most essential questions in psychiatric diagnosis: a pluralogue part 1: conceptual and definitional issues in psychiatric diagnosis. Philos Ethics Humanit Med. 2012;7:3.
5. Phillips J, Frances A, Cerullo MA, Chardavoyne J, Decker HS, First MB, Ghaemi N, Greenberg G, Hinderliter AC, Kinghorn WA, LoBello SG, Martin EB, Mishara AL, Paris J, Pierre JM, Pies RW, Pincus HA, Porter D, Pouncey C, Schwartz MA, Szasz T, Wakefield JC, Waterman GS, Whooley O, Zachar P. The six most essential questions in psychiatric diagnosis: a pluralogue. Part 4: general conclusion. Philos Ethics Humanit Med. 2012;7:14.
6. Maj M. Are we able to differentiate between true mental disorders and homeostatic reactions to adverse life events? Psychother Psychosom. 2007;76:257–9.
7. Zachar P. The practical kinds model as a pragmatist theory of classification. Phil Psychiatry Psychol. 2003;9:219–27.
8. Zachar P. Real kinds but not true taxonomy. In: Kendler KS, Parnas J, editors. Philosophical issues in psychiatry. Baltimore, MD: Johns Hopkins University Press; 2008.
9. Wakefield JC. The concept of mental disorder. Am Psychol. 1992;47:373–88.
10. Jablensky A. Does psychiatry need an overarching concept of "mental disorder"? World Psychiatry. 2007;6:157–8.
11. Berrios GE. Classifications in psychiatry: a conceptual history. Aust N Z J Psychiatry. 1999;33:145–60.

12. Jablensky A. The nature of psychiatric classification: issues beyond ICD-10 and DSM-IV. Aust N Z J Psychiatry. 1999;33:137–44.
13. Kendler KS, Zachar P. The incredible insecurity of psychiatric nosology. In: Kendler KS, Parnas J, editors. Philosophical issues in psychiatry. Baltimore, MD: Johns Hopkins University Press; 2008.
14. Kendler KS. The phenomenology of major depression and the representativeness and nature of DSM criteria. Am J Psychiatry. 2016;173:771–80.
15. Robins E, Guze SB. Establishment of diagnostic validity in psychiatric illness: its application to schizophrenia. Am J Psychiatry. 1970;126:983–7.
16. Kendler K. Toward a scientific psychiatric nosology: strengths and limitations. Arch Gen Psychiatry. 1990;47:969–73.
17. Cohen BM. Embracing complexity in psychiatric diagnosis, treatment, and research. JAMA Psychiat. 2016;73:1211–2.
18. Kendell R, Jablensky A. Distinguishing between the validity and utility of psychiatric diagnoses. Am J Psychiatry. 2003;160:4–12.
19. Zachar P, Kendler KS. The removal of Pluto from the class of planets and homosexuality from the class of psychiatric disorders: a comparison. Philos Ethics Humanit Med. 2012;7:4.
20. Schaffner KFA. philosophical overview of the problems of validity for psychiatric disorders. In: Kendler KS, Parnas J, editors. Philosophical issues in psychiatry. Baltimore, MD: Johns Hopkins University Press; 2008.
21. Schwartz MA, Wiggins OP. Typifications. The first step for clinical diagnosis in psychiatry. J Nerv Ment Dis. 1987;175:65–77.
22. Husserl E. Erfahrung und Urteil: Untersuchungen zur Genealogie der Logik. Prague: Academia Verlagsbuchhandlung, 1939. English translation: experience and judgment. Evanston, IL: Northwestern University Press; 1973.
23. Westen D. Prototype diagnosis of psychiatric syndromes. World Psychiatry. 2012;11:16–21.
24. Morey LC, Ochoa ES. An investigation of adherence to diagnostic criteria: clinical diagnosis of the DSM-III personality disorders. J Pers Disord. 1989;3:180–92.
25. Schaffner KS. A tail of a tiger. In: Pendler KS, Parnas J, editors. Philosophical issues in psychiatry. Baltimore, MD: Johns Hopkins University Press; 2008.
26. Westen D, Shedler J, Bradley R. A prototype approach to personality disorder diagnosis. Am J Psychiatry. 2006;163:846–56.
27. DeFife JA, Peart J, Bradley B, Ressler K, Drill R, Westen D. Validity of prototype diagnosis for mood and anxiety disorders. JAMA Psychiat. 2013;70:140–8.
28. Maj M. Psychiatric diagnosis: pros and cons of prototypes vs. operational criteria. World Psychiatry. 2011;10:81–2.
29. Goldberg D. Plato versus Aristotle: categorical and dimensional models for common mental disorders. Compr Psychiatry. 2000;41:8–13.
30. Kraemer HC, Noda A, O'Hara R. Categorical versus dimensional approaches to diagnosis: methodological challenges. J Psychiatr Res. 2004;38:17–25.
31. Zachar P, Kendler KS. Psychiatric disorders: a conceptual taxonomy. Am J Psychiatry. 2007;164:557–65.
32. Van Praag HM, Leijnse B. Neubewertung des syndroms. Skizze einerfunktionellen Pathologie. Psychiatr Neurol Neurochir. 1965;68:50–66.
33. Mundt C. Psychotic continuum or distinct entities: perspectives from psychopathology. In: Marneros A, Andreasen NC, Tsuang MT, editors. Psychotic continuum. Berlin, Heidelberg: Springer; 1995. p. 7–15.
34. van Praag HM, Korf J, Lakke JPWF, Schut T. Dopamine metabolism in depressions, psychoses, and Parkinson's disease: the problem of the specificity of biological variables in behaviour disorders. Psychol Med. 1975;5:138–46.
35. Van Praag HM, Asnis GM, Kahn RS, Brown SL, Korn M, Friedman JM, Wetzler S. Nosological tunnel vision in biological psychiatry. A plea for a functional psychopathology. Ann N Y Acad Sci. 1990;600:501–10.

36. Widiger TA, Simonsen E. Alternative dimensional models of personality disorder: finding a common ground. J Pers Disord. 2005;19:110–30.
37. van Os J, Fahy TA, Jones P, Harvey I, Sham P, Lewis S, Bebbington P, Toone B, Williams M, Murray R. Psychopathological syndromes in the functional psychoses: associations with course and outcome. Psychol Med. 1996;26:161–76.
38. Peralta V, Cuesta MJ. Exploring the borders of the schizoaffective spectrum: a categorical and dimensional approach. J Affect Disord. 2008;108:71–86.
39. Potuzak M, Ravichandran C, Lewandowski KE, Ongür D, Cohen BM. Categorical vs dimensional classifications of psychotic disorders. Compr Psychiatry. 2012;53:1118–29.
40. Picardi A, Viroli C, Tarsitani L, Miglio R, de Girolamo G, Dell'Acqua G, Biondi M. Heterogeneity and symptom structure of schizophrenia. Psychiatry Res. 2012;198:386–94.
41. Helmes E, Landmark J. Subtypes of schizophrenia: a cluster analytic approach. Can J Psychiatry. 2003;48:702–8.
42. Lykouras L, Oulis P, Daskalopoulou E, Psarros K, Christodoulou GN. Clinical subtypes of schizophrenic disorders: a cluster analytic study. Psychopathology. 2001;34:23–8.
43. Tandon R, Gaebel W, Barch DM, Bustillo J, Gur RE, Heckers S, Malaspina D, Owen MJ, Schultz S, Tsuang M, Van Os J, Carpenter W. Definition and description of schizophrenia in the DSM-5. Schizophr Res. 2013;150:3–10.
44. Insel T, Cuthbert B, Garvey M, Heinssen R, Pine DS, Quinn K, Sanislow C, Wang P. Research domain criteria (RDoC): toward a new classification framework for research on mental disorders. Am J Psychiatry. 2010;167:748–51.
45. Fava GA. Road to nowhere. World Psychiatry. 2014;13:49–50.
46. Kraemer HC. Research Domain Criteria (RDoC) and the DSM – two methodological approaches to mental health diagnosis. JAMA Psychiat. 2015;72:1163–4.
47. Weinberger DR, Glick ID, Klein DF. Whither Research Domain Criteria (RDoC)?: the good, the bad, and the ugly. JAMA Psychiat. 2015;72:1161–2.
48. Reed GM, Correia JM, Esparza P, Saxena S, Maj M. The WPA-WHO global survey of psychiatrists' attitudes towards mental disorders classification. World Psychiatry. 2011;10:118–31.
49. Phillips J, Frances A, Cerullo MA, Chardavoyne J, Decker HS, First MB, Ghaemi N, Greenberg G, Hinderliter AC, Kinghorn WA, LoBello SG, Martin EB, Mishara AL, Paris J, Pierre JM, Pies RW, Pincus HA, Porter D, Pouncey C, Schwartz MA, Szasz T, Wakefield JC, Waterman GS, Whooley O, Zachar P. The six most essential questions in psychiatric diagnosis: a pluralogue part 3: issues of utility and alternative approaches in psychiatric diagnosis. Philos Ethics Humanit Med. 2012;7:9.
50. Pancheri P, Biondi M, Gaetano P, Picardi A. Costruzione della SVARAD (Scala di Valutazione Rapida Dimensionale). Riv Psichiatr. 1999;34:72–83.
51. Pancheri P, Picardi A, Gaetano P, Pasquini M, Puzella A, Di Fabio F, Biondi M. Validazione della scala per la valutazione rapida dimensionale SVARAD. Riv Psichiatr. 1999;34:84–93.
52. Aiken LR. Rating scales and checklists. New York, NY: John Wiley & Sons; 1996.
53. Picardi A, Martinotti G, Paci M, Simi C, Caroppo E. Reliability of self-reported attachment style in patients with severe psychopathology. J Pers Assess. 2011;93:491–9.
54. Picardi A, Caroppo E, Porcelli P, Di Maria G, Munittola G, Martinotti G. Alexithymia and severe psychopathology: a study on psychiatric inpatients. Psychopathology. 2012;45:159–66.
55. Picardi A, Caroppo E, Fabi E, Proietti S, Di Gennaro G, Meldolesi GN, Martinotti G. Attachment and parenting in adult patients with anxiety disorders. Clin Prac Epidemiol Mental Health. 2013;9:157–63.
56. Pancheri P, Biondi M, Gaetano P, Picardi A, Pasquini M. Impiego della scala per la valutazione rapida dimensionale SVARAD in una popolazione clinica ambulatoriale di 1124 soggetti. Riv Psichiatr. 2001;36:204–16.
57. Biondi M, Picardi A, Pasquini M, Gaetano P, Pancheri P. Dimensional psychopathology of depression: detection of an "activation" dimension in unipolar depressed outpatients. J Affect Disord. 2005;84:133–9.
58. Pasquini M, Picardi A, Biondi M, Gaetano P, Morosini P. Relevance of anger and irritability in outpatients with major depressive disorder. Psychopathology. 2004;37:155–60.

59. Picardi A, Morosini P, Gaetano P, Pasquini M, Biondi M. Higher levels of anger and aggressiveness in major depressive disorder than in anxiety and somatoform disorders. J Clin Psychiatry. 2004;65:442–3.
60. Abraham K. Notes on the psychoanalytical investigation of manic-depressive insanity and allied conditions; in Abraham K: Selected papers of Karl Abraham. London: Hogarth Press; 1948.
61. Freud S. Mourning and melancholia (1917). In: Strachey J, editor. Standard edition of the complete psychological works of Sigmund Freud, vol. 14. London: Hogarth Press; 1957. p. 239–58.
62. Guidano VF. Complexity of the self. New York, NY: Guilford Press; 1987.
63. Van Praag HM. Serotonergic dysfunction and aggression control. Psychol Med. 1991;21:15–9.
64. Bowlby J. Attachment and loss. New York, NY: Basic Books; 1980. vol 3: Loss.
65. Pasquini M, Biondi M, Costantini A, Cairoli F, Ferrarese G, Picardi A, Sternberg C. Detection and treatment of depressive and anxiety disorders among cancer patients: feasibility and preliminary findings from a liaison service in an oncology division. Depress Anxiety. 2006;23:441–8.
66. Pasquini M, Picardi A, Speca A, Orlandi V, Tarsitani L, Morosini P, Cascavilla I, Biondi M. Combining an SSRI with an anticonvulsant in depressed patients with dysphoric mood: an open study. Clin Pract Epidemiol Ment Health. 2007;3:3.
67. Piacentino D, Pasquini M, Tarsitani L, Berardelli I, Roselli V, Maraone A, Biondi M. The association of anger with symptom subtypes in severe obsessive-compulsive disorder outpatients. Psychopathology. 2016;49:40–6.
68. Rozin P, Lowery L, Imada S, Haidt J. The CAD triad hypothesis: a mapping between three moral emotions (contempt, anger, disgust) and three moral codes (community, autonomy, divinity). J Pers Soc Psychol. 1999;76:574–86.
69. Freud S. Notes upon a case of obsessional neurosis (1909). In: Strachey J, editor. Standard edition of the complete psychological works of Sigmund Freud, vol. 10. London: Hogarth Press; 1955. p. 153–249.
70. Pasquini M, Picardi A, Biondi M, Scarciglia P, Pancheri P. Analisi psicopatologica dimensionale del Disturbo Ossessivo-Compulsivo: appartiene ai Disturbi d'Ansia? Riv Psichiatr. 2003;38:86–91.
71. Dazzi F, Picardi A, Orso L, Biondi M. Predictors of inpatient psychiatric admission in patients presenting to the emergency department: the role of dimensional assessment. Gen Hosp Psychiatry. 2015;37:587–94.
72. Suozzo AC, Pasquini M, Biondi M. Importance of perspective dimensional psychopathology of depression: a proposal of evaluation for Brazilian patients indifferent settings. Arq Med Hosp Fac Cienc Med Santa Casa São Paulo. 2015;60:44–5.
73. Guillemin F, Bombardier C, Beaton D. Cross-cultural adaptation of health-related quality of life measures: literature review and proposed guidelines. J Clin Epidemiol. 1993;46:1417–32.
74. Harkness J. In pursuit of quality: issues for cross-National Survey Research. Soc Res Methodol. 1999;2:125–40.
75. Mohatlane EJ. Back-translation as a quality control mechanism in Sesotho translation. J Soc Sci. 2014;41:167–75.
76. Bulmer M. The problem of exporting social survey research. Am Behav Sci. 1998;42:153–67.
77. Biondi M, Pasquini M. Dimensional psychopharmacology in somatising patients. Adv Psychosom Med. 2015;34:24–35.

2 Dimensional Assessment with SVARAD in Clinical Practice

Massimo Biondi, Martina Valentini, Corinna Pancheri, Daria Piacentino, Massimo Pasquini, and Angelo Picardi

The SVARAD has been routinely used for many years in clinical practice at the Psychiatric Clinic of the Policlinico Umberto I, Sapienza University Hospital in Rome. Policlinico Umberto I, located in downtown Rome, is the largest hospital in central Italy, with 1200 beds and an emergency department (ED) visited by about 140,000 patients each year (one every 7 min). Every 24 h, 8–12 patients visit the emergency department with severe psychopathological symptoms, requiring specific psychiatric consultation. Patients who come to the ED are from different areas: about one third are from the downtown area of Rome, about one third are from the rest of the city or from the regional area (Lazio), and a final third are from elsewhere in Italy or other countries. About one or two of them a day are admitted to the psychiatric inpatient service (Servizio Psichiatrico di Diagnosi e Cura—SPDC), while the others are referred to other psychiatric services. The SPDC has 13 beds and about 380 acute admissions each year. Diagnostic categories for admission are mainly acute psychosis, schizophrenia, bipolar disorders, mood disorders with attempted suicide, and decompensated borderline personality disorders. The mean stay in SPDC is about 9 days, and ranges from 2 to 20 days according to several variables, including

M. Biondi (✉)
Department of Human Neurosciences, Policlinico Umberto I Hospital, Sapienza University of Rome, Rome, Italy
e-mail: massimo.biondi@uniroma1.it

M. Valentini · C. Pancheri · M. Pasquini
Department of Human Neurosciences, Sapienza University of Rome, Rome, Italy
e-mail: massimo.pasquini@uniroma1.it

D. Piacentino
Department of Neuroscience, Mental Health, and Sensory Organs (NESMOS), Sapienza University of Rome, Rome, Italy

A. Picardi
Centre for Behavioural Sciences and Mental Health, Italian National Institute of Health, Rome, Italy
e-mail: angelo.picardi@iss.it

M. Biondi et al. (eds.), *Dimensional Psychopathology*,
https://doi.org/10.1007/978-3-319-78202-7_2

severity and complexity of psychopathological and somatic conditions, response to treatment, and organisational and social intervention made before the discharge. For less severe cases, patients are invited to contact the outpatient service (OPS) of the psychiatric clinic or the psychiatric day hospital (DH) service located in the same structure, where about 120 patients are treated each year. The OPS treats about 5500 patients each year, on an appointment and walk-in basis, providing treatment to adult outpatients who come from Rome and the surrounding area. It is open Monday through Friday, from 8:00 a.m. to the early afternoon, and about 20–25 psychiatric visits take place there each day. Each visit lasts about 45–60 min and is carried out in a quiet room, with a semi-structured interview conducted by a senior psychiatrist and a psychiatric resident, and, in some instances, with a medical student in training. The visit includes both a general medical examination and laboratory assessments, if needed. Each visit results in a medical record, regularly supervised by a third senior psychiatrist with at least 20 years of experience in clinical diagnosis, as well as in the supervision of individual resident/student projects.

The present chapter describes the findings of two naturalistic studies. The first study was conducted on a sample of patients from the outpatient psychiatric service (OPS), the second on a sample of acute psychiatric inpatients of the SPDC. The aims are several: (a) to gather data on the usefulness and feasibility of SVARAD in a busy clinical setting with patients affected by common psychiatric conditions; (b) to describe mean dimensional profiles of several psychiatric diagnostic categories; (c) to explore the different components of psychopathological suffering within a single diagnostic category according to a dimensional perspective, that is, to investigate the true diversity of cases satisfying the criteria for a single diagnostic category; and (d) to give preliminary findings that suggest how recognising different dimensional profiles with SVARAD could permit personalised—not standardised—choices of treatment, i.e. precision psychiatric treatment. Further and more detailed analysis and discussion of these findings can be found later in the book, in specific chapters dedicated to individual SVARAD dimensions.

2.1 Study Methods

2.1.1 Design

Outpatients. This study has a naturalistic design, with transversal assessment of cases.

Inpatients. This study was done with a naturalistic, retrospective cross-sectional design.

2.1.2 Sample

Outpatients. A total of 1174 outpatients were enrolled from January 1996 to January 2001. Of these, 46 subjects were excluded from the study due to missed data or data that could not be evaluated due to confusion or uncooperativeness. Selection bias was unlikely to have affected results because key characteristics of excluded patients, including age, sex, marital status, nationality, clinical characteristics, and substance use status, were similar in excluded and included patients. The final sample included 1124 outpatients.

Inpatients. A total of 960 patients with a severe acute psychopathological state, admitted and treated in the SPDC, were enrolled from January 2011 to June 2014. Of these, 93 subjects were excluded from the study due to missed data or data that could not be evaluated due to confusion, excessive sedation, or uncooperativeness. The final sample included 846 patients. Selection bias was unlikely to have affected results, because key characteristics of excluded patients, including age, sex, marital status, nationality, clinical characteristics, compulsory vs. voluntary admission status, and substance use status, were similar in excluded and included patients. Patients were admitted with voluntary or, in a minority of cases, compulsory admission. Hospitalisation lasted a mean of 8 days, ranging from a few days to 2–3 weeks in a small number of cases.

2.1.3 Procedure

Each patient was asked for consent for the use of personal and treatment data, and gave informed consent for treatment, except in the case of compulsory treatment. The clinical data underwent a daily review (supervised by the chief psychiatrist or a senior psychiatrist with at least 10 years of experience) for testing the accuracy and coherence of the assessment of the clinical global picture. Diagnosis was performed at the end of the visit by a resident and a senior psychiatrist with at least 10 years of clinical experience and further validated in a weekly clinical meeting by the department head (MB). Outpatients were evaluated during a standard psychiatric visit lasting 45–60 min. Inpatients were assessed during the psychiatric ward admission (within 24 h).

2.1.4 Instruments

Outpatients. Psychopathological assessment instruments included the SVARAD [1] and the Minnesota Multiphasic Personality Inventory. The SVARAD was administered at the end of the visit, with ratings reported in the patient's clinical data sheet together with those from other scales. SVARAD procedures and characteristics were described in Chap. 1.

Inpatients. Psychopathological assessment was performed utilising the SVARAD and the Brief Psychiatric Rating Scale (BPRS). The SVARAD is part of the routine clinical psychopathological assessment at the initial evaluation, and at the time of discharge after treatment, with ratings reported in the patient's clinical data sheet.

2.1.5 Data Analysis

Distribution, mean, and standard deviation were calculated for all ten SVARAD dimensions. A 10-dimension "multiparametric" analysis was then done for the sample as a whole and for each DSM-IV diagnostic group. A further descriptive analysis detailed the patients' level of each specific dimension according to the following scores: 0 = absent, 1 = mild, 2 = moderate, 3 = severe, 4 = extreme. We introduced the "code type" analysis of SVARAD, which at first sight determines the three highest SVARAD peaks, leading to a rapid and easy characterisation of the principal components of a psychopathological description.

2.2 Main Findings and Discussion

We present in subsequent order the general characteristics of the outpatient and inpatient samples, the mean values of each SVARAD dimension, and the SVARAD dimensional profile of each DSM-IV category.

2.2.1 DSM-IV Diagnoses and Characteristics of the Whole Sample

Outpatients. The mean age of the outpatient sample (n = 1124) was 41.5 years (SD 15.2); 52% were of female gender; and 95% were patients with no reported psychopharmacological treatment in the previous 3 months. All the patients were residents of the Lazio region, mainly in the Rome district. From the larger sample, we selected a smaller sample of patients with sufficiently numerous and/or more relevant diagnoses. According to DSM-IV diagnosis, we included the following groups: borderline personality disorder (n = 31); major depressive disorder (n = 172); depressive disorder NOS (n = 63); dysthymia (n = 158); bipolar disorder (n = 18) with the following subtypes: bipolar disorder, depressive episode (n = 11); bipolar disorder, manic/hypomanic episode (n = 7); obsessive-compulsive disorder (n = 31); anxiety disorder NOS (n = 51); panic disorder (n = 92); generalised anxiety disorder (n = 11); somatic symptom disorder (n = 31); eating disorders (n = 37); schizophrenia, chronic (n = 31); psychotic disorder not otherwise specified (n = 29); delusional disorder (n = 12) (Table 2.1).

In the clinical characterisation of the sample, we also included mean values of the MMPI of 810 patients, although these data are not discussed in the present chapter. As a whole, the mean MMPI profile shows a "2-3" code type (D-Hy), with Hs (mean 73.32) and Pt (mean 71.75) as high peaks, suggesting the preponderance of a

Table 2.1 Sociodemographic characteristics of outpatients (N = 1124)

Variable		Mean (SD) or N (%)
Age in years, mean (SD)		41.5 (15.2)
Female gender, N (%)		588 (52.3)
Psychopharmacological treatment in the previous 3 months, N (%)	Yes	55 (4.9)
	No	1069 (95.1)
Response to psychopharmacological treatment, N (%)	Yes	36/55 (65.5)
	No	19/55 (34.5)

SD standard deviation

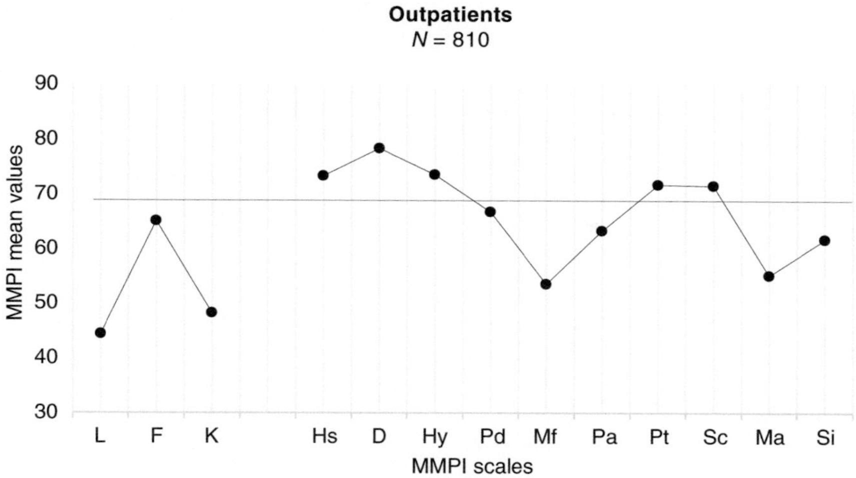

Fig. 2.1 MMPI profile of the psychiatric outpatient sample as a whole ($n = 810$). *MMPI* Minnesota Multiphasic Personality Inventory, *D* depression, *Hs* hypochondriasis, *Hy* hysteria, *K* key of correction, *L* lie, *F* infrequency, *Ma* hypomania, *Mf* masculinity/femininity, *Pa* paranoia, *Pd* psychopathic deviation, *Pt* psychasthenia, *Sc* schizophrenia, *Si* social introversion

"neurotic" type of psychopathological suffering, as expected in an outpatient sample; the mean *F* scale score was 65.09 (Fig. 2.1).

Inpatients. Sociodemographic characteristics of the inpatient sample ($n = 846$) are reported in Table 2.2. From the whole sample, patients with sufficiently numerous and/or more relevant diagnoses were selected for further analysis. According to DSM-IV diagnosis, we included the following groups: major depressive disorder ($n = 47$); depressive disorder NOS (104); bipolar disorder ($n = 186$) with the following subtypes: bipolar disorder, depressive episode ($n = 74$); bipolar disorder, manic/hypomanic episode ($n = 78$); bipolar disorder, mixed episode ($n = 34$); obsessive-compulsive disorder ($n = 8$); schizophrenia ($n = 82$); schizoaffective disorder ($n = 53$); psychotic disorder not otherwise specified ($n = 213$); borderline personality disorder ($n = 22$). The relatively high number of "not otherwise specified" diagnoses could be due to several factors: acute symptomatology, a short stay by the patient, the need for immediate treatment due to the severity of the clinical conditions, the scarcity of anamnestic information in some cases, and a high number of non-Italian-speaking patients. All of these elements contributed to a less accurate specification of disorders. All major and common psychopathological conditions leading to acute admission were well represented in this sample.

With regard to clinical data, 140 patients had had a severe suicide attempt requiring hospital admission. The mean length of the hospital stay was 11.8 days (SD = 9.7). Of all of the inpatients, 138 were admitted involuntarily, compared with 708 who were admitted voluntarily. Regarding global assessment of functioning (GAF), 39.3 was the mean value (SD = 12.5). The mean score on the BPRS at the initial assessment was 52.2 (SD = 13.5), while the mean score on the 21-item Hamilton depression rating scale (HDRS-21) was 17.3 (SD = 6.1) (Table 2.3). The BPRS profile for the inpatient sample is shown in Fig. 2.2.

Table 2.2 Sociodemographic characteristics of inpatients (*N* = 846)

Variable		Mean (SD) or *N* (%)
Age in years, mean (SD)		42 (14.3)
Female gender, *N* (%)		407 (48.1)
Educational level, *N* (%)	Primary school	60 (7.1)
	Middle school	222 (26.2)
	Secondary school	308 (36.4)
	University	124 (14.7)
	Other	132 (15.6)
Occupation, *N* (%)	Student	51 (6)
	Employee	221 (25.5)
	Self-employed	96 (11.2)
	Unemployed	282 (33.3)
	Invalid	12 (1.4)
	Retired	65 (7.7)
	Other	119 (14.1)
Marital status, *N* (%)	Single	476 (56.2)
	Married/cohabitant	166 (19.6)
	Separated/divorced	105 (12.4)
	Widowed	23 (2.7)
	Other	76 (9)
Housing, *N* (%)	Homeless	17 (2)
	Alone	205 (24.2)
	Family of origin	227 (26.8)
	Family of creation	205 (24.2)
	Psychiatric community	10 (1.2)
	Nonpsychiatric community	6 (0.7)
	Nursing home/foster home	11 (1.3)
	Other	165 (19.5)
Civil ability, *N* (%)	Capable	687 (81.2)
	Incapable (with public guardian/trustee)	41 (4.8)
	Other	118 (13.9)

Table 2.3 Clinical characteristics of inpatients (*N* = 846)

Variable		Mean (SD) or *N* (%)
Type of admission to psychiatric ward, *N* (%)	Voluntary	708 (83.7)
	Involuntary	138 (16.3)
Length of stay in psychiatric ward in days, mean (SD)		11.8 (9.6)
Attempted suicide requiring admission in psychiatric ward, *N* (%)		140 (16.5)
BPRS at admission, mean (SD)		52.2 (13.5)

BPRS Brief Psychiatric Rating Scale, *SD* standard deviation

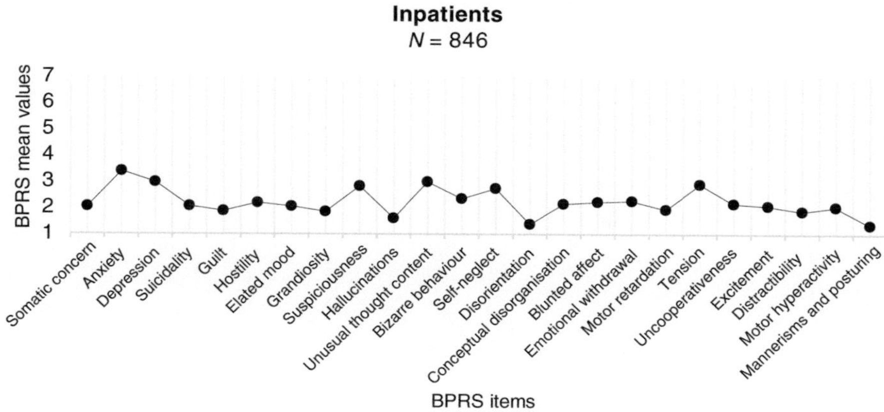

Fig. 2.2 BPRS profile of the psychiatric inpatient sample as a whole ($N = 846$)

2.2.2 Analysis of Each SVARAD Psychopathological Dimension in the Whole Sample

Mean SVARAD values in the whole sample are shown in Fig. 2.3. For the outpatient sample, the three dimensions with the highest values were Apprehension/Fear, Sadness/Demoralisation, Apathy, and Anger/Aggressiveness. The top scores for the inpatient sample were Apprehension/Fear as the highest score (above 2), followed by Reality Distortion, and then Sadness/Demoralisation, Impulsivity, and Apathy. This seems to suggest that the dominant, diffuse psychopathological suffering of the inpatient sample is characterised by moderate to intense feelings of tension, vulnerability, threat, fear, and anxiety associated with experiences of delusions, psychoticism, and hallucinations but also to a significant degree, demoralisation and dysregulation of behaviour and emotions.

The mean SVARAD profile suggested less acute and severe symptomatology for outpatients than for patients admitted to the inpatient service.

2.2.2.1 Apprehension/Fear

Outpatients. The SVARAD dimensions showed mean values that were moderately high (2 score) in almost every diagnostic group (Fig. 2.4). Generalised anxiety, obsessive-compulsive, and panic disorders were the three highest ones (above 2).

Inpatients. The SVARAD dimensions displayed a mean value around 2 score throughout the inpatient group. This appears to be a "trans-diagnostic" dimension that was present to a significant degree in *all* the diagnostic groups (Fig. 2.5).

This finding suggests how Apprehension/Fear (with its components of tension, anxiety, and nervousness) might be better viewed as a trans-diagnostic dimension rather than one typical only of anxiety disorders. SVARAD analysis suggests that this dimension actually represents ubiquitous psychopathological phenomena of suffering, common in diverse psychiatric patients. For instance, mean scores near 2 (moderate

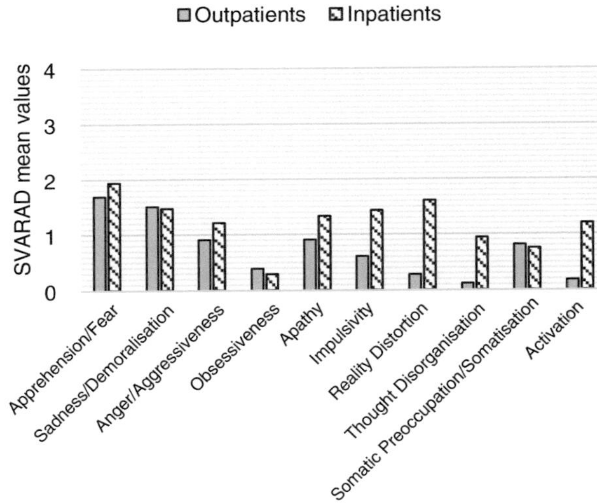

Fig. 2.3 SVARAD profile of the whole psychiatric sample: mean scores and standard deviations (outpatients $N = 1124$; inpatients $N = 846$)

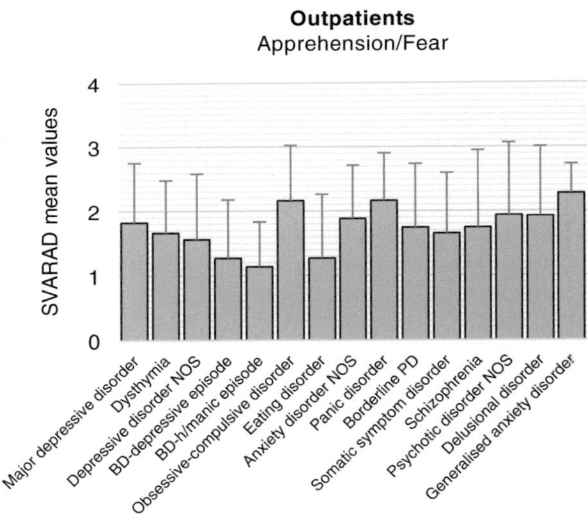

Fig. 2.4 SVARAD Apprehension/Fear dimension across outpatients' diagnostic categories: mean scores and standard deviation

Apprehension/Fear) actually suggest a consistent component of anxiety/tension in major depression, dysthymia, depressive disorder NOS, and psychotic disorders.

Standardised diagnostic criteria of the DSM-IV and ICD-10 do not describe or include these components in the clinical picture. These findings underscore the need to pay attention to and recognise anxiety in many diagnostic groups, allowing the clinician to correctly address and treat them with specific drugs or psychosocial interventions [2]. In our experience, this is essential for psychiatric residents and clinical psychologists, as well as psychiatrists.

The finding of high Apprehension/Fear in every diagnostic group might also explain the common use of benzodiazepines in clinical practice for many

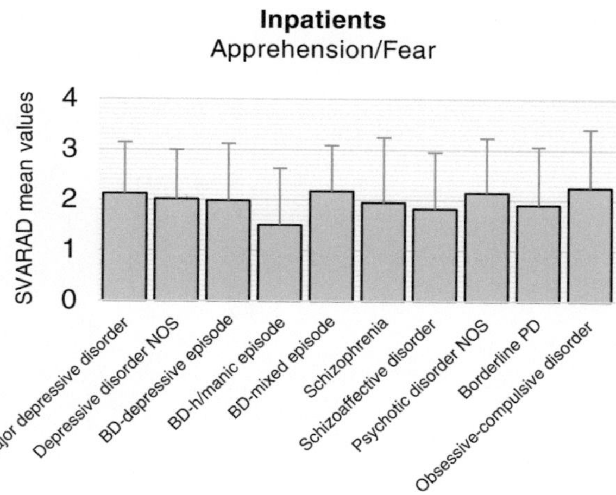

Fig. 2.5 SVARAD Apprehension/Fear dimension across inpatients' diagnostic categories: mean values and standard deviations

disorders and not just in the anxiety disorders area (e.g. major depression and dysthymia, bipolar disorder, obsessive-compulsive disorder, somatoform disorders, borderline personality disorder, and schizophrenia and other psychoses) [3, 4]. Benzodiazepines give the patient a prompt benefit with the reduction of painful psychic tension, somatic symptoms of anxiety, worries, and nervousness. It might also explain the frequent unwanted activation, with increased tension, anxiety, and insomnia, up to agitation and suicidal thoughts, after the use of selective serotonin reuptake inhibitors (SSRIs) or other antidepressants in major depression or dysthymia (if there is a significant level of undetected anxiety/nervousness). It might also help to explain the variability in response to the same psychotherapeutic intervention for the same disorder (e.g. cognitive or psychodynamic treatment in major depression, panic disorder, or borderline personality disorder, to give some examples).

2.2.2.2 Sadness/Demoralisation

Outpatients. The Sadness/Demoralisation dimension showed relatively high scores throughout all the diagnostic groups. Sadness/Demoralisation showed high mean values for bipolar disorder (depressive episode) (score 2.5), major depression (score 2.3), and moderate scores for dysthymia (mean score 1.9), depressive disorder NOS, and borderline personality disorder (mean score 1.7 for each). Low to moderate scores (mean scores 1.5–1.6) were found in the psychotic disorder NOS, obsessive-compulsive disorder, and eating disorder groups. Schizophrenia, panic disorder, and generalised anxiety disorder showed mean scores in the lower range (mean score around 1, mild). The lowest values were for delusional disorders and bipolar disorder—manic episode (mean score below 0.5) (Fig. 2.6).

Inpatients. The Sadness/Demoralisation dimension showed the highest mean scores for major depression (2.7), bipolar disorder (depressive episode) (2.5), and depressive disorder NOS (2.3), as expected. Borderline personality disorder showed a moderate Sadness/Demoralisation mean (1.9); this value was significant from a

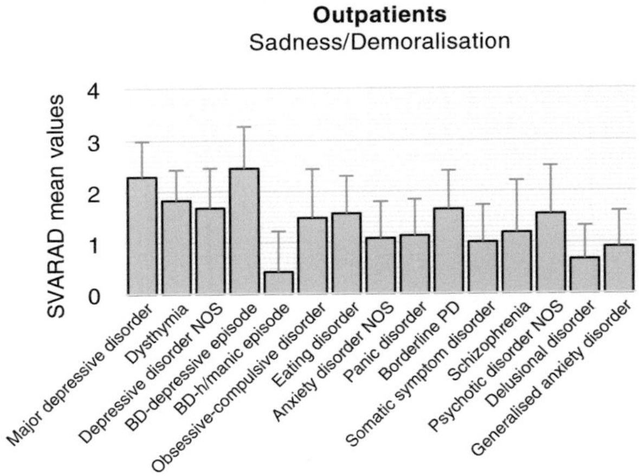

Fig. 2.6 SVARAD Sadness/Demoralisation dimension across outpatients' diagnostic categories: mean scores and standard deviations

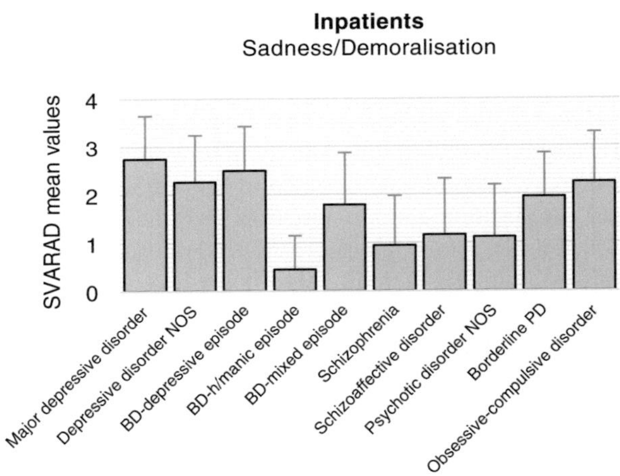

Fig. 2.7 SVARAD Sadness/Demoralisation dimension across inpatients' diagnostic categories: mean values and standard deviations

clinical viewpoint, and it probably explains the major reason for admission for this group of patients. The lowest mean scores were for bipolar disorder (manic episode) (0.4) and schizophrenia (0.9), together with schizoaffective disorder (1.2) and psychotic disorder NOS groups (1.1) (Fig. 2.7).

Severe and moderate degrees of Sadness/Demoralisation are certainly expected in disorders of the mood spectrum. The low to moderate values found in several other disorders, however, suggest that diagnosis according to standardised criteria might fit well for the aim of the coherence and concordance among clinicians for that specific disorder, but might lead to underestimation of significant components of suffering. As concerns antidepressant treatment, a similar view can be applied to several other diagnostic groups: within a single diagnostic category, an antidepressant can be indicated or not indicated, according to the psychopathological profile of the single patient.

2.2.2.3 Anger/Aggressiveness

Outpatients. Borderline personality disorder showed the highest mean value, with a mean score near 2, followed by the delusional disorder group (mean score 1.5) (Fig. 2.8).

Inpatients. The Anger/Aggressiveness dimension in this group showed higher mean SVARAD scores for bipolar disorder (manic and mixed episodes) (1.7 and 1.3) and borderline personality disorder (1.4). Low mean values were present in all the other groups (Fig. 2.9).

Anger/Aggressiveness is among the more interesting dimensions, because no DSM-IV or ICD-10 categories fully represent it, or correspond to it, to the extent that the SVARAD dimension does. It is interesting to point out that these mean values actually include two to three subgroups, satisfying criteria for the same diagnostic category that are quite different according to Anger/Aggressiveness

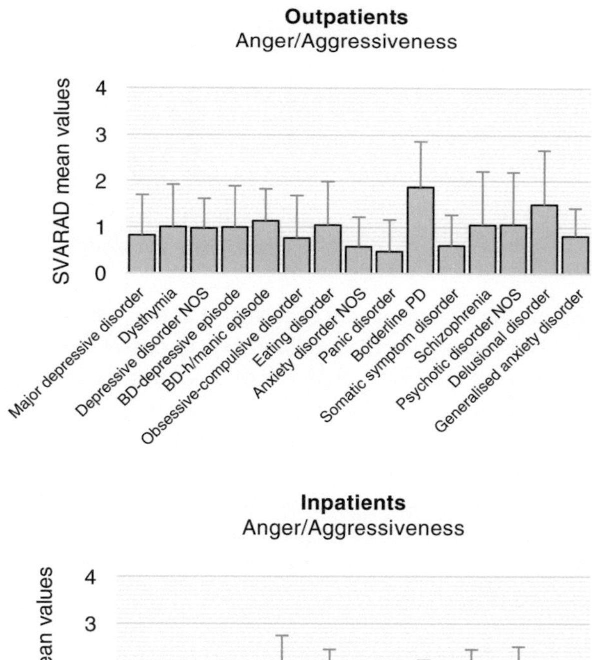

Fig. 2.8 SVARAD Anger/Aggressiveness dimension across outpatients' diagnostic categories: mean scores and standard deviations

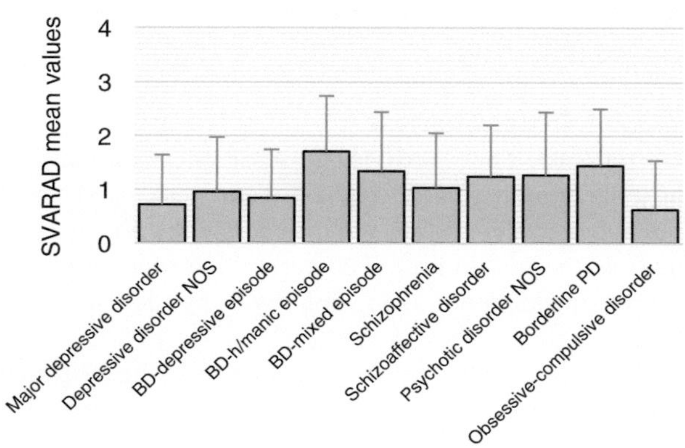

Fig. 2.9 SVARAD Anger/Aggressiveness dimension across inpatients' diagnostic categories: mean values and standard deviations

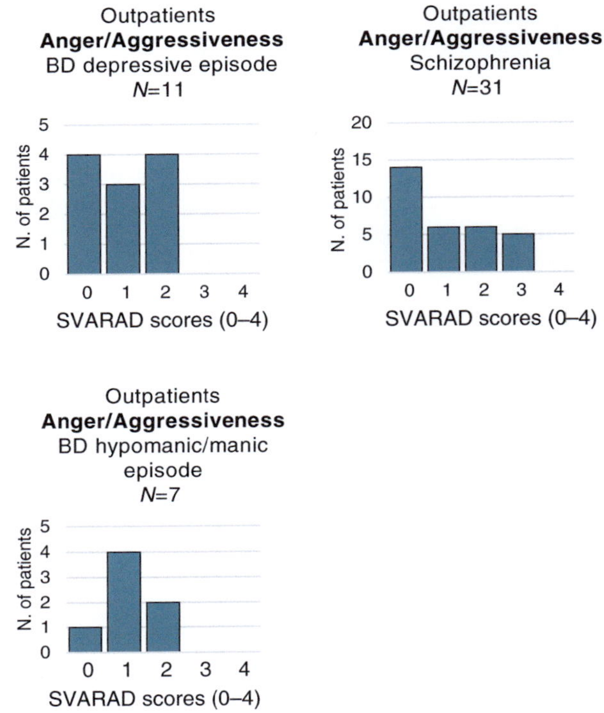

Fig. 2.10 Number of outpatients scoring 0-1-2-3-4 at the SVARAD Anger/Aggressiveness dimension divided by diagnostic category: bipolar disorder-depressive episode, bipolar disorder-hypomanic/manic episode and schizophrenia

dimension scores: absent or low (0–1 scores), moderate (2), severe (3), and extreme (4). Both psychotherapeutic and psychopharmacological treatment should take into account the dimensional profile of this category. Divergence in eligibility, outcomes, and side effects might well be explained by dimensional variability, as the figure suggests. In the outpatient sample, the delusional disorder group showed mean values above 1, while the schizophrenia and bipolar disorder groups showed mean values slightly under 1. Cases with scores higher than 2 were rare (schizophrenia $N = 5$; bipolar disorder (depressive episode) $N = 0$; bipolar disorder (manic episode) $N = 0$). However, if they are present, this is clinically significant, as Chap. 7 will discuss (Fig. 2.10).

2.2.2.4 Obsessiveness

Outpatients. The SVARAD Obsessiveness dimension (including compulsiveness as a descriptor) has the highest mean peak—as expected—in the obsessive-compulsive disorder group, with mean values above 2 and ranging at the top end to near 4 (the maximum). The eating disorder group had the second highest mean score (mean score 0.7), followed by delusional disorder (mean score 0.5) (Fig. 2.11).

Inpatients. In the inpatient sample, as in the outpatient sample, the Obsessiveness dimension had very high mean values in the obsessive-compulsive group. All the other groups show very low mean scores (<0.5) (Fig. 2.12).

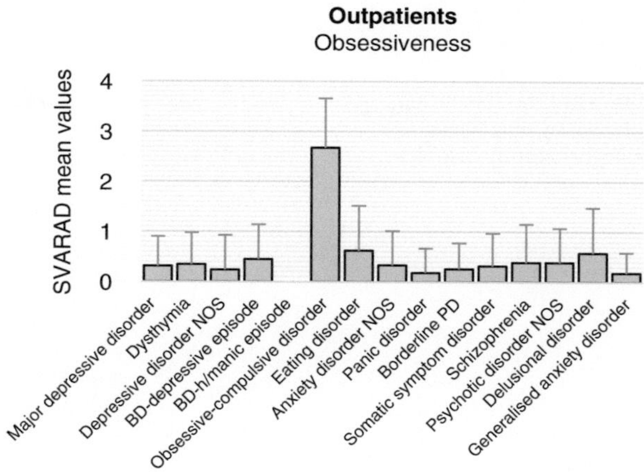

Fig. 2.11 SVARAD Obsessiveness dimension across outpatients' diagnostic categories: mean scores and standard deviations

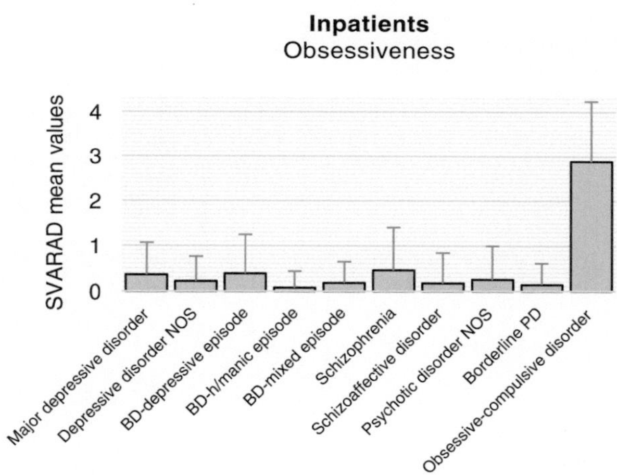

Fig. 2.12 SVARAD Obsessiveness dimension across inpatients' diagnostic categories: mean values and standard deviations

The relationship between Obsessiveness—and particularly compulsiveness—and anorexia or bulimia has been often discussed both in the literature about eating disorders and in obsessive-compulsive spectrum research [5]. A small but clinically significant subgroup of eating disorders within our outpatient sample had Obsessiveness values above 2 (eating disorders, $N = 2$) (Fig. 2.13). This suggests that the eating disorders group as a whole does not fall within the obsessive spectrum per se, but Obsessiveness might be present in some subjects. It is interesting to remember that extreme (score 4) values were found in a few eating disorders patients admitted to our ward as acute inpatients.

2.2.2.5 Apathy

Outpatients. The SVARAD Apathy dimension showed the highest mean score in the bipolar disorder—depressive episode group (mean score above 2, moderate to severe)—followed by the schizophrenia group (mean score at 1.9, moderate) and

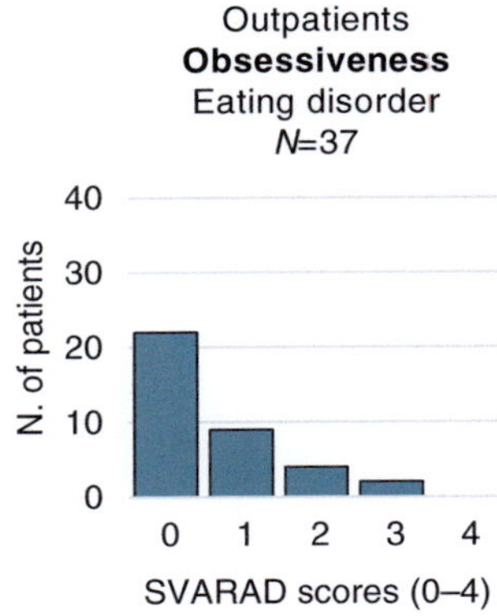

Fig. 2.13 Number of outpatients scoring 0-1-2-3-4 on the SVARAD obsessiveness dimension divided by diagnostic category: eating disorder

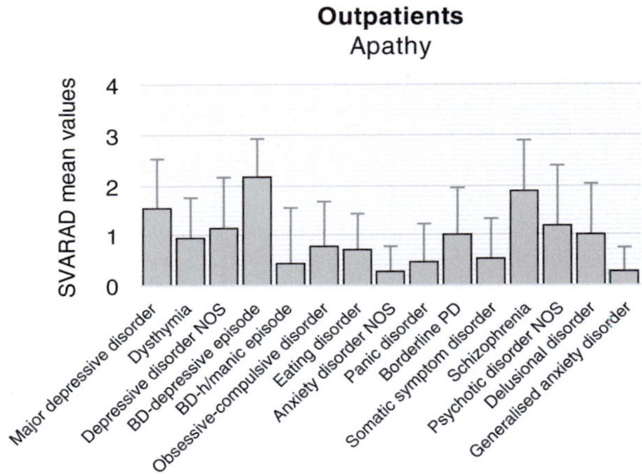

Fig. 2.14 SVARAD apathy dimension across outpatients' diagnostic categories: mean scores and standard deviations

major depressive disorder group (mean score 1.6, low to moderate). Low but clinically significant mean scores were found in the depressive disorder NOS (mean score 1.2), psychotic disorder NOS (mean score 1.2), and borderline personality disorder (mean score 1) groups. The lowest scores were seen in the anxiety disorder, bipolar disorder manic/hypomanic episode, and somatic symptom disorder groups (Fig. 2.14).

Inpatients. Apathy in inpatients showed the highest mean scores in the major depression (2.1, moderate), bipolar disorder (depressive episode) (1.8), schizophrenia (1.8), and depressive disorder NOS (1.5) groups (Fig. 2.15).

Apathy thus is present in the two different broad clinical areas of depressive and psychotic disorders.

2.2.2.6 Impulsivity

Outpatients. In outpatients, the SVARAD Impulsivity dimension showed a peak for borderline personality disorder, with a mean score of 1.8, moderate degree. The eating disorders group showed a mean score of 1.5, intermediate between low and moderate degree, and the delusional disorder mean score was 1.3, slightly above a

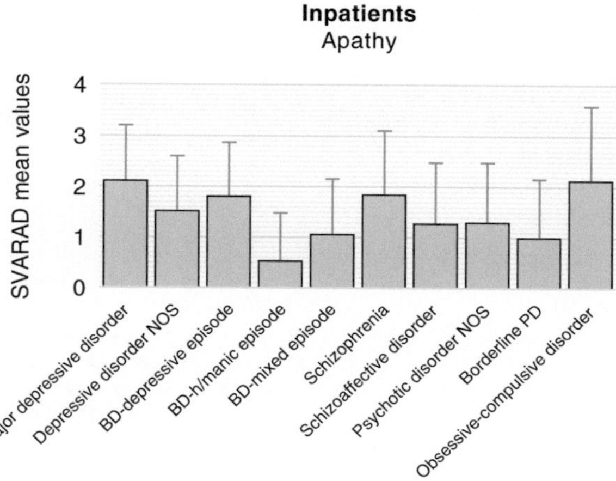

Fig. 2.15 SVARAD apathy dimension across inpatients' diagnostic categories: mean values and standard deviations

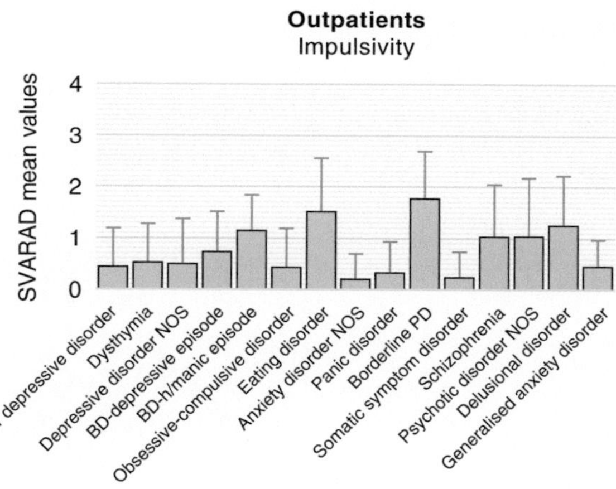

Fig. 2.16 SVARAD impulsivity dimension across outpatients' diagnostic categories: mean scores and standard deviations

low value. It might seem strange to find relatively low values for bipolar disorder (manic episode), psychotic disorder NOS, and schizophrenia, with mean scores close to 1, just above a low degree. This finding could be explained by the characteristics of the sample, mainly composed by non-acute patients in an ambulatory setting. The several other diagnostic groups had very low mean scores (near or below 0.5) (Fig. 2.16).

Inpatients. In inpatients, impulsivity showed the highest mean scores for borderline personality disorder (1.9, moderate). Scores near 2 were also seen for bipolar disorder—manic/hypomanic and mixed episodes—as expected (1.8 and 1.6) (Fig. 2.17).

2.2.2.7 Reality Distortion

Outpatients. In the outpatients, the SVARAD Reality Distortion dimension showed the highest values for schizophrenia and delusional disorder (mean values of 2.3 and 2.2, just above a moderate degree of severity), followed by psychosis NOS (mean

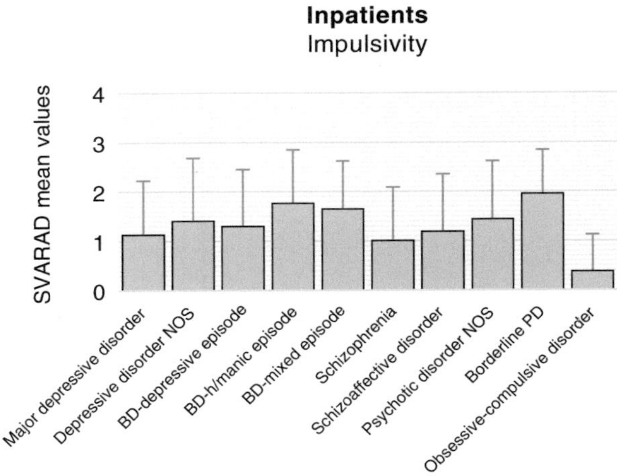

Fig. 2.17 SVARAD impulsivity dimension across inpatients' diagnostic categories: mean values and standard deviations

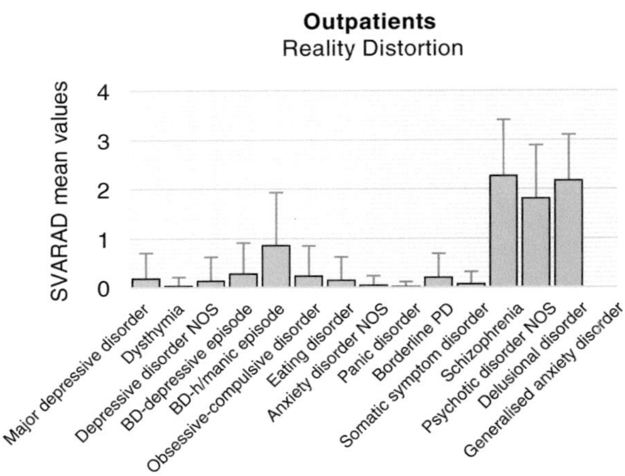

Fig. 2.18 SVARAD Reality Distortion dimension across outpatients' diagnostic categories: mean scores and standard deviations

score 1.8, near a moderate degree), as expected. Bipolar disorder—manic episode—showed a low mean score (0.9); however, this score was higher than all the other diagnostic groups. The low values for this latter category might reflect the characteristics of the outpatient setting, with less severe patients. Scores near zero were common in other diagnostic groups, as expected (Fig. 2.18).

Inpatients. In the inpatient sample, the highest mean values for the Reality Distortion dimension were in schizophrenia (near 3, severe), and schizoaffective disorder and psychotic disorder NOS (both 2.4), followed by bipolar disorder—manic episode (1.9). All the other groups were in a low range (Fig. 2.19).

2.2.2.8 Thought Disorganisation

Outpatients. In outpatients, the SVARAD Thought Disorganisation dimension showed the highest mean value in the psychotic disorders groups, as expected: the group with schizophrenia had the highest mean score (1.2, mild degree), followed

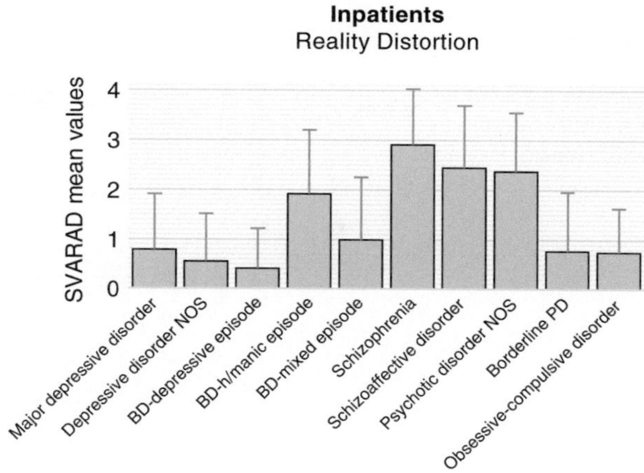

Fig. 2.19 SVARAD Reality Distortion dimension across inpatients' diagnostic categories: mean values and standard deviations

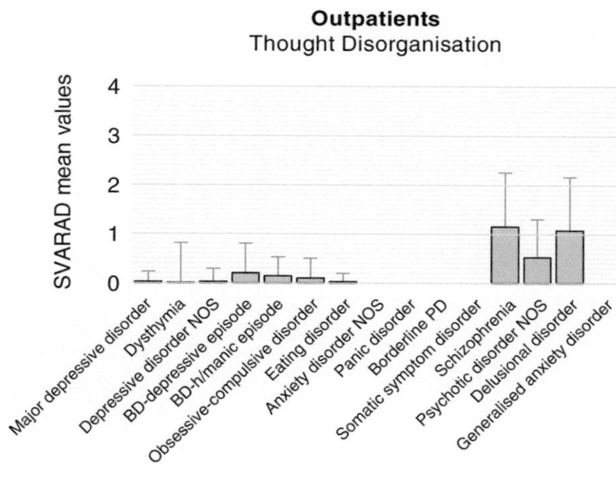

Fig. 2.20 SVARAD Thought Disorganisation dimension across outpatients' diagnostic categories: mean scores and standard deviations

by the delusional disorder group, with similar mean values of 1.1, and then the psychotic disorder NOS group, with a mean value of 0.5. The other diagnostic groups were near zero (Fig. 2.20).

Inpatients. In inpatients, the Thought Disorganisation dimension displayed the highest mean values for schizophrenia (near value 2), bipolar disorder (manic episode) (1.5), schizoaffective disorder (1.4), and psychotic disorder NOS (1.2). All the other groups were in a low range (Fig. 2.21).

2.2.2.9 Somatic Preoccupation/Somatisation

Outpatients. In outpatients, the SVARAD Somatic Preoccupation/Somatisation dimension displayed the highest mean values for somatic symptom disorders (mean group value, 2.2; moderate to severe), followed by panic disorder (1.2), eating disorders (1.1), anxiety disorder NOS (1.0), and generalised anxiety disorder (1.0) (Fig. 2.22).

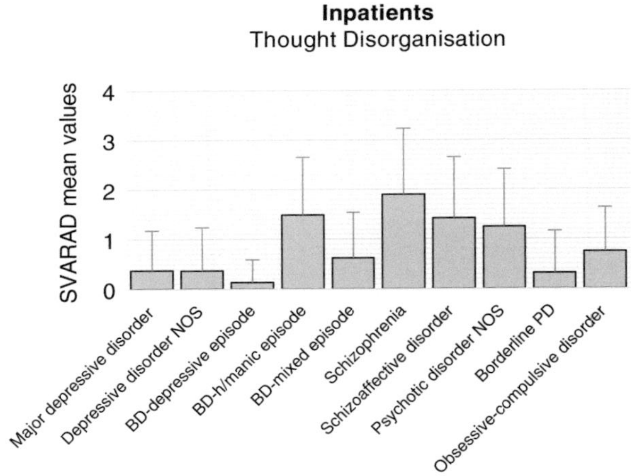

Fig. 2.21 SVARAD Thought Disorganisation dimension across inpatients' diagnostic categories: mean values and standard deviations

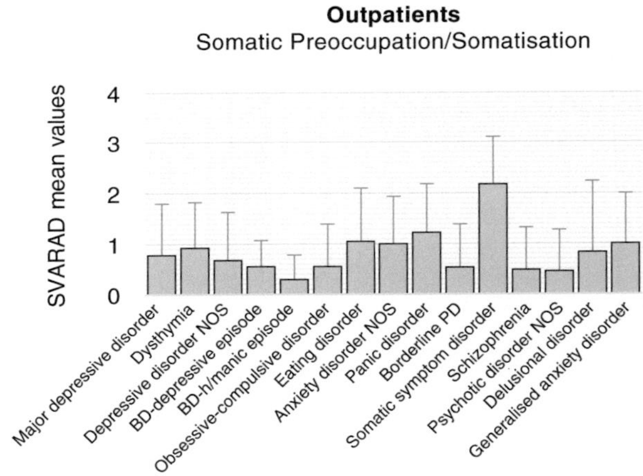

Fig. 2.22 SVARAD Somatic Preoccupation/Somatisation dimension across outpatients' diagnostic categories: mean scores and standard deviations

Inpatients. In inpatients, the Somatic Preoccupation/Somatisation dimension had the highest mean values in the major depressive groups (1.2, just above the mild score), with low values in all the other groups (Fig. 2.23).

A low mean degree of Somatic Preoccupation/Somatisation was commonly seen among several diagnostic groups.

2.2.2.10 Activation

Outpatients. In the outpatients, the SVARAD Activation dimension reached the highest peak in the bipolar disorder (manic/hypomanic episode) (mean value 1.3) and borderline personality disorder (0.6) groups (Fig. 2.24).

Inpatients. In the inpatients, the Activation dimension showed the highest mean value in the bipolar disorder (manic episode) (2.5) group, followed by the bipolar disorder (mixed episode) (1.6) group. Schizophrenia and other psychotic disorders, as well as borderline personality disorder, showed mild mean values (Fig. 2.25).

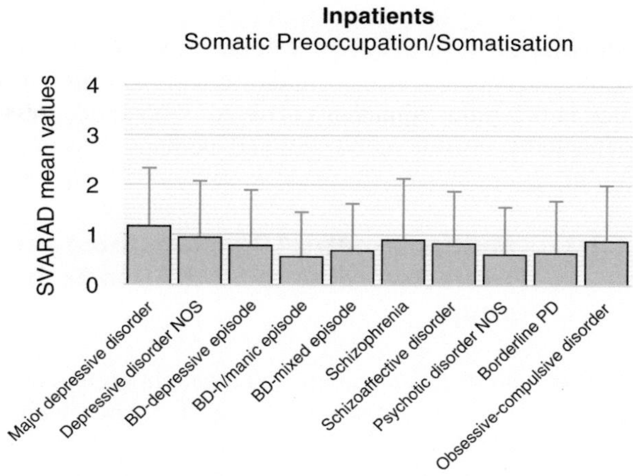

Fig. 2.23 SVARAD Somatic Preoccupation/Somatisation dimension across inpatients' diagnostic categories: mean values and standard deviations

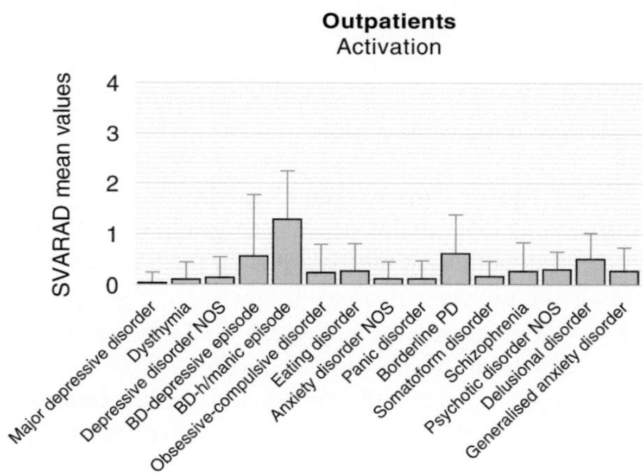

Fig. 2.24 SVARAD Activation dimension across outpatients' diagnostic categories: mean scores and standard deviations

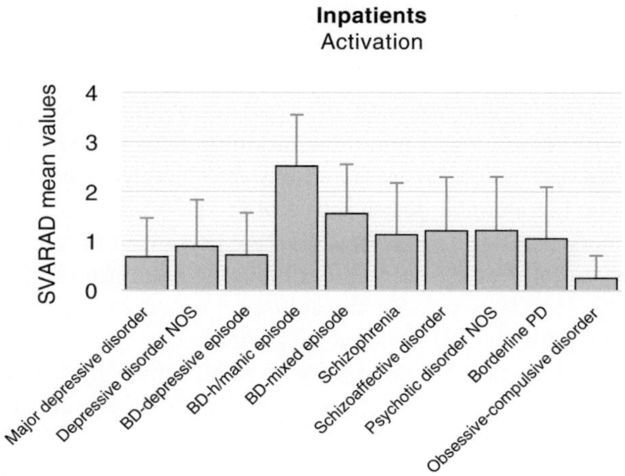

Fig. 2.25 SVARAD Activation dimension across inpatients' diagnostic categories: mean values and standard deviations

A past study from our research group found an Activation factor (as assessed by SVARAD; MMPI-2) to be one of the three most significant components in unipolar depressed patients [6]. This finding could be of significant clinical interest if depression in this group of patients is treated with antidepressants, considering the risk of a manic/hypomanic mood shift and, perhaps, self-damaging behaviours.

2.2.3 "Multiparametric" Psychopathological Dimensional Profile According to SVARAD in Each DSM-IV Diagnostic Group

The analysis of the outpatient and inpatient groups was performed according to the primary diagnosis (thus including all the cases with Axis I and II comorbidity). Each dimensional profile was presented as a code type, which expresses a SVARAD profile as the numbers of the two or three highest items. For example, a 2-1 code type would indicate an individual scoring high in the Sadness/Demoralisation and Apprehension/Fear dimensions. We also performed the same analysis excluding any comorbidity. Differences in scores across dimensions in every diagnosis were limited. The presence of comorbidities seems not to compromise the validity of the code type dimension model.

2.2.3.1 Major Depressive Disorder

Outpatients. The major depression group (n = 172) showed, as expected, a 2-1-5 code type, with high mean scores on the SVARAD Sadness/Demoralisation of 2.3 (moderate to severe degree), followed by high mean scores for Apprehension/Fear (1.8, near moderate degree), Apathy (1.6), Anger/Aggressiveness (0.8), and Somatic Preoccupation/Somatisation (0.8) (Fig. 2.26).

Inpatients. As with the outpatients, the inpatients with major depression (n = 47) showed a SVARAD mean profile characterised by a 2-1-5 code type, with a triad of dimensions: a very high value (severe) for Sadness/Demoralisation (2.7) and a moderate value for Apprehension/Fear and Apathy (both 2.1). Somatic Preoccupation/Somatisation showed mild mean value (1.2), suggesting the presence of a mild component of somatic suffering in depression, together with a mild component of impulsivity (1.1) (Fig. 2.27).

The significant peak of Apprehension/Fear in both inpatients and outpatients with major depression certainly introduces the issue of the coexistence of anxiety and depression, an area of extensive research due to the presence of comorbidity [7]. The hierarchical decision tree of the DSM-IV leads to the identification of the primary diagnosis as the best one; the presence of comorbidities will be determined if symptoms of the secondary clinical condition fulfil the set of criteria of an anxiety disorder. If they don't, any anxiety component disappears from the categorical diagnosis, although the patient may be suffering to some extent because of it. This might be a very relevant issue for treatment, especially for partially improving and resistant major depression cases. This will be further detailed in the next section about anxiety disorders.

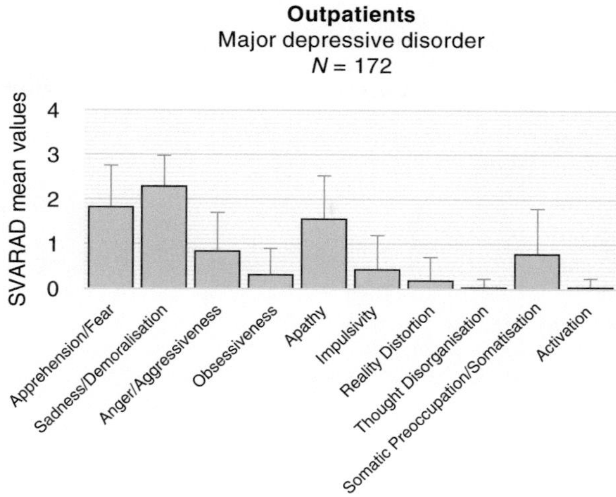

Fig. 2.26 SVARAD profile of outpatients with major depressive disorder: mean scores and standard deviations. Code type: 2-1-5 ($n = 172$)

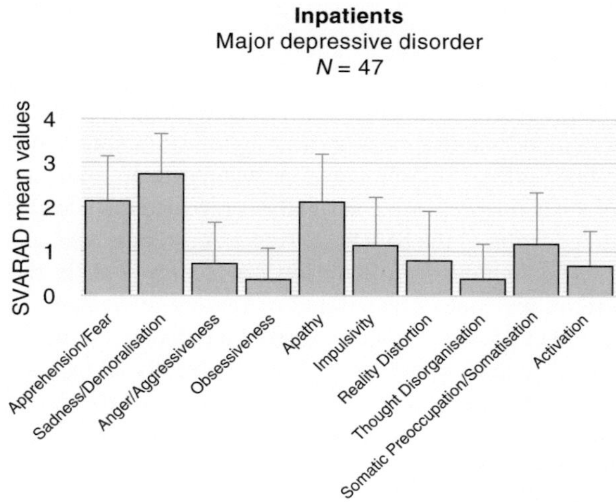

Fig. 2.27 SVARAD profile of inpatients with major depressive disorder: mean scores and standard deviations. Code type: 2-1-5 ($n = 47$)

The SVARAD also showed the presence, especially in the outpatient group, of a significant degree of Anger/Aggressiveness, a component of psychopathological suffering that is not reported in the DSM-IV and ICD-10 criteria for inclusion in this diagnostic group. A previous study from our research group, based on factor analysis, found that unipolar depressive patients display three main components of SVARAD dimensions: (1) Sadness/Demoralisation + Apathy, (2) Apprehension/Fear + Somatic Preoccupation/Somatisation, and (3) Anger/Aggressiveness + Impulsivity + Activati on, confirmed also by factor analysis of MMPI-2 [6]. Figure 2.25 shows that 40 outpatients (23%) of the major depression diagnostic group had a score of 2 (moderate) or 3 (severe) in Anger/Aggressiveness. The fact that about ¼ of the outpatient sample had these scores could be relevant both for understanding the complexity of the

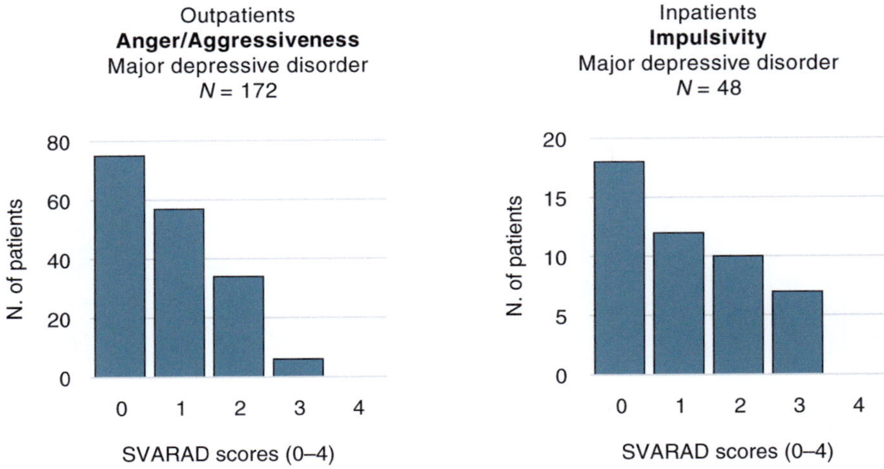

Fig. 2.28 Number of outpatients and inpatients scoring 0-1-2-3-4 on the SVARAD anger/aggressiveness and impulsivity dimensions, divided by diagnostic category: major depressive disorder

psychopathology of major depression and particularly for constructing treatment plans. A recent study from our research group suggested that the add-on of gabapentin as a mood stabiliser resulted in a significant improvement of the Hamilton depression rating scale scores of major depression patients with partial improvement after the use of the antidepressant. Data analysis revealed that the clinical improvement was linked to reduction of the SVARAD Anger/Aggressiveness score in the outpatients [8].

The Impulsivity dimension, the fourth peak in the SVARAD inpatient profile group, was not expected and extremely interesting. Particularly, there was a subgroup of 35.4% of major depression inpatients scoring 2 and 3 (20.8% as moderate, 14.6% as severe), which should be carefully taken into account for the management and treatment of these patients (Fig. 2.28). Its role could be significant in some clinical cases, due to the risk of suicide attempts and the activation response after antidepressant initiation.

Finally, the SVARAD Somatic Preoccupation/Somatisation dimension was high for both groups, suggesting a role for somatic suffering and preoccupation, frequently found in patients with depression.

2.2.3.2 Dysthymia and Depressive Disorder NOS

Outpatients. The dysthymia ($n = 158$) and depressive disorder NOS ($n = 63$) diagnostic groups both showed SVARAD profiles with a 2-1 code type, similar to that of the major depression group (Figs. 2.29 and 2.30). The "multiparametric" analysis, however, showed that dysthymia has Anger/Aggressiveness as the third component of suffering (30.4% of the outpatients scored at 2 or above), whereas the depressive NOS group showed Apathy as the third component of psychopathological suffering (1.1). However, 33.3% of the patients in this group also had Anger/Aggressiveness scores at or above 2 (Fig. 2.31).

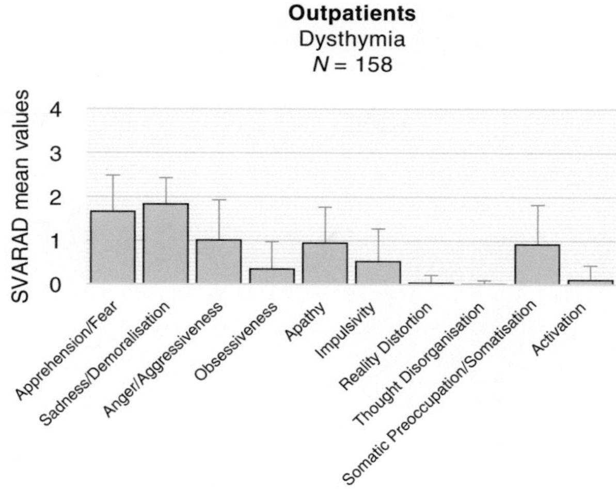

Fig. 2.29 SVARAD profile of outpatients with dysthymia: mean scores and standard deviations. Code type: 2-1-3 ($n = 158$)

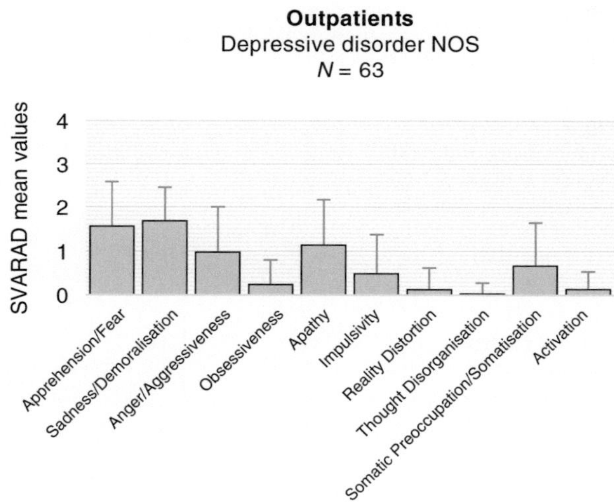

Fig. 2.30 SVARAD profile of outpatients with depressive disorder NOS: mean scores and standard deviations. Code type: 2-1-5 ($n = 63$)

Inpatients. Inpatients with depressive disorder NOS (n = 104) showed a similar SVARAD 2-1-5 code type, with a high mean of Sadness/Demoralisation (2.3), as expected, as well as Apprehension/Fear (2). Apathy showed an intermediate value between mild and moderate (1.5), and, very interestingly, Impulsivity showed a mean score of 1.4, indicating that in the depressive disorder NOS group, there was a more pronounced impulsivity component than in the major depressive disorder group (Fig. 2.32).

These findings suggest that anger and impulsivity play a more significant role in dysthymic and depressive NOS patients than in major depression patients. These findings might suggest that antidepressant treatment for these kinds of patients should be more cautious than that for major depression, because of the risks of the onset of dysphoric mood, anxiety, irritability, insomnia, and impulsive self-damaging behaviours.

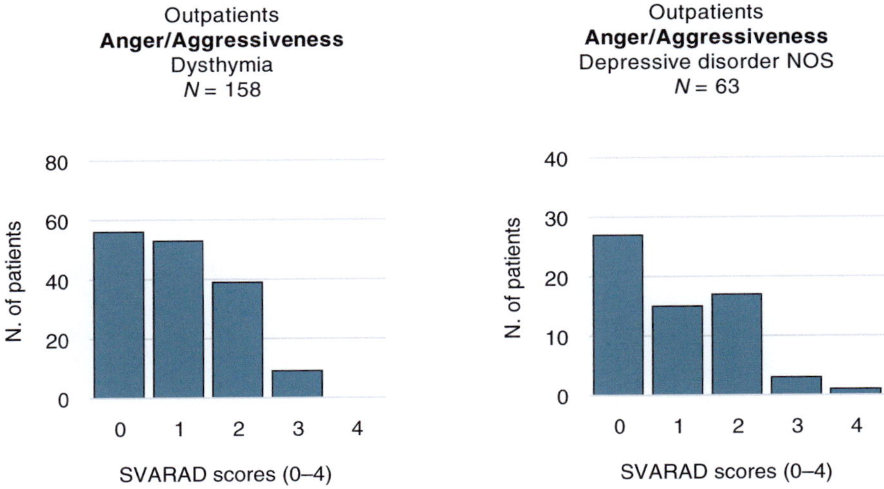

Fig. 2.31 Number of outpatients scoring 0-1-2-3-4 on the SVARAD Anger/Aggressiveness dimension divided by diagnostic category: dysthymia and depressive disorder NOS

Fig. 2.32 SVARAD profile of inpatients with depressive disorder NOS: mean scores and standard deviations. Code type: 2-1-5 ($n = 104$)

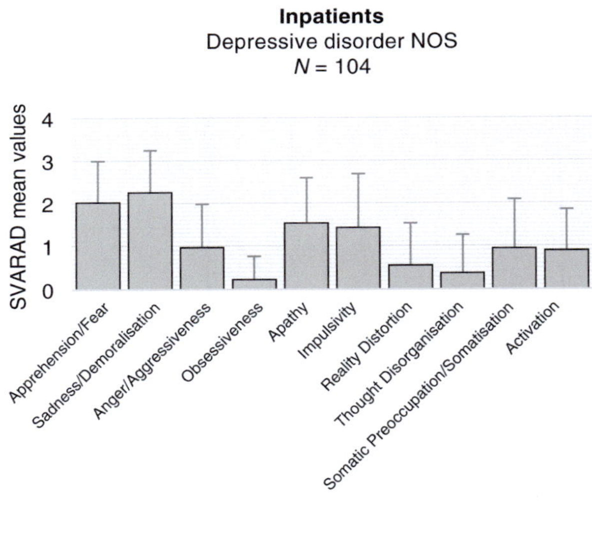

Mood stabilisers or low-dose atypical antipsychotics could be useful in these dysthymia and depressive subgroups. SVARAD "multiparametric" assessment outlines this characteristic and might be a better guide for use in choosing a treatment.

2.2.3.3 Bipolar Disorder: Depressive Episode

Outpatients. Outpatients with a diagnosis of bipolar disorder—depressive episode ($n = 11$)—displayed a 2-5-1 code type (Sadness/Demoralisation = 2.5; Apathy = 2.2;

Apprehension/Fear = 1.3), followed by Anger/Aggressiveness (1). While depressive mood and apathy certainly predominated, it should be highlighted that patients experienced impulsivity and irritability, at mild (one out of three patients) and moderate (one out of four patients) levels, respectively, as well as symptoms of activation (at mild to moderate levels) (Fig. 2.33).

Inpatients. Inpatients with a diagnosis of bipolar disorder—depressive episode ($n = 74$)—also showed a SVARAD mean profile with a 2-1-5 code type, with moderate to severe Sadness/Demoralisation (2.5), moderate Apprehension/Fear (2), and moderate Apathy (1.8), while Impulsivity that was just above mild values (1.3) (Fig. 2.34). Impulsivity values at or above 2 were seen in 45.2% of the group. One patient out of five had a severe or extreme degree of Impulsivity (Fig. 2.35).

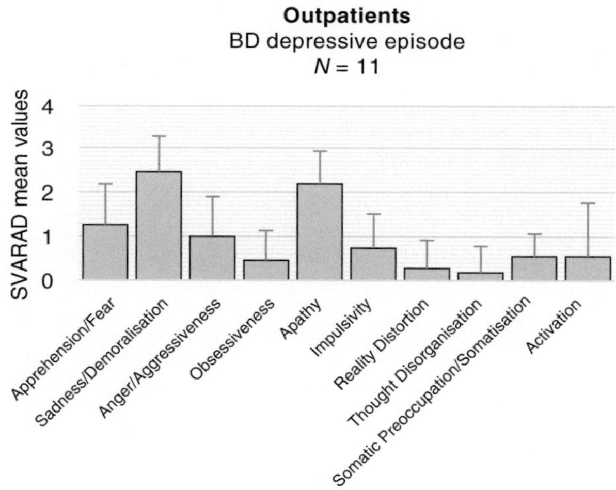

Fig. 2.33 SVARAD profile of outpatients with bipolar disorder-depressive episode: mean scores and standard deviations. Code type: 2-5-1 ($n = 11$)

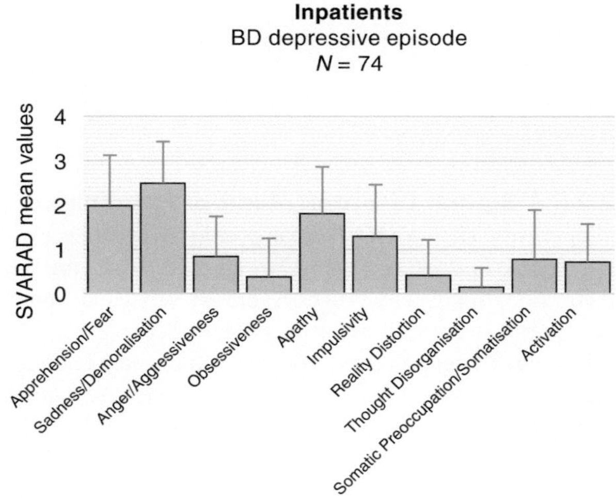

Fig. 2.34 SVARAD profile of inpatients with bipolar disorder-depressive episode: mean values and standard deviations. Code type: 2-1-5 ($n = 74$)

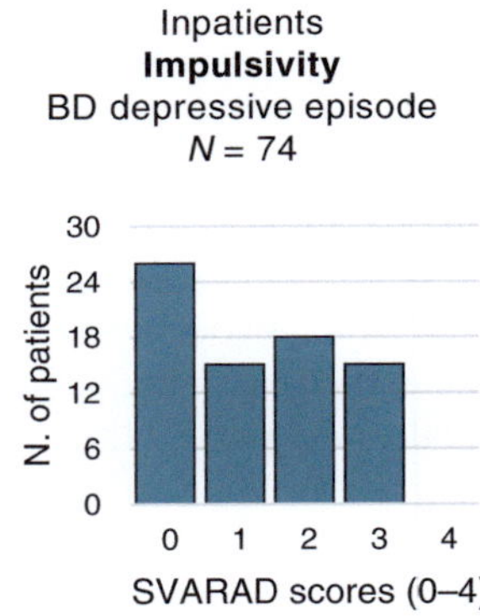

Fig. 2.35 Number of inpatients scoring 0-1-2-3-4 on the SVARAD Impulsivity dimension divided by diagnostic category: bipolar disorder-depressive episode

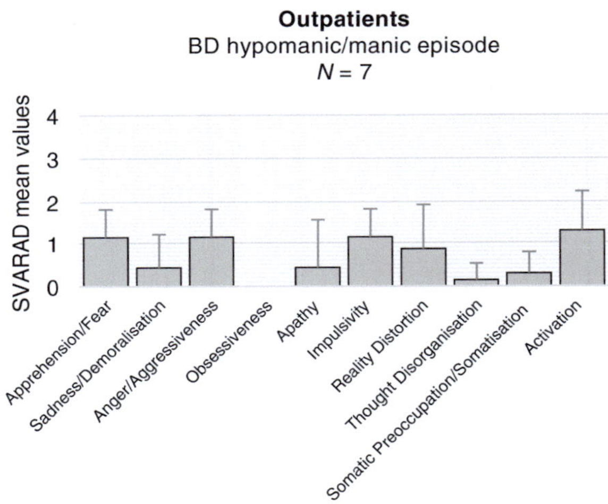

Fig. 2.36 SVARAD profile of outpatients with bipolar disorder-hypomanic/manic episode: mean scores and standard deviations. Code type: 10-1-3-6 ($n = 7$)

These findings have led to considerations about the risks of antidepressant treatment use in this group. Treatment should be very cautious and should always to be done in association with mood stabilisers, as previously discussed (see the major depression section).

2.2.3.4 Bipolar Disorder: Manic or Hypomanic Episode

Outpatients. The SVARAD mean profile of outpatients with bipolar disorder—manic or hypomanic episode ($n = 7$)—showed a 10-6-1-3 code type, with the highest peaks in Activation, Impulsivity, Anger/Aggressiveness, and Apprehension/Fear. However, these components were still only at the mild level (Fig. 2.36). More than one of the patients had mild Anger/Aggressiveness, and one out of three had moderate Anger/Aggressiveness, and the same results were seen for Impulsivity. Activation was moderate in 57% of the cases. Reality Distortion was mildly present in 43% of the sample (Fig. 2.37).

Inpatients. Inpatients with manic or hypomanic episodes ($n = 78$) displayed a 10-7-6 code type, with high mean peaks of moderate to severe Activation (2.5),

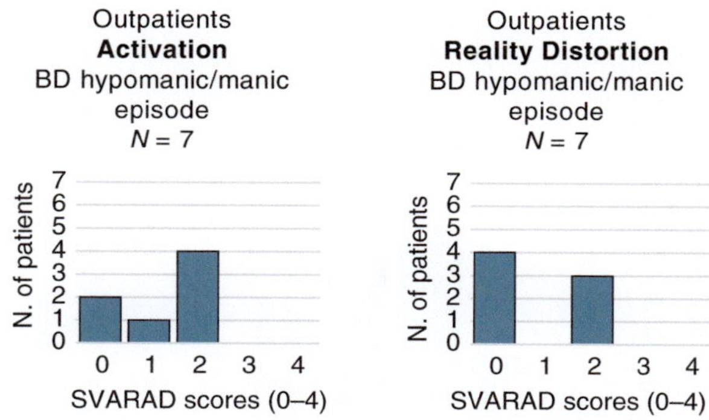

Fig. 2.37 Number of outpatients scoring 0-1-2-3-4 on the SVARAD Activation and Reality Distortion dimensions divided by diagnostic category: bipolar disorder-hypomanic/manic episode

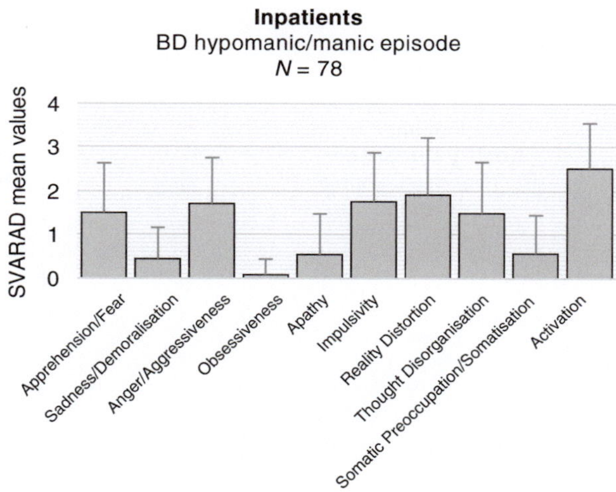

Fig. 2.38 SVARAD profile of inpatients with bipolar disorder-hypomanic/manic episode: mean values and standard deviations. Code type: 10-7-6 (n = 78)

Reality Distortion (1.9), and Impulsivity (1.8). The mean Anger/Aggressiveness score is also high (1.7) (Fig. 2.38). Analysis of the frequency of the SVARAD scores highlights a very critical and psychopathologically severe group: 19.2% had extreme and 30.8% severe Activation; 60.2% had an Impulsivity score above 2; 57.7% had a Reality Distortion score above 2; and 56.4% of the group had moderate to extreme Anger/Aggressiveness (Fig. 2.39).

These descriptive SVARAD findings could suggest the use of classical mood stabilisers (including lithium) for both groups of patients, while atypical antipsychotics could be specified for those presenting with Reality Distortion (which appeared to a mild degree in outpatients, where the relatively low values could be explained by the outpatient setting of the service, which excludes severe or very acute patients).

2.2.3.5 Bipolar Disorder: Mixed Episode

The mixed episode group is made up solely of inpatients (n = 34) and presents a 1-2-6 code type, with the Apprehension/Fear dimension as the primary one (2.2), followed by Sadness/Demoralisation (1.8) and Impulsivity (1.65). Patients also

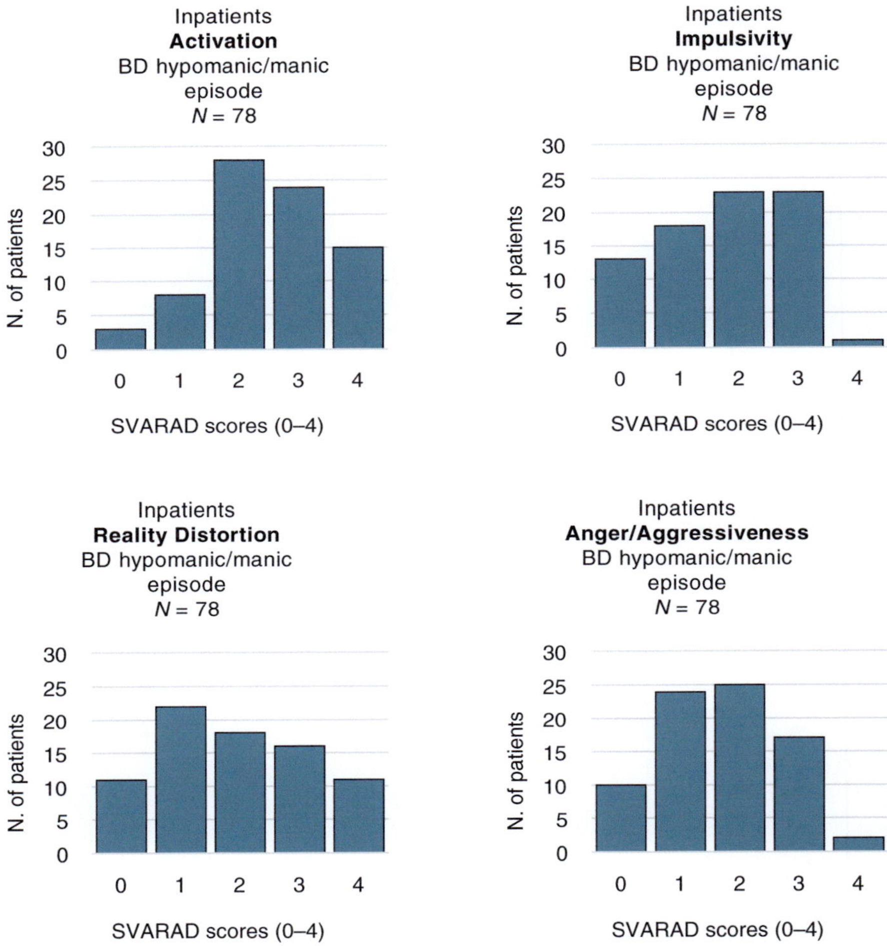

Fig. 2.39 Number of inpatients scoring 0-1-2-3-4 on the SVARAD Activation, Impulsivity, Reality Distortion, and Anger/Aggressiveness dimensions divided by diagnostic category: bipolar disorder-hypomanic/manic episode

presented mean values that fell between mild and moderate for Activation (1.56) (Fig. 2.40). Forty-seven percent had a moderate score for Impulsivity, while 38.2% had moderate, and about 15% had severe to extreme scores on the SVARAD. For Anger/Aggressiveness, 20.6% of this group had moderate, and 20.6% had severe scores (Fig. 2.41).

These findings seem to better characterise the clinical picture of the typical patient with a mixed episode, which might be interpreted solely as anxiousness and depression, when in fact the patient is in a mixed suffering episode, with different components of feelings. From a therapeutic perspective, antidepressants are clearly contraindicated in the majority of cases, while mood stabilisers are the first choice.

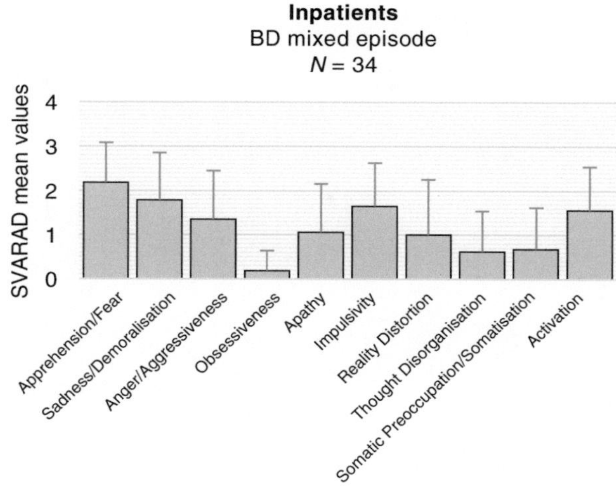

Fig. 2.40 SVARAD profile of inpatients with bipolar disorder-mixed episode: mean values and standard deviations. Code type: 1-2-6 ($n = 34$)

Reality distortion is mild in 14.7%, moderate in 14.7%, and severe in another 14.7% of patients, supporting the appropriateness of atypical antipsychotics in medium-low dosages (Fig. 2.41).

2.2.3.6 Panic Disorder, Generalised Anxiety Disorder, and Anxiety Disorder NOS

The anxiety disorder group consisted solely of outpatients. The panic disorder group ($n = 92$) presented a SVARAD mean profile with a 1-9-2 code type, with high mean values for Apprehension/Fear (2.2) and Somatic Preoccupation/Somatisation (1.2), followed by Sadness/Demoralisation (1.1). The generalised anxiety disorder group ($n = 11$) showed a similar SVARAD profile with a 1-9-2 code type, with the highest mean values for Apprehension/Fear (2.3) and Somatic Preoccupation/Somatisation (1), followed by Sadness/Demoralisation (0.9). Interestingly, this group reports a mean score of 0.8 for Anger/Aggressiveness, a value significantly higher than that of panic disorder patients (Fig. 2.42). The anxiety disorder NOS group ($n = 51$) showed a SVARAD profile with a 1-2-9 code type, with Somatic Preoccupation/Somatisation as the third peak and, interestingly, Anger/Aggressiveness as the fourth peak of the dimensional profile, with 43% of the patients scoring mild and 7% moderate (Figs. 2.42 and 2.43). This profile resembles the one in the generalised anxiety disorder group, rather than the panic disorder group. This finding could be of clinical interest for drug treatment, because standard recommended SSRIs or noradrenalin reuptake inhibitors (NARIs) not only might not resolve it but could perhaps induce a worsening of irritability and an increase in symptoms of activation.

As concerns the "multiparametric" analysis of the DSM-IV anxiety groups, quite apart from the expected Apprehension/Fear and Somatic Preoccupation/Somatisation peaks, the SVARAD presented a significant component of Sadness/Demoralisation for these groups. This finding certainly raises the old issue of the coexistence of anxiety and depression. It is interesting to underline that the shape of the dimensional

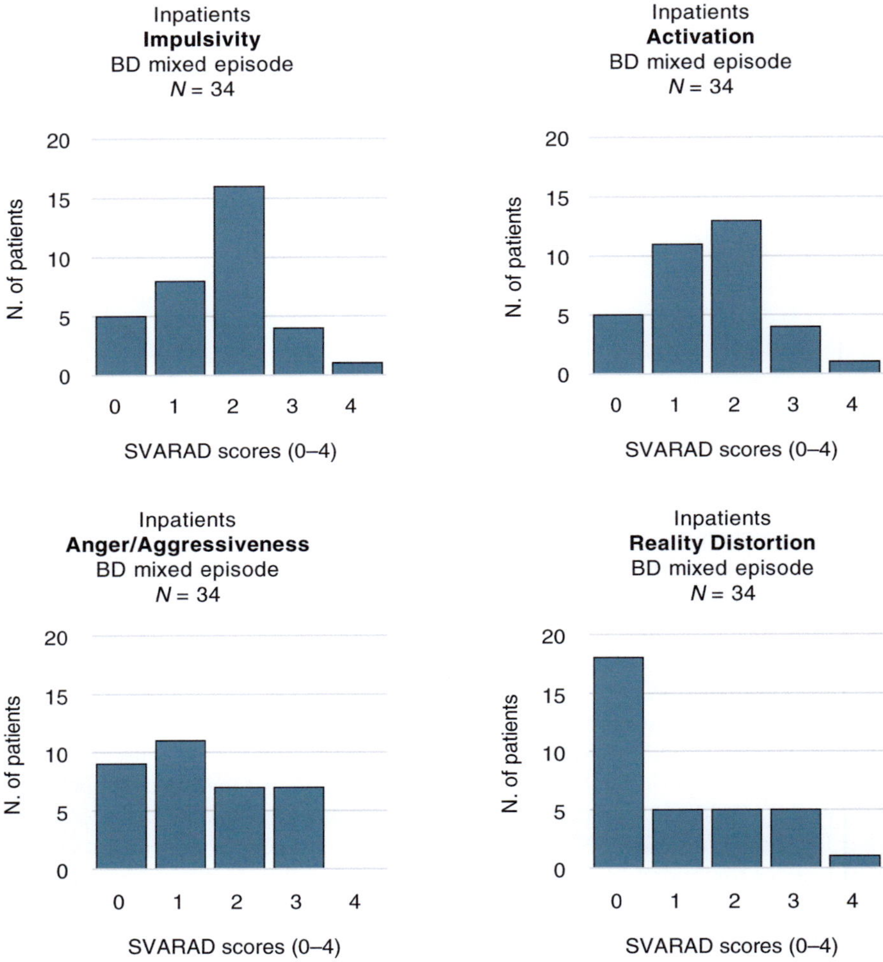

Fig. 2.41 Number of inpatients scoring 0-1-2-3-4 on the SVARAD Impulsivity, Activation, Anger/Aggressiveness, and Reality Distortion dimensions divided by diagnostic category: bipolar disorder-mixed episode

profile, with a significant peak for Sadness/Demoralisation, does not change in a further analysis excluding all the cases with comorbidity: that is, the patients without any axis 1 comorbidity also have a significant Sadness/Demoralisation component. We can suppose that these cases are *subthreshold* ones for an associated mood disorder diagnosis or, better, *that there is a common, natural coexistence of anxiety and depression that the categorical diagnosis does not permit the recognition of*, if symptoms do not fulfil predetermined criteria for severity, length, and number. However, the Sadness/Demoralisation suffering for the patient does exist, and in these cases where it is present, it needs to be treated. The possible coexistence of anxiety and

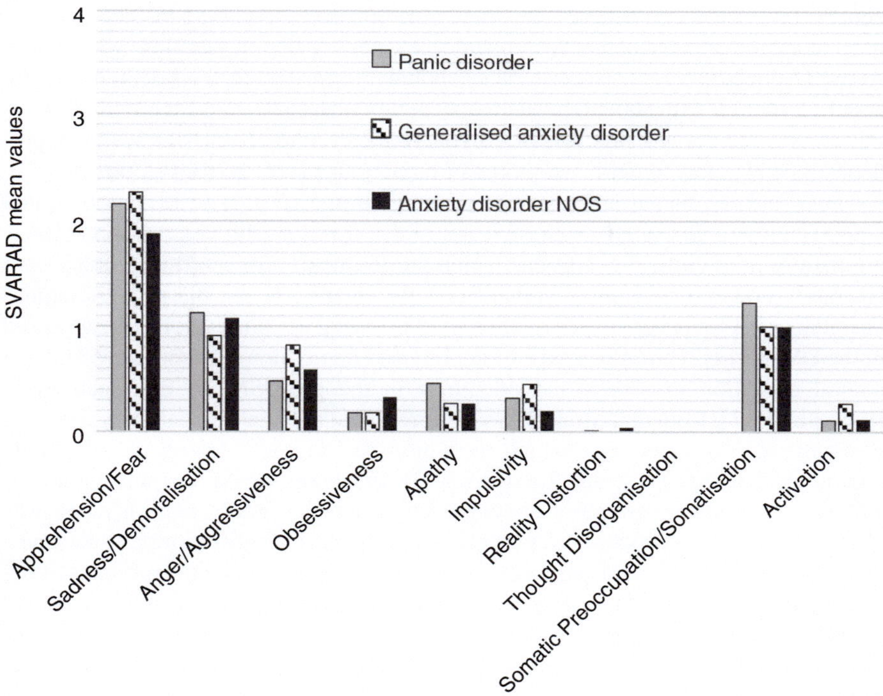

Fig. 2.42 SVARAD profile of outpatients with panic disorder ($n = 92$), generalised anxiety disorder ($n = 11$) and anxiety disorder NOS ($n = 51$): mean scores and standard deviations

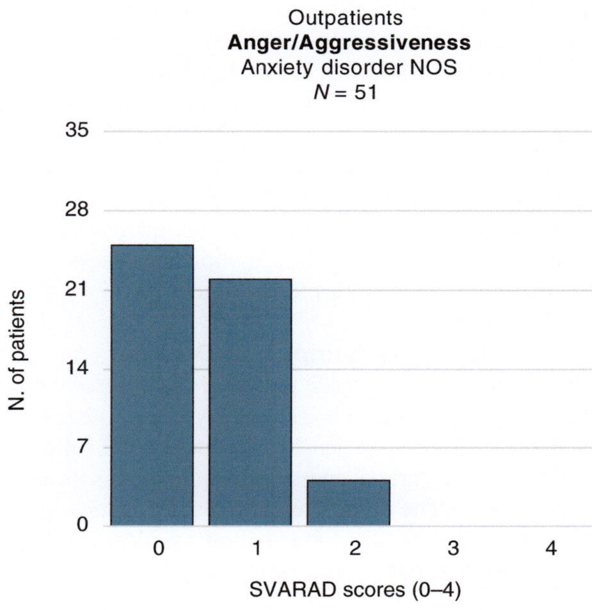

Fig. 2.43 Number of outpatients scoring 0-1-2-3-4 on the SVARAD Anger/Aggressiveness dimension divided by diagnostic category: anxiety disorder NOS

depression in the same patient was already recognised in the DSM-IV-TR as a category of mixed anxiety-depressive disorder, a category that at that time was "under study" (Appendix B) [9]. The criteria set included persistent dysphoric mood lasting at least 1 month, with some typical anxiety symptoms, such as: difficulty in concentrating or mind going blank, worries, hypervigilance, anticipating the worst, apprehension, angst, and somatic symptoms of anxiety or panic. Moreover, this putative category and the mixed anxiety-depressive syndrome (F41.2) of the International Classification of Diseases—10th Revision of the World Health Organization [10]—are mainly mild clinical conditions which are frequently observed in primary care but rarely undergo psychiatric consultation for whether or not the patients require hospitalisation for severe psychopathology. This category, however, has disappeared in the DSM-5 [4]. The coexistence of anxiety and depression was considered to be a natural condition in several classifications in the pre-DSM III era, such as in Kielholtz's [11], Roth and Mountjoy's (they reserved a diagnostic category for *anxiety/depression*, a mixed state which recognises that the two clusters of complaints commonly coexist) [12], and Langen's [13]. Furthermore, Maser and Cloninger edited a systematic investigation about anxiety and depression comorbidity, suggesting that although psychopathology involves a complex array of comorbid syndromes, this comorbidity has a stable structure [7]. Johansson investigated the presence of the comorbidity of anxiety and depression symptoms in a sample of 3001 randomly selected Swedish adults; he found that among participants with either clinically significant depression or anxiety, nearly 50% had comorbid disorders [14]. Furthermore, subthreshold issues about anxiety and depression have been repeatedly raised in the literature [15]. Even if a preponderance of data suggests that in clinical practice anxiety and depression often coexist, the DSM-5 and ICD-10 diagnostic classifications nevertheless define clear criteria sets for both major depression and GAD or panic disorder. The classifications are made according to hierarchical (between depression and anxiety) and mutually exclusive criteria, leading to clear-cut differentiation that is useful for epidemiological and statistical use [4, 10]. If anxiety and depression are both present in a given patient, the diagnostic categories are poorly suited to represent this phenomenon, unless both the GAD and MD criteria are fulfilled and there is recognised comorbidity. The SVARAD dimensional approach might be useful in giving a finer recognition of the coexistence of anxiety and depression in depressive and anxiety disorders. The dimensional recognition of an important anxiety component in MDD patients and of a significant sadness component in GAD patients thus has important implications for treatment.

2.2.3.7 Obsessive-Compulsive Disorder

Outpatients. The SVARAD profile of the obsessive-compulsive disorder patients in the outpatient group ($n = 31$) displayed a 4-1-2 code type, with an Obsessiveness peak at 2.7 and an Apprehension/Fear peak at 2.2, followed by Sadness/Demoralisation at 1.5. Anger/Aggressiveness and Apathy both showed a value of 0.8 (Fig. 2.44).

Inpatients. The obsessive-compulsive inpatient group ($n = 8$) showed a similar 4-1-2 code type, with an Obsessiveness peak at 2.9, Apprehension/Fear and Sadness/Demoralisation both at 2.2, and Apathy as the fourth peak at 1.3 (Fig. 2.45).

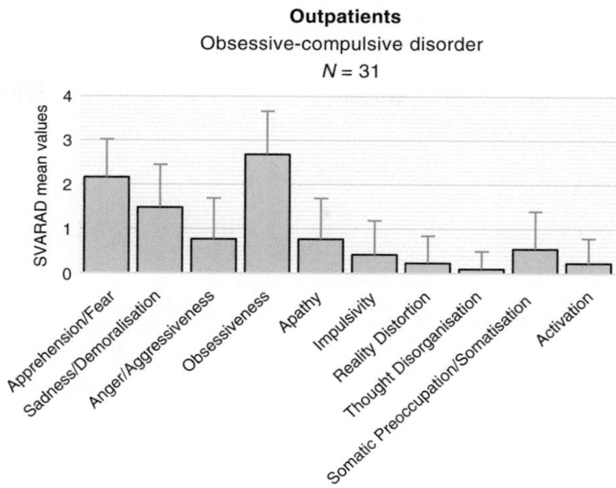

Fig. 2.44 SVARAD profile of outpatients with obsessive-compulsive disorder: mean scores and standard deviations. Code type: 4-1-2 ($n = 31$)

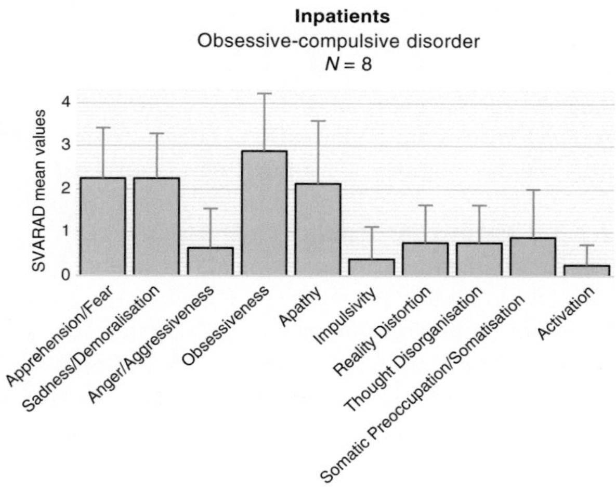

Fig. 2.45 SVARAD profile of inpatients with obsessive-compulsive disorder: mean values and standard deviations. Code type: 4-1-2 ($n = 8$)

Anxiety and depressive components thus seem to play a significant role in obsessive-compulsive suffering. The SVARAD profile, however, reveals clinical differences in individual patients that are potentially relevant for treatment: a subgroup of patients have moderate to extreme anger feelings, moderate impulsivity, and some activation as well. They are not suitable candidates for a high dosage of antidepressant 5HT drugs, particularly if they are "resistant" to a progressive increase of dosages, as guidelines for obsessive-compulsive disorder treatment impartially suggest. It could be better in these cases to supplement with D2/5HT2 drugs (such as risperidone, aripiprazole, and others). Valproic acid or gabapentin could be added to address this "activated" component. The subgroup in which apprehension, fear, apathy, and somatic preoccupations predominate seems a better candidate for 5HT antidepressant drugs, and for higher dosages of these drugs,

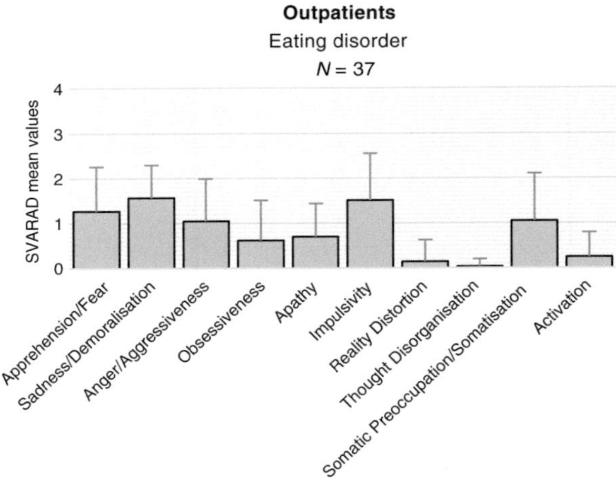

Fig. 2.46 SVARAD profile of outpatients with eating disorder: mean scores and standard deviations. Code type: 2-6-1 ($n = 37$)

together with benzodiazepines. Finally, the identification of different dimension profiles in these types of patients could be useful in detecting specific obsession subtypes. For example, obsessive-compulsive patients with a predominance of Anger/Aggressiveness may express aggressive and sexual obsessions, whereas patients with more Sadness/Demoralisation and Apprehension/Fear components may express religious and somatic obsessions [16].

2.2.3.8 Eating Disorders

The eating disorders group was made up solely of outpatients ($n = 37$). This group showed a mean SVARAD profile with a 2-6-1 code type, with Sadness/Demoralisation and Impulsivity at similar levels (both scored around 1.5), Apprehension/Fear as a third peak (1.3), and Somatic Preoccupation/Somatisation near 1. The profile appears to be significantly characterised by Impulsivity (Fig. 2.46). About 60% of eating disorder outpatients had an Impulsivity score from moderate to extreme, as well as a Sadness/Demoralisation score at the same approximate level. Anger/Aggressiveness, together with Apprehension/Fear, also characterised the sample, with 40% of patients ranging from 2 to 4 (Fig. 2.47).

Overall, this group appears to have a multifaceted psychopathological profile, with a mix of depression, anger, and anxiety. A deeper analysis reveals the lack of a single, unitary pattern. There are different subgroups of this group: one dominated by impulsivity/anger, another by anxiety/depression/impulsivity, and another by extreme impulsivity. This SVARAD distinction could be of significant interest for treatment, both for choosing medication and for establishing the psychotherapeutic alliance and defining treatment targets.

2.2.3.9 Somatic Symptom Disorder

The somatic symptom disorder group was also solely composed of outpatients ($n = 31$). It showed a SVARAD mean profile with a 9-1-2 code type (Somatic Preoccupation/Somatisation = 2.1; Apprehension/Fear = 1.6), with Sadness/

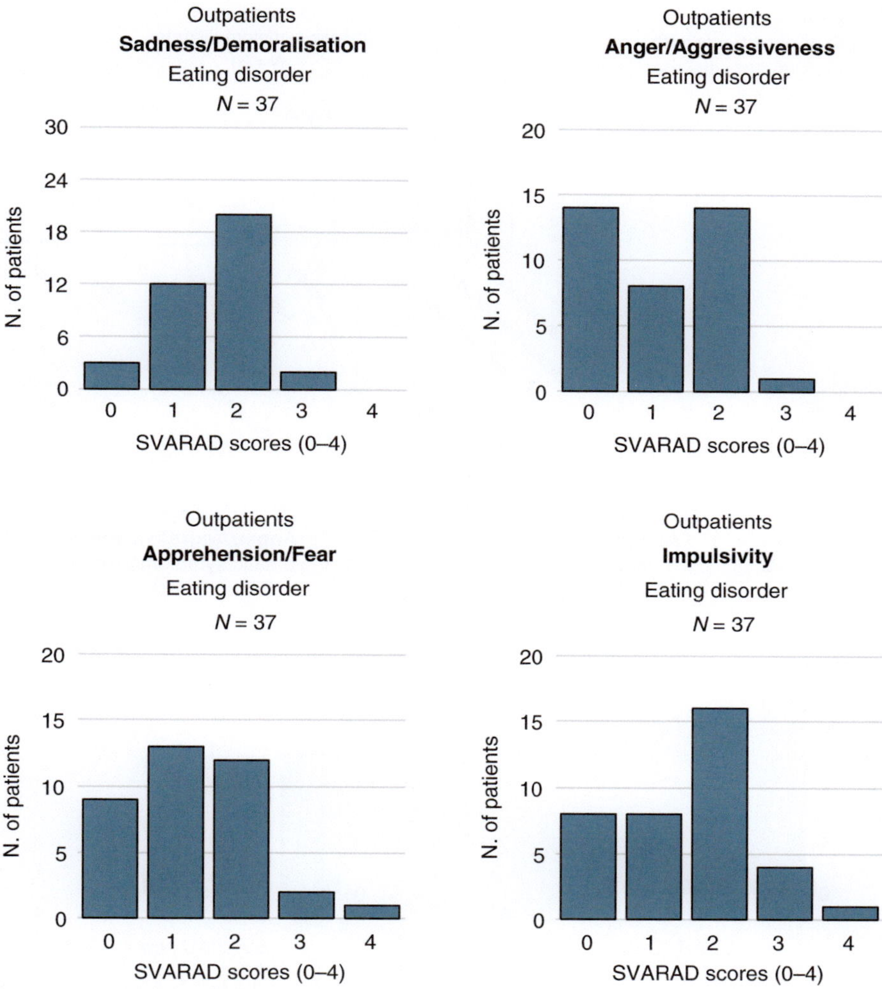

Fig. 2.47 Number of outpatients scoring 0-1-2-3-4 on the SVARAD Sadness/Demoralisation, Anger/Aggressiveness, Apprehension/Fear, and Impulsivity dimensions divided by diagnostic category: eating disorder

Demoralisation (1) as the third component of suffering (Fig. 2.48). One out of three patients showed moderate to mild Apathy, with common mild (13%) or moderate (19%) anger feelings (Fig. 2.49).

This diagnostic group seems to have a multifaceted psychopathological profile rather than a unified one. As an in-depth tool, the Diagnostic Criteria for Psychosomatic Research (DCPR), developed by Fava and colleagues, may be a useful instrument to further investigate the somatic spectrum and to personalise treatments according to specific symptom manifestation [17].

Fig. 2.48 SVARAD profile of outpatients with somatic symptom disorder: mean scores and standard deviations. Code type: 9-1-2 ($n = 31$)

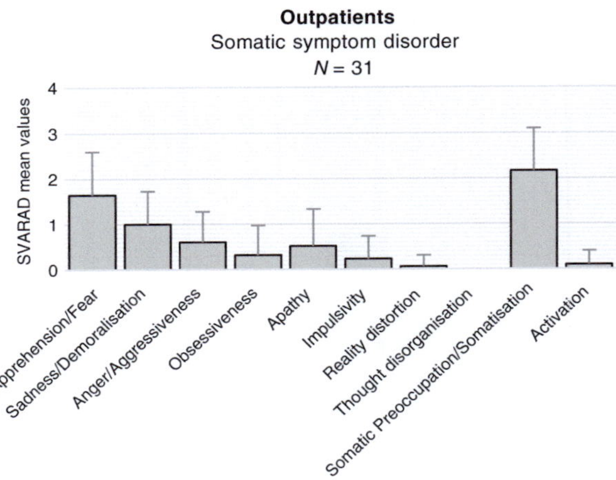

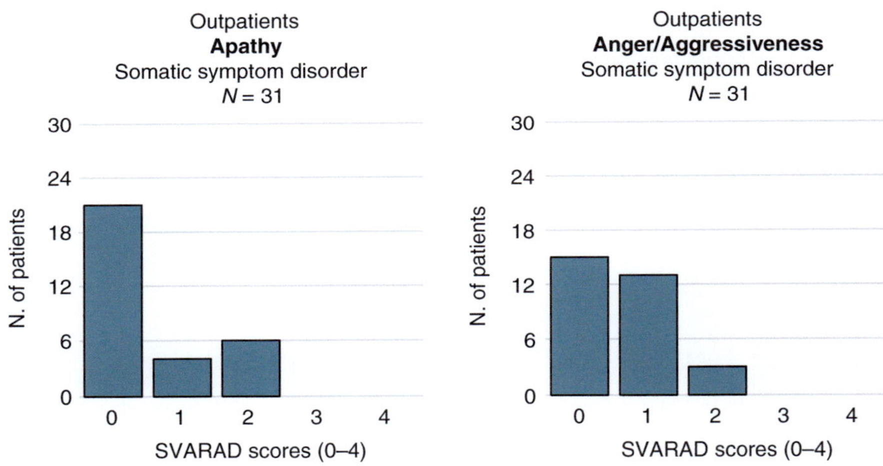

Fig. 2.49 Number of outpatients scoring 0-1-2-3-4 on the SVARAD Apathy and Anger/Aggressiveness dimensions divided by diagnostic category: somatic symptom disorder

2.2.3.10 Schizophrenia

Outpatients. The schizophrenia outpatient group (n = 31) displayed a SVARAD profile with a 7-5-1 code type (Reality Distortion = 2.3; Apathy = 1.9; Apprehension/Fear = 1.7). Mild scores of Sadness/Demoralisation, Anger/Aggressiveness, and Impulsivity also characterised this outpatient group (Fig. 2.50). A subgroup (35%) of patients presented with moderate to severe feelings of anger, while almost one out of three had severe to extreme fear feelings (Fig. 2.51).

Inpatients. The mean profile of the schizophrenia inpatient group (n = 82) showed a 7-1-8 code type with high Reality Distortion (2.9) and Apprehension/Fear (1.9) and the appearance of Thought Disorganisation with a SVARAD mean

Fig. 2.50 SVARAD profile of outpatients with schizophrenia: mean scores and standard deviations. Code type: 7-5-1 ($n = 31$)

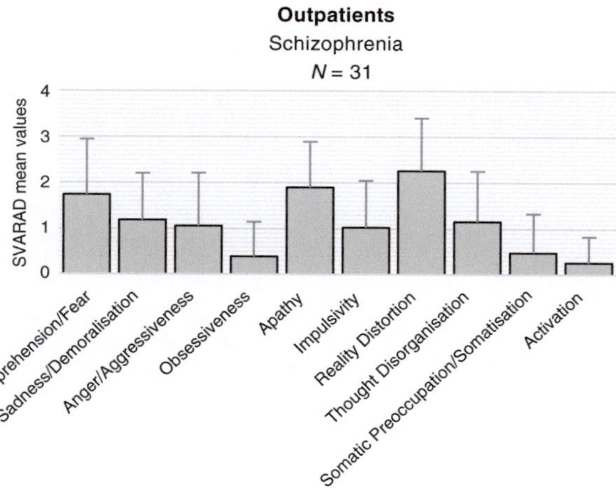

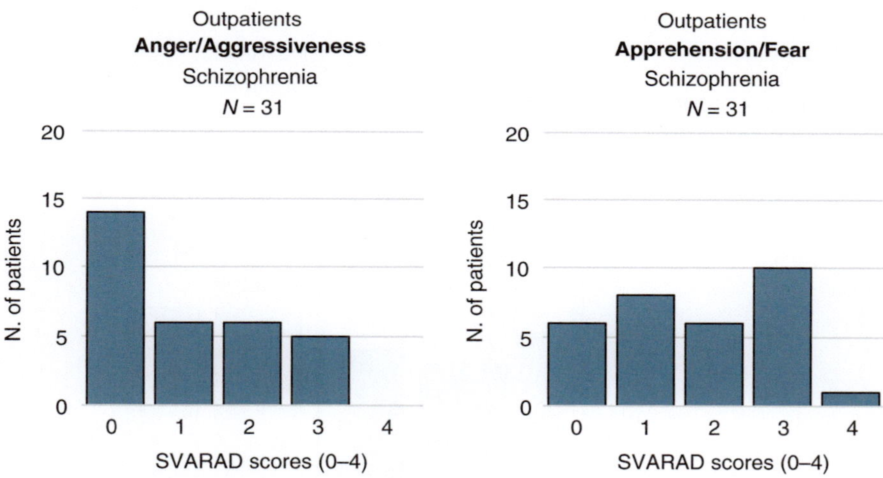

Fig. 2.51 Number of outpatients scoring 0-1-2-3-4 on the SVARAD Anger/Aggressiveness and Apprehension/Fear dimensions divided by diagnostic category: schizophrenia

of 1.9. Apathy was the fourth main component (1.8) (Fig. 2.52). The analysis of frequencies showed a subgroup with mild or no Apathy (19.5% scoring 0 and 20.7% scoring 1), while a larger group (61%) scored 2 or above (Fig. 2.53).

Reality Distortion and Apathy characterised the typical schizophrenic outpatient and inpatient but with a multifaceted psychopathology that the standard DSM-IV criteria probably do not fully represent. Apathy and sadness—which probably represent the main "negative" symptoms—were present in about half of the sample. The SVARAD descriptive analysis seemed to suggest two main subtypes of the schizophrenia group, which might be an important consideration for treatment. The typical or atypical D2 blockers are the first-line drugs for the first subgroup (high in SVARAD Reality Distortion), but anticonvulsants could be added for targeting

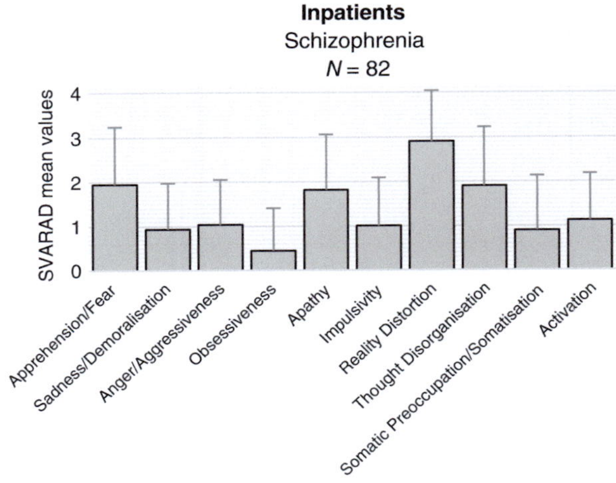

Fig. 2.52 SVARAD profile of inpatients with schizophrenia: mean values and standard deviations. Code type: 7-1-8 ($n = 82$)

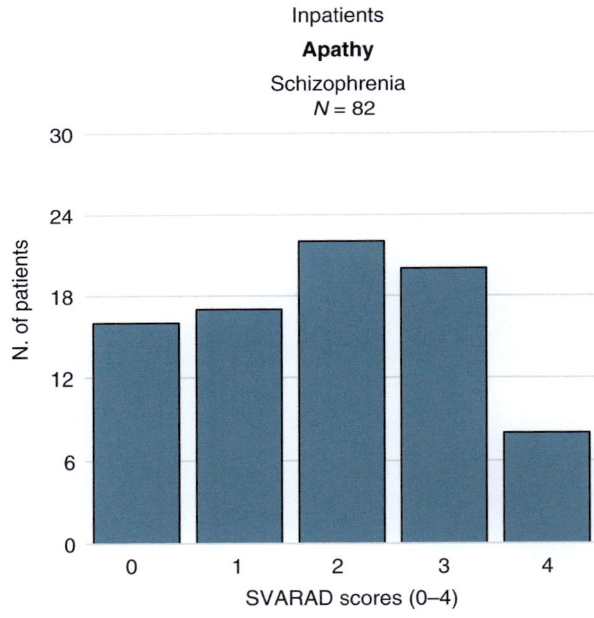

Fig. 2.53 Number of inpatients scoring 0-1-2-3-4 on the SVARAD Apathy dimension divided by diagnostic category: schizophrenia

Anger/Aggressiveness and Impulsivity in more severe cases. On the other hand, typical D2 blockers (and to some extent the atypical D2/5H2 drugs) might not be indicated for patients low in Reality Distortion and with prominent negative symptoms (high Apathy and Sadness/Demoralisation). An alternative choice might be the prescription of low-dose antidepressants, while the role of benzodiazepines might be considered for patients high in Apprehension/Fear. Especially with regard to treatment choice, the psychopathological dimensions display much greater discriminatory power between schizophrenia groups, as has been shown in a previous study by our research group [18].

2.2.3.11 Schizoaffective Disorder

Schizoaffective disorder inpatients ($n = 53$) showed a 7-1-8 code type, with Reality Distortion (2.4) as the first psychopathological dimension, followed by Apprehension/Fear (1.8), Thought Disorganisation (1.4), and Apathy (1.3) (Fig. 2.54). Although the same code type, the profile differs from the one of the schizophrenia group.

2.2.3.12 Delusional Disorder

Similar to other psychotic groups, the outpatient group with delusional disorder ($n = 12$) had a 7-1-2 code type, with Anger/Aggressiveness as the fourth relevant component (1.5) and Impulsivity (1.2) as the fifth (Fig. 2.55). The mean profile painted the typical patient as a delirious, fearful, potentially aggressive, and irritable

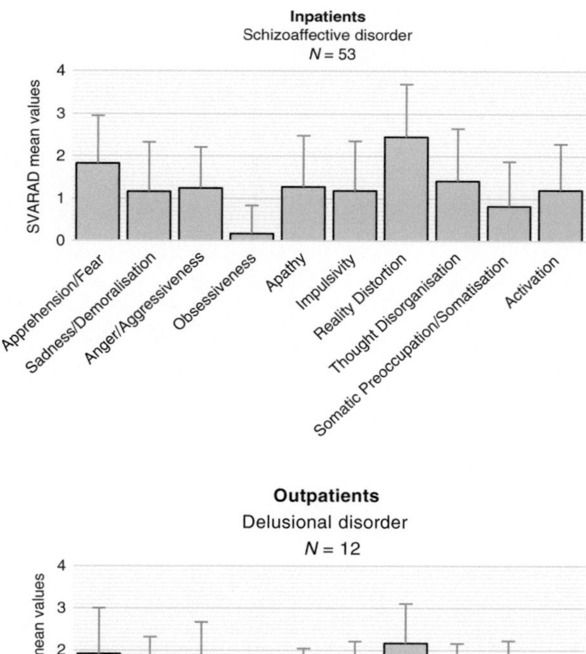

Fig. 2.54 SVARAD profile of inpatients with schizoaffective disorder: mean values and standard deviations. Code type: 7-1-8 ($n = 53$)

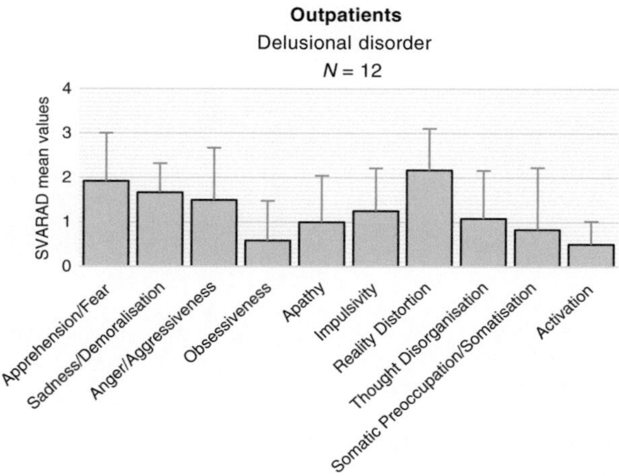

Fig. 2.55 SVARAD profile of outpatients with delusional disorder: mean scores and standard deviations. Code type: 7-1-2 ($n = 12$)

subject, prone to some impulsive reactions. The small size of this sample does not permit any further analysis.

2.2.3.13 Psychotic Disorder NOS

Outpatients. The psychotic NOS outpatient group included 29 subjects and showed a 1-7-2 code type (Apprehension/Fear = 1.9; Reality Distortion = 1.8; Sadness/Demoralisation = 1.5), suggesting that the emotional components of fear, threat, and anxiety dominate the clinical picture, accompanying delusional thoughts. Apathy (1.2) is more relevant than Anger/Aggressiveness (1.1) or Impulsivity (1), which were both lower than in schizophrenia (Fig. 2.56).

Inpatients. The sample of inpatients with psychotic disorder NOS (n = 213) showed a 7-1-6 code type, with peaks in Reality Distortion (2.4), Apprehension/Fear (2.1), and Apathy (1.4) (Fig. 2.57). The mean profile was more similar to the schizoaffective profile than to the schizophrenic one, with lower Reality Distortion, Apathy, and Thought Disorganisation (Fig. 2.58).

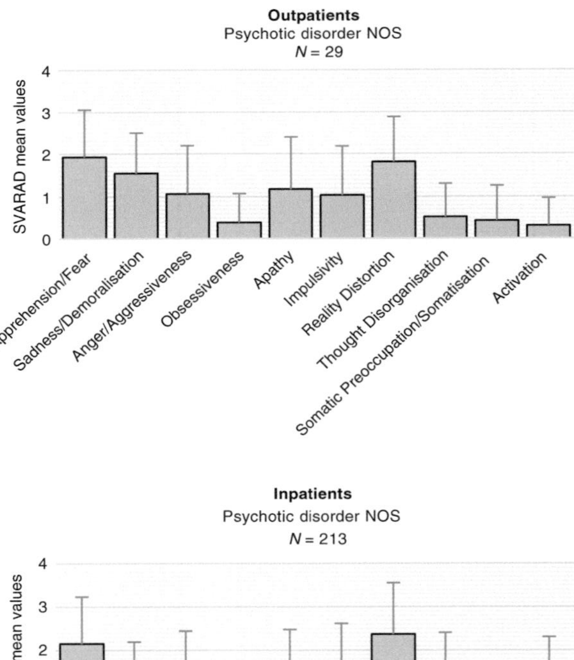

Fig. 2.56 SVARAD profile of outpatients with psychotic disorder NOS: mean scores and standard deviations. Code type: 1-7-2 (n = 29)

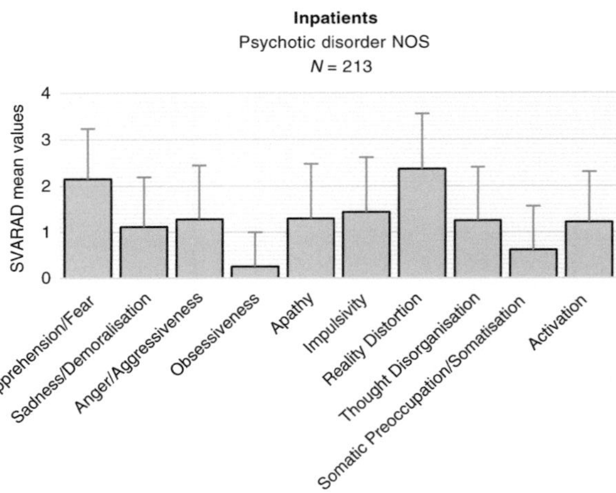

Fig. 2.57 SVARAD profile of inpatients with psychotic disorder NOS: mean values and standard deviations. Code type: 7-1-6 (n = 213)

2.2.3.14 Borderline Personality Disorder

Outpatients. The borderline personality disorder outpatient group ($n = 30$) showed a 3-6-1 code type, with high Anger/Aggressiveness (1.9), Impulsivity (1.8), and Apprehension/Fear (1.7) scores. Also, the Sadness/Demoralisation dimension was consistently high (1.6) (Fig. 2.59). This group presented a multifaceted psychopathological profile, with relevant within-group differences: 29% of patients showed severe Anger/Aggressiveness, and 42% showed moderate levels of this dimension; 23% presented with severe and 42% with moderate Impulsivity; 52% showed moderate Sadness/Demoralisation, suggesting significant depressive symptoms; and 16% showed mild or moderate Reality Distortion, suggesting possible hallucinations, paranoid ideation, psychoticism, and delusional thoughts (Fig. 2.60).

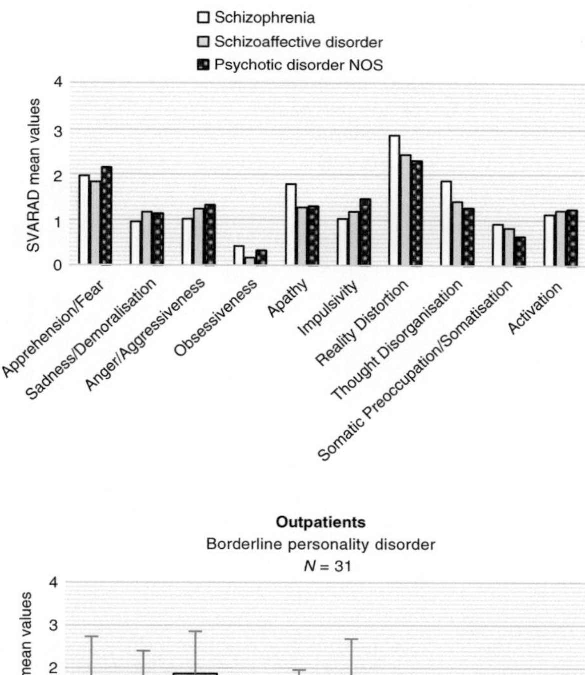

Fig. 2.58 SVARAD profile of inpatients with schizophrenia ($n = 85$), schizoaffective disorder ($n = 53$), psychotic disorder NOS ($n = 226$)

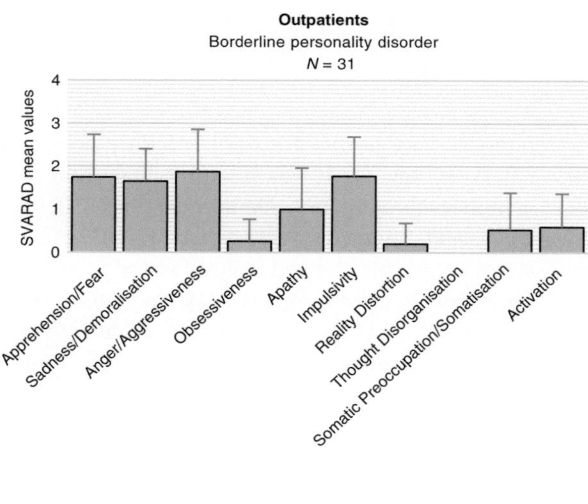

Fig. 2.59 SVARAD profile of outpatients with borderline personality disorder: mean scores and standard deviations. Code type: 3-6-1 ($n = 31$)

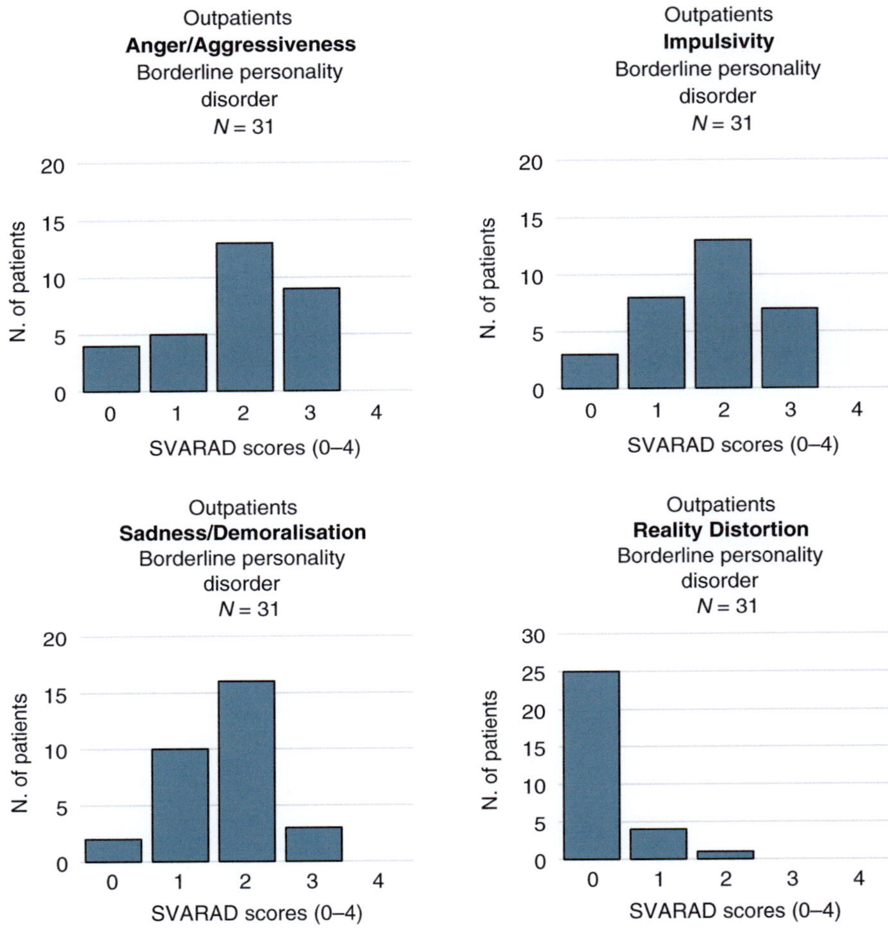

Fig. 2.60 Number of outpatients scoring 0-1-2-3-4 on the SVARAD Anger/Aggressiveness, Impulsivity, Sadness/Demoralisation, and Reality Distortion dimensions divided by diagnostic category: borderline personality disorder

Inpatients. The profile for the borderline personality disorder inpatient group ($n = 22$), code type 2-6-1, was similar to that of the outpatient group, with the presence of high levels of Sadness/Demoralisation (1.9), Impulsivity (1.9), Apprehension/Fear (1.9), and Anger/Aggressiveness (1.4) (Fig. 2.61). The profile of borderline personality inpatients showed a multifaceted mixed emotional profile characterised by impulsiveness, depression, anxiety, and anger. With respect to the frequency of these dimensions, 77.3% showed high Impulsivity; 50% and 27.3% of patients, respectively, showed moderate (score 2) or severe (score 3) Sadness/Demoralisation;

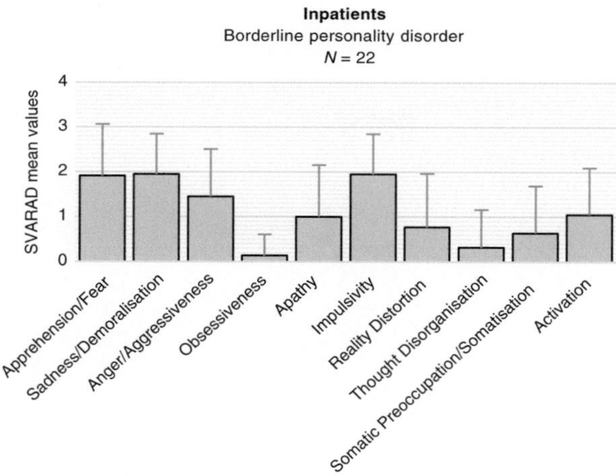

Fig. 2.61 SVARAD profile of inpatients with borderline personality disorder: mean values and standard deviations. Code type: 2-6-1 ($N = 22$)

50% showed moderate or severe Anger/Aggressiveness; 63.6% showed moderate, severe, or extreme Apprehension/Fear; and 22.7% showed a score of 2 or 3 in Reality Distortion (Fig. 2.62).

These different psychopathological dimensions suggest the necessity for carefully designing a personalised drug treatment regimen for borderline personality patients, to target the different prominent psychopathological components. For instance, DSM-IV criteria for borderline personality disorder do not describe depressive symptoms as a criterion for inclusion. However, the mean value of 1.9 in our inpatient sample suggested that a large subgroup was suffering from depression, with score values over 2. This led to the recognition of the importance of depressive symptoms in these patients and the need to specifically treat them with antidepressants. Meta-analyses and reviews discuss psychopharmacological treatment of borderline personality disorder, suggesting mood stabilisers, antidepressants, and atypical antipsychotics, often with conflicting results from controlled trials, which serves to bring about conflicting recommendations [19]. Often, suggestions are presented on the basis of *mean* therapeutic responses to different drugs for borderline personality disorder, but much of the research in this area does not distinguish *which* psychopathological components improve in *which* patients. These inconsistent views in the literature might well be explained by the *variance* of psychopathological dimensional components within a sample of the same diagnostic category, as the SVARAD findings suggest. An antidepressant treatment could be needed in borderline personality patients with a Sadness/Demoralisation mean score near 2, whereas that same treatment could be detrimental in patients with low Sadness/Demoralisation mean scores (under 1), who also had high scores in Activation, Anger/Aggressiveness, or Reality Distortion.

In summary, borderline personality disorder is an example of how SVARAD-based precision in diagnosis can provide practical guidance to clinicians in terms of

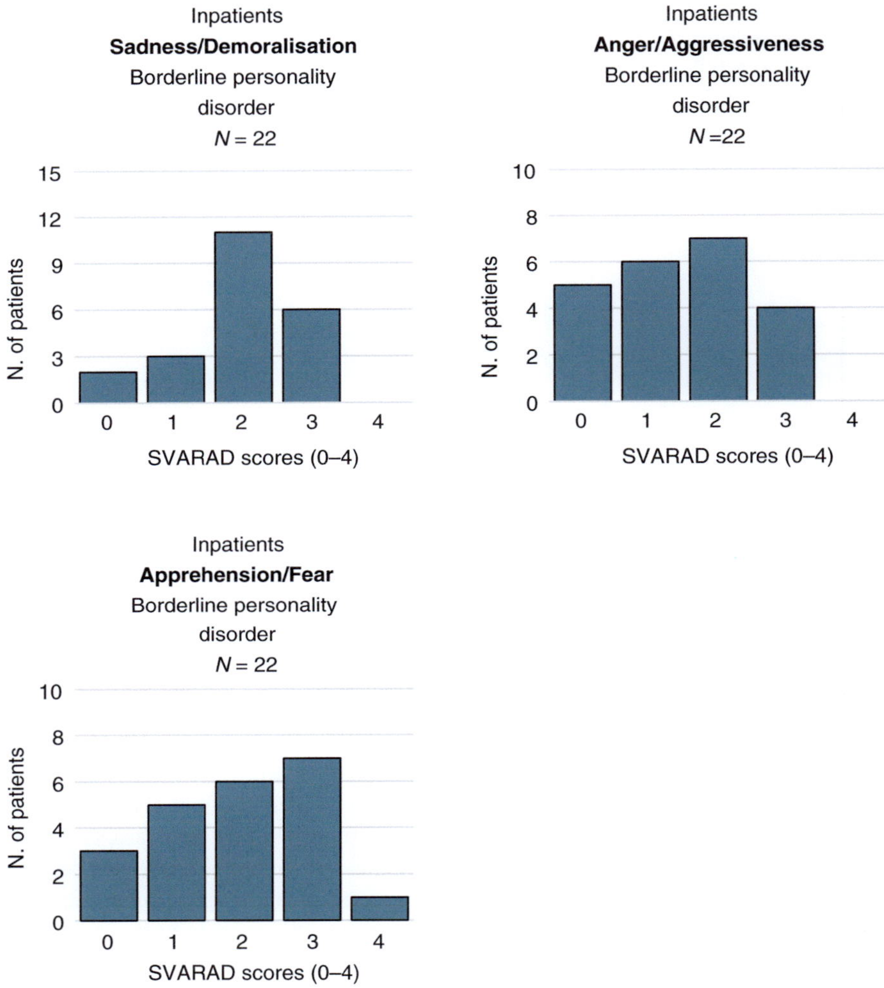

Fig. 2.62 Number of inpatients scoring 0-1-2-3-4 on the SVARAD Sadness/Demoralisation, Anger/Aggressiveness, and Apprehension/Fear dimensions divided by diagnostic category: borderline personality disorder

personalised treatment. High scores for Sadness/Demoralisation in borderline patients suggest the use of low-dose 5HT antidepressants; high scores for Anger/Aggressiveness and Impulsivity suggest the use of mood stabilisers; mild or moderate scores for Reality Distortion suggest the use of atypical antipsychotics. It can also serve as a warning to prevent harm; for example, high Impulsivity together with Anger/Aggressiveness and Sadness/Demoralisation might suggest a significant risk of self-harm behaviours and attempted suicide, so patients with these SVARAD parameters should be given the highest attention and careful treatment, with possible hospitalisation.

2.2.4 Contribution of SVARAD "Multiparametric" Profiles to the Treatment of Patients

As our findings show, several SVARAD dimensions are present in many diagnostic DSM-IV categories. A single diagnostic group often shows high variability in psychopathological characteristics, with subgroups that differ along SVARAD dimensions. This can be highly relevant for the understanding of cases and the optimisation of treatment.

Our findings suggest that DSM-IV diagnostic categories do respond to a standard criteria set, but they also differ according to some relevant dimensions that are not always represented in the criteria set. While this difference might not be relevant from a statistical and epidemiological viewpoint, it can make a substantial difference in terms of the therapeutic relationship and the tailoring of drug treatments to specific psychopathological components within a single diagnostic category.

For instance, antidepressant drugs are, as a class, all authorised for the treatment of major depressive disorder or generalised anxiety disorder. However, the variance within these two diagnostic categories might profoundly affect the response to an antidepressant, according to the dimensional profile of the patient [20]. Several molecules have been authorised for major depression or generalised anxiety disorder, leading to the misleading notion that *all* the cases of those disorders will be appropriately and correctly treated if they are given those approved molecules. Although psychotherapies do not have authorisation and labels, something similar to drug treatment probably happens. When a controlled study finds that a psychotherapy is better than no treatment or a waiting list condition, just the *mean* treatment effect for the group as a whole could be valuable. The nonresponders subgroup might have a dimensional profile of their psychopathological suffering that is not addressed by that psychotherapeutic intervention (for instance, anger, fear, sadness, somatisation, apathy, impulsivity, and so on). Taking into consideration the psychopathological dimensional profile of each patient might be useful for both the patient-therapist relationship and the choice of a specific psychotherapy, including its planning and combination with drug therapy. SVARAD assessment distinguishes among dimensional subgroups, allowing the clinician to better address the individual needs and problems of each patient while at the same time driving the clinician's attention to all the relevant dimensions in each case.

In some cases, a given drug may not produce the expected response for a given diagnostic category. Mean responses, effect sizes, and evidence of effectiveness for a drug or psychotherapeutic treatment vary according to several variables. One of these, the psychopathological dimensions of the individual case, could play a major role in the effectiveness of the approach. Variability of psychopathological dimensions, in other words, might explain variability in response to treatments.

SVARAD "multiparametric" analysis might be useful in composing a specific psychopathological profile for each patient, according to the different dimensions present in the patient. At present, anecdotal and unpublished data—but not controlled or systematic data—support this suggestion. Further research is needed on the topic, but we think that treatments will be more effective if they are planned using the patient's dimensional profile within a diagnostic category, in other words, *personalised* treatment. The same considerations apply to both drug and psychotherapeutic interventions; different techniques and therapeutic programmes may be shaped according to the different psychopathological components of each individual patient.

Conclusions

The findings of these studies give rise to several points of discussion.

First, the findings from our inpatient and outpatient studies confirm the feasibility and usefulness of the SVARAD. Dimensional assessment using the SVARAD can be done in a busy clinical setting, during an ordinary psychiatric visit, and in an acute clinical context, with limited effort by a minimally trained clinician, providing interesting and useful information without additional costs. Although quick and easy to use, requiring only a few minutes for completion, our data indicate that this ease of use has not come at the price of decreased validity; the psychopathological dimensions of the SVARAD were well represented in our samples. Furthermore, these assessments resulted in interesting characterisations, quite apart from DSM-IV categorical diagnoses. The SVARAD was used daily in a ward of severely acute inpatients, admitted by both voluntary and compulsory procedures, as well as in the assessment of patients in the emergency department [21] and in the outpatient clinic. Mean time for assessment and compilation was around 4 min for each patient, and fulfilled the requirement for providing basic, essential descriptions of the principal component of suffering (according to ten psychopathological dimensions), with minimal pressure on the clinician to complete records. Another important quality, the repeatability of this basic assessment in subsequent visits, was also seen, although this was not primarily tested in the present report. The SVARAD seems thus to satisfy the need of the clinician for a valid, easy, and very short instrument, fitted for use in a busy clinical setting.

Second, SVARAD dimensions clearly show the multifaceted composition of the psychopathology of severe, acute mental disorders, suggesting the existence of "trans-diagnostic" descriptors across DSM-IV and ICD-10 categories. A "trans-diagnostic role" can be envisioned for certain SVARAD dimensions that consistently contribute to the acute psychopathological suffering of several categories of DSM-IV mental disorders. For instance, the SVARAD offers concise and clear descriptions of how patients suffer highly significant rates of Apprehension/Fear in disorders outside the anxiety disorders area. Such disorders (for which the DSM or ICD criteria do not formally require description of anxiety or fear in the criteria set for categorical diagnosis) include major depression, depressive disorder NOS, schizophrenia, psychotic disorder NOS, bipolar disorders both in manic/hypomanic and depressive episodes, and borderline personality disorder, with the highest peak in bipolar disorder-mixed episode. The SVARAD Sadness/Demoralisation dimension is also high in several mental disorders categories, outside the mood disorders area. Again, our study found that the SVARAD Anger/Aggressiveness dimension (which does not have a definite DSM or ICD diagnostic category except in antisocial personality disorder) was well represented in several diagnostic groups, including borderline personality disorder, bipolar disorder (manic/hypomanic and mixed episodes), and psychotic disorder NOS. Interestingly, it is also represented in a significant subgroup of depressed patients (see Chap. 6 for a more detailed examination of this issue). Few SVARAD dimensions, e.g. Disorganisation, Reality Distortion, and Obsessiveness, were involved in only a few diagnostic groups; most spanned many diagnostic groups.

Third, an interesting finding is the ability of the SVARAD to describe the psychopathological variability *within a single diagnostic* DSM-IV (and now DSM-5) category of mental disorders in a large sample. The findings suggest how SVARAD assessment could give a first glance at how *variable and multifaceted* the psychopathological components within a single diagnostic category can be. This constitutes one major contribution of dimensional analysis in enhancing the present categorical approach to mental disorders. The case of anxiety and depression in our sample is a fine example of this. Patients who fit the diagnostic criteria set for major depressive disorder report a mean SVARAD profile consisting of the triad Sadness/Demoralisation, Apathy, and Apprehension/Fear SVARAD dimensions, followed by the Somatic Preoccupation/Somatisation dimension. While the first two SVARAD dimensions are well recognised in DSM-IV and ICD criteria as components of major depressive disorder, Apprehension/Fear is not well recognised by the DSM-IV or the ICD. In our inpatient sample, it was the third most prominent dimension in major depressive disorder. Another undervalued dimension is Somatic Preoccupation/Somatisation, which was the fourth most prominent clinical presentation in our major depressive disorder inpatient sample. This might be related to the comorbidity of anxiety and depressive disorders [15, 22], as well as to the debate about the choice of hierarchical and exclusion criteria followed by DSM criteria [23]. The natural coexistence of anxiety and depression was a commonly recognised presentation in classifications from the pre-DSM III era [24]. The Feighner criteria [25] provided a first attempt to cope with the problem, while the research diagnostic criteria (RDC) [26], with the introduction of a set of criteria for many disorders, cleared the way for the dichotomous-tree decision rules of the algorithmic procedure of the DSM-III, which assign a patient to a depressive disorder or an anxiety disorder according to his/her main psychopathological symptoms.

Another example is the relevance of the Impulsivity and Anger/Aggressiveness psychopathological dimensions in the major depressive disorder group. In accordance with several previous studies by our research group, they confirm these components as the fourth and fifth significant dimensions, respectively, after Sadness/Demoralisation, Apprehension/Fear, and Apathy. A subgroup of major depressive disorder and depressive NOS patients seem to display substantial levels of anger, irritability, and impulsivity. This finding underscores the need to recognise this area of suffering in major depressive disorder patients—an area that is not openly represented in DSM and ICD criteria—and the need to tailor specific psychopharmacological and psychotherapeutic treatments to address these dimensions, in addition to the dimensions classically associated with depression. Recognising and addressing these two psychopathological dimensions could reduce suffering, the risk of activation, and suicide risk in these patients, especially during the first phase of antidepressant treatment. Patients with certain dimensional profiles might require D2-blockers and especially mood stabilisers, together with appropriate psychological interventions and communication strategies (such as specific attention to de-escalation techniques by the healthcare personnel). The clinically routine use of SVARAD could give the psychiatrist and the clinical psychologist the ability to explore and "see" these significant components of suffering—currently overlooked by standardised diagnostic criteria—and to treat them appropriately.

Finally, the findings from a severe, acute inpatient sample, together with those from the outpatient sample, led to the consideration that the categorical approach to diagnosis is, while without a doubt valid, not the only one available. The supplemental use of the dimensional approach can better capture the complexity and multifaceted psychopathology underlying current categories of mental illness and thereby optimise diagnosis and, moreover, treatment choices. The categorical approach can represent a valid choice as an efficient diagnostic system, but it is accompanied by two costs: one is the underrepresentation of components present in the mental disorder (e.g. anxiety in depressives and depression in anxious patients), which can lead to untreated needs. Another is the possibility of an unintentional misconception—implied by the DSM classification—that anxiety/fear is not a component in major depressive disorder (because the criteria do not include anxiety or fear in the characterisation of major depressive disorder). Conversely, depression components are not significant in anxiety disorders (at least they do not fulfil the criteria for a full comorbidity with two separate diagnoses). One could argue that the DSM era might have led to a generation of psychiatrists well equipped to diagnose through dichotomous decisions with mutually exclusive categories, but unable to recognise the complex components and sub-components of psychopathological suffering. Andreasen, in 2007, said: "Since the publication of the DSM-III in 1980, there has been a steady decline in the teaching of careful clinical evaluation that is targeted to the individual person's problems and social context and that is enriched by a good general knowledge of psychopathology. Students are taught to memorize DSM rather than to learn complexities from the great psychopathologists of the past. By 2005, the decline has become so severe that it could be referred to as the death of phenomenology in the United States" [27]. A dimensional approach such as that facilitated by the SVARAD could mitigate this negative drift. The categorical approach might well be integrated with a dimensional one, to reach a full representation of the distinct components of suffering in each diagnostic category of mental disorder.

On the basis of our experience, we suggest the need to integrate the categorical diagnosis process with a dimensional approach. We propose a *three-step* procedure:

1. Perform a preliminary categorical diagnosis (for instance, according to DSM 5 or ICD-10).
2. Perform a dimensional assessment using the SVARAD instrument.
3. Construct a pharmacological and psychotherapeutic treatment regimen on the basis of the dimensional profile.

This three-step procedure could be very useful in addressing variability within the same category of mental disorder and in working toward the aim of *personalised treatment* [28]. This sequential, categorical-dimensional approach may help train young psychiatrists and clinical psychologists to think in a more flexible way that allows for the recognition of all the actual components of a patient's psychopathological suffering. Such recognition can then form that basis for tailored, personalised treatments. The development of optimal treatments that consider the additional information provided by dimensional assessment is currently a work in progress that requires further study.

Table 2.4 shows a possible overview of the dimensional approach, including the link between each SVARAD dimension, the neurotransmitters implicated in their alteration, the underlying brain networks, and suggested pharmacological and psychotherapeutic treatments (Table 2.4).

The present study has several limitations. First, patients were not diagnosed with a structured diagnostic interview. However, the diagnoses were made after a professional psychiatric examination, and were confirmed for accuracy by an experienced faculty psychiatrist, who carefully reviewed all clinical records. Although the presence of a comorbid disorder may have been missed in some cases, this should not have had a substantial influence on the results. Second, our analysis does not include psychometric data, such as the MMPI-2 and BPRS scores (we presented only mean group values of these as a description of the samples). Further analysis is expected to enrich our preliminary findings. Third, we based our evaluation on only a cross-sectional dimensional assessment; we did not report longitudinal data or pre- and post-treatment data. We also did not present data coupling different SVARAD profiles with different psychopharmacological choices within the same diagnostic category. Further, carefully designed, prospective studies are needed to confirm our preliminary results and to further the aim of our work: the development of enhanced psychiatric methods for personalised diagnosis and treatment of mental disorders.

Table 2.4 SVARAD dimensional approach: pathophysiology of neurotransmitters, connectivity networks, and drug target involved and suggested drug and psychotherapy treatments

Code	Dimensions	Pathophysiology (neurotransmitters)	Connectivity (brain neural networks)	Main drug targets	Suggested drugs	Suggested psychotherapy
1	APPREHENSION/FEAR	Gaba↓ 5HT↓	*Fear circuits* – Raphe nucleus – Amygdala – Locus coeruleus – ANS – ENS	Gaba 5HT Ach	– Benzodiazepines – Serotonergic and noradrenergic compounds (SSRIs, TCAs) – Dual-acting drugs (serotonergic and noradrenergic)	RIL CBT BFB
2	SADNESS/ DEMORALISATION	NA↓ 5HT↓ DA↓	– Emotion-regulation circuitry – Reward-processing circuitry – HPA – ANS	5HT NA	– Serotonergic and noradrenergic drugs – Dual-acting drugs (serotonergic and noradrenergic) – MAOIs – NSAIDs[a] – High-dose omega-3 supplements[a] – Ketamine[a] – Triiodothyronine[a]	CBT IPT BT SPP[a]
3	ANGER/ AGGRESSIVENESS	DA↑ Glu↑	Testosterone	Glu	– Mood stabilisers – Low-dose serotonergic drugs (TCAs or SSRIs)	CBT
4	OBSESSIVENESS	5HT↓/Glu	– Right caudate nucleus – Orbitofrontal cortex	5HT DA Glu	Serotonergic drugs (SSRIs, clomipramine)	CBT
5	APATHY	DA↓ NA↓	Prefrontal-basal ganglia system	DA NA	– Dopaminergic drugs – Noradrenergic drugs – MAOIs – Low-dose sulpiride or amisulpride[a] – Acetylcholinesterase inhibitors[a] – Modafinil[a] – Nefiracetam[a]	BT CBT[a] CRT[a]

6	IMPULSIVITY	5HT↓ DA↑	– Prefrontal cortex – Orbitofrontal cortex	5HT GLU	– Mood stabilisers – D_2 or D_2/5HT antagonists (low-dose typical or atypical antipsychotics) – BDZ or non-BDZ hypnoinducers – Antihistaminics and antidepressants with high antihistaminic sedative properties	CBT DBT MBT
7	REALITY DISTORTION	DA↑ Glu↑	*Salience circuits* – DA mesocortical	DA $5HT_2$	– D_1 and D_2 antagonists – $D_2/5HT_2$ antagonists	CBT
8	THOUGHT DISORGANISATION		Cognitive Network		– D_1 and D_2 antagonists – $D_2/5HT_2$ antagonists	CRT[a] MERIT[a]
9	SOMATIC PREOCCUPATION/ SOMATISATION		– Limbic system – ANS – ENS – Immune system	Gaba	Serotonergic drugs (SSRIs, Clomipramine)	CBT BFB SOM
10	ACTIVATION	DA↑		Glu DA $5HT_2$	– Mood stabilisers – D_2 or D_2/5HT antagonists (low-dose typical or atypical antipsychotics) – BDZ or non-BDZ hypnoinducers – Antihistaminics and antidepressants with high antihistaminic sedative properties	

[a]Possibly effective, but more evidence is needed, *5HT* serotonin, *NA* noradrenalin, *Ach* acetylcholine, *DA* dopamine, *Glu* glutamate, *ANS* autonomic nervous system, *ENS* enteric nervous system, *HPA* hypothalamic-pituitary-adrenal axis, *BDZ* benzodiazepines, *SSRI* selective serotonin reuptake inhibitor, *TCA* tricyclic antidepressant, *MAOI* monoamine oxidase inhibitor, *NSAIDs* nonsteroidal anti-inflammatory drugs, *RIL* relaxation treatment/mindfulness, *CBT* cognitive-behavioural psychotherapy, *BFB* biofeedback therapy, *IPT* interpersonal psychotherapy, *BT* behavioural activation, *SPP* short-term psychodynamic psychotherapy, *DBT* dialectical behavioural therapy, *MBT* mentalisation-based treatment, *CRT* cognitive remediation therapy, *MERIT* metacognitive reflection and insight therapy, *SOM* somatic treatments

References

1. Pancheri P, Biondi M, Gaetano P, Picardi A, Pasquini M. Use of the scale for the rapid dimensional assessment SVARAD in a sample of 1,124 psychiatric outpatients. Riv Psichiatr. 2001;36:4. Available from: http://www.rivistadipsichiatria.it/allegati/00186_2001_04/fulltext/204-216.pdf.
2. Andrews G, Charney D, Sirovatka PJ, Regier D. Stress-induced and fear circuitry disorders. Arlington VA: American Psychiatric Association; 2009.
3. Wolkowitz OM, Pickar D. Benzodiazepines in the treatment of schizophrenia: a review and reappraisal. Am J Psychiatry. 1991;148(6):714–26. Available from: http://www.ncbi.nlm.nih.gov/pubmed/1674645.
4. APA. Diagnostic and statistical manual of mental disorders, Fifth Edition (DSM-5). Arlington VA: APA; 2013.
5. Lavender JM, Goodman EL, Culbert KM, Wonderlich SA, Crosby RD, Engel SG, et al. Facets of impulsivity and compulsivity in women with anorexia nervosa. Eur Eat Disord Rev. 2017;25(4):309–13. Available from: http://www.ncbi.nlm.nih.gov/pubmed/28387426.
6. Biondi M, Picardi A, Pasquini M, Gaetano P, Pancheri P. Dimensional psychopathology of depression: detection of an "activation" dimension in unipolar depressed outpatients. J Affect Disord. 2005;84(2–3):133–9. Available from: http://linkinghub.elsevier.com/retrieve/pii/S0165032702001039.
7. Maser JD, Cloninger RD. Comorbidity of mood and anxiety disorders. London: American Psychiatric Press, Inc.; 1990. 888 p.
8. Pasquini M, Picardi A, Speca A, Orlandi V, Tarsitani L, Morosini P, et al. Combining an SSRI with an anticonvulsant in depressed patients with dysphoric mood: an open study. Clin Pract Epidemiol Ment Heal. 2007;3(1):3. Available from: http://www.ncbi.nlm.nih.gov/pubmed/17288584.
9. APA. Diagnostic and Statistical manual of mental disorders - 4th ed. Text Rev. (DSM IV-TR). Washington, DC: APA; 2000.
10. WHO. International statistical classification of diseases and related health problems, 10th Revision (ICD-10). Geneva: WHO; 1992.
11. Sartorius N, Andreoli V, Cassano G, Eisenberg L, Kielholz P, Pancheri P, et al. Anxiety psychobiological and clinical perspective. New York, NY: Hemisphere Publishing Corporation; 1990.
12. Roth M, Mountjoy C. The distinction between anxiety states and depressive disorders. In: Paykel E, editor. Handbook of affective disorders. Edinburgh: Churchill Livingstone; 1982. p. 70–92.
13. Jablensky A. Approaches to the definition and classification of anxiety and related disorders in European psychiatry. In: Anxiety and the anxiety disorders. Hillsdale, NJ: Lawrence Erlbaum; 1985. p. 735–58.
14. Johansson R, Carlbring P, Heedman Å, Paxling B, Andersson G. Depression, anxiety and their comorbidity in the Swedish general population: point prevalence and the effect on health-related quality of life. PeerJ. 2013;1:e98. Available from: http://www.ncbi.nlm.nih.gov/pubmed/23862109
15. Möller H-J, Bandelow B, Volz H-P, Barnikol UB, Seifritz E, Kasper S. The relevance of "mixed anxiety and depression" as a diagnostic category in clinical practice. Eur Arch Psychiatry Clin Neurosci. 2016;266(8):725–36. Available from: http://www.ncbi.nlm.nih.gov/pubmed/27002521.
16. Piacentino D, Pasquini M, Tarsitani L, Berardelli I, Roselli V, Maraone A, et al. The association of anger with symptom subtypes in severe obsessive-compulsive disorder outpatients. Psychopathology. 2016;49(1):40–6.
17. Fava GA, Freyberger HJ, Bech P, Christodoulou G, Sensky T, Theorell T, et al. Diagnostic criteria for use in psychosomatic research. Psychother Psychosom. 1995;63(1):1–8. Available from: http://www.ncbi.nlm.nih.gov/pubmed/7740096.

18. Picardi A, Viroli C, Tarsitani L, Miglio R, de Girolamo G, Dell'Acqua G, et al. Heterogeneity and symptom structure of schizophrenia. Psychiatry Res. 2012;198(3):386–94. Available from: http://www.ncbi.nlm.nih.gov/pubmed/22424892
19. Stoffers JM, Lieb K. Pharmacotherapy for borderline personality disorder—current evidence and recent trends. Curr Psychiatry Rep. 2015;17(1):534. Available from: http://www.ncbi.nlm.nih.gov/pubmed/25413640.
20. Goldberg D, Kendler K, Sirovatka P, DA R. Diagnostic issues in depression and generalised anxiety disorder. Arlington VA: American Psychiatric Association; 2010.
21. Dazzi F, Picardi A, Orso L, Biondi M. Predictors of inpatient psychiatric admission in patients presenting to the emergency department: the role of dimensional assessment. Gen Hosp Psychiatry. 2015;37(6):587–94.
22. Pincus HA, Tew JD, First MB. Psychiatric comorbidity: is more less? World Psychiatry. 2004;3(1):18–23. Available from: http://www.ncbi.nlm.nih.gov/pubmed/16633444.
23. Clarke DE, Narrow WE, Regier DA, Kuramoto SJ, Kupfer DJ, Kuhl EA, et al. DSM-5 field trials in the United States and Canada, Part I: study design, sampling strategy, implementation, and analytic approaches. Am J Psychiatry. 2013;170(1):43–58. Available from: http://psychiatryonline.org/doi/abs/10.1176/appi.ajp.2012.12070998.
24. Gorman JM. Comorbid depression and anxiety spectrum disorders. Depress Anxiety. 1996;4(4):160–8. Available from: http://www.ncbi.nlm.nih.gov/pubmed/9166648.
25. Feighner JP, Robins E, Guze SB, Woodruff RA, Winokur G, Munoz R. Diagnostic criteria for use in psychiatric research. Arch Gen Psychiatry. 1972;26:57–63.
26. Spitzer RL, Robins E. Research diagnostic criteria: rationale and reliability. Arch Gen Psychiatry. 1978;35(6):773–82.
27. Andreasen NCDSM. the death of phenomenology in america: an example of unintended consequences. Schizophr Bull. 2006;33(1):108–12. Available from: http://www.ncbi.nlm.nih.gov/pubmed/17158191.
28. Pancheri P. Approccio dimensionale ed approccio categoriale alla diagnosi psichiatrica. G Ital di Psicopatol. 1999;38:1–14.

3 The Apprehension/Fear and Somatic Preoccupation/Somatisation Dimensions

Roberto Delle Chiaie and Amedeo Minichino

3.1 The Apprehension/Fear Dimension: General Considerations

As extensively reviewed by Crocq in a recent work on the history of anxiety [1], the word anxiety derives from the Latin substantive *angor* and the corresponding verb *ango* (to constrict). A cognate word is *angustus* (narrow). These words derive from an Indo-European root that has produced *Angst* in modern German (and related words in Dutch, Danish, Norwegian, and Swedish). Interestingly, the same relationship between the idea of narrowness and anxiety is attested in Biblical Hebrew. In fact, Job expresses his anguish (Job 7:10) literally with the Hebrew expression "the *narrowness* (tsar) of my spirit". In French, as well as in other Romance languages, *anxiété* (anxiety, from the Latin anxietas) is often differentiated from *angoisse* (anguish, from the Latin angustıa). Sometimes, the two terms are considered synonymous by some authors. *Anxiété* was described as including the psychological and cognitive aspects of worrying. In contrast, *angoisse* was defined as the experience of spastic constriction of voluntary or involuntary muscle fibres. *Angoisse* could be experienced as a constriction affecting the muscles of all systems: bronchial spasm, shortness of breath, intestinal cramps, vaginismus, urinary urgency, pseudo-angina pectoris, and headache.

Between classical antiquity and modern psychiatry, there was an interval of centuries when the concept of anxiety as an illness seems to have disappeared from written records. Patients with anxiety did exist, but they were diagnosed with other diagnostic terms. The last and most successful of these new diagnoses was Beard's neurasthenia. In 1621, Robert Burton published his treatise *The Anatomy of*

R. Delle Chiaie (✉) · A. Minichino
Department of Human Neurosciences, Sapienza University of Rome, Rome, Italy
e-mail: r.dellechiaie@centrokahlbaum.it

M. Biondi et al. (eds.), *Dimensional Psychopathology*,
https://doi.org/10.1007/978-3-319-78202-7_3

Melancholy. Burton's work is generally quoted in the context of depression. However, Burton was also concerned with anxiety. At that time, the meaning of melancholia was not limited to depression but encompassed anxiety as well. Generally, the diagnosis of melancholia could be applied to a variety of clinical pictures with negative affect or internalising symptoms. A key criterion of melancholia was the fact that the patient would remain quiet, an agitated patient qualified for a diagnosis of mania, in Greek, or furor, in Latin. For Burton, fear and sorrow were intimately linked. These concepts influenced clinicians until the eighteenth century, since medical authors published clinical descriptions of panic attacks, but they did not label them as a separate illness. Rather, symptoms of panic attacks were often considered to be symptoms of melancholia. In the late nineteenth and early twentieth century, anxiety was a key component of various new diagnostic categories, from neurasthenia to neuroses. George Miller Beard first described neurasthenia in 1869. Its symptoms were manifold, ranging from general malaise, neuralgic pains, hysteria, and hypochondriasis to symptoms of anxiety and chronic depression. Pierre Janet (born 1859) developed the idea that anxious manifestations could be triggered by subconscious "fixed ideas". He coined the term "psychastenia" for what was supposed to be one of the two major neuroses, along with hysteria. Freud separated anxiety neurosis from neurasthenia. He coined many of the terms that are used today for various anxiety disorders, even though these terms have by now largely shaken off their original psychoanalytical connotations. In the same period, Emil Kraepelin gave much attention to anxiety as a symptom associated with other diagnoses, but wrote less extensively on anxiety as a separate diagnosis.

In the DSM-I (1952), anxiety was almost synonymous with "psychoneurotic disorders". The DSM-I states that the chief characteristic of neurotic disorders is anxiety. According to the apparent manifestations, the diagnosis might be "anxiety reaction" (when the anxiety is diffuse and not restricted to specific situations or objects), "dissociative reaction" or "conversion reaction" (when the impulse causing the anxiety is "converted" into functional symptoms in organs or parts of the body), "phobic reaction" (when the patient's anxiety becomes detached from a specific idea, object, or situation in his or her daily life and is displaced to some symbolic idea in the form of a specific neurotic fear), "obsessive–compulsive reaction" (when the anxiety is associated with the persistence of unwanted ideas and of repetitive impulses to perform acts), and "depressive reaction" (when the anxiety is allayed and partially relieved by depression and self-deprecation). In the DSM-II, the overarching category for anxiety symptomatology was called "neuroses". Panic disorder is a relatively new diagnostic category. Its conception arose out of clinical observations by Donald Klein (1964) that a subgroup of anxious patients who had panic attacks tended to do well when treated with imipramine in contrast to those with other anxiety disorders. Panic disorder was first formally recognised by the Feighner Criteria (1972) and Research Diagnostic Criteria (1978), before finally entering the official American Psychiatric Association nomenclature in the DSM-III (1980).

Anxiety- and fear-related disorders are extremely common in the general population, being among the most frequent psychiatric illnesses, with a lifetime prevalence of 29% [2]. Anxiety and fear are adaptive evolutionary states that can guard against environmental (external) or self-triggered (internal) threats, but maladaptive reactions have also emerged in human evolution [3, 4]. Thus, anxiety- and fear-related disorders may be considered as maladaptive conditions in which disproportionate responses to stress, or even self-evoked responses, are displayed and associated with impaired real-world functioning [5, 6]. Anxiety- and fear-related disorders have a deep nosological history. Greek and Latin physicians and philosophers distinguished anxiety and fear from other types of negative affect and identified them as medical disorders; this has been extensively reviewed elsewhere [1].

The current classification of anxiety disorders has become increasingly specific since the late 1970s [7]. In 1968, in the second edition of the *Diagnostic and Statistical Manual of Mental Disorders* (DSM-II), the overarching category for anxious symptomatology was called neuroses [8]. It was stated that anxiety was the chief characteristic of the neuroses, which established anxiety and neurosis as quasi synonyms [7, 8]. The DSM-II included three anxiety disorders (anxiety, phobic, and obsessional neuroses). In the DSM-III (1980), the DSM-II's anxiety neurosis was split into two newly created disorders, panic disorder (PD) and generalised anxiety disorder (GAD); later, post-traumatic stress disorder (PTSD) was introduced [6, 9]. Thus, the DSM-III's chapter on anxiety disorders was reorganised as follows: (1) phobic disorders, subdivided into agoraphobia (with or without panic attacks), social phobia, and simple phobia; (2) anxiety states, subdivided into PD, GAD, and obsessive–compulsive disorder (OCD); and (3) PTSD [6–8]. The increasing knowledge about neurobiological, genetic, and psychological features of anxiety- and fear-related disorders led to the most recent DSM classification [10, 11]. In the DSM-V, OCD and PTSD have been separated from the category of anxiety spectrum disorders, which still includes (among others) the diagnoses of PD, GAD, and phobias. These diagnostic categories share the core essential feature of being characterised by an excessive fear response and/or worry that interferes with functioning or causes significant distress (see ref. [11] for an extensive review on diagnostic classification of anxiety disorders).

Increasingly, the scientific community is recognising that such a diagnostic classification and the resulting diagnostic categories are not optimal for classifying and comparing individuals for the purposes of understanding pathophysiology and, ultimately, improving therapeutics [12–15]. Starting from evidence of epidemiological and clinical studies consistently reporting high rates of co-occurrence between anxiety and other major psychiatric disorders, many investigators are advocating taking a dimensional approach to anxiety- and fear-related symptoms (see ref. [15] for a review). This co-occurrence, indeed, seems to relate to shared symptom descriptors that suggest underlying dimensions that have been arbitrarily divided into discrete categories. Of note, these latent variables demonstrate clearer patterns of heritability and relationships to early psychosocial stressors than do disorders formally classified through a categorical approach [16].

Starting from these considerations, we tried to classify the anxiety- and fear-related symptom descriptors into a unique dimension, called "Apprehension/Fear" (A/F). A clear definition of the A/F dimension has been provided by our research group: "State of anxiety and worry; sense of constriction; perception of imminent threat; feelings of worry, fear, and anguish" [17]. Individuals with high levels of A/F are plagued by unpleasant, fearful thoughts and images and concerns about physiological symptoms. They invariably report persistent fear reactions that include pounding heart, palpitations, exaggerated startle, breathing irregularities, sweaty palms, and tense muscles [18]. These symptoms are associated with impaired workplace performance and hefty economic costs [19], as well as an increased risk of cardiovascular morbidity and mortality [20]. As conceptualised before by Yerkes–Dodson [21], and later redefined by Hans Selye [22], performance shows an inverted-U relation with anxiety and fear, with increasing A/F improving performance until a maximum is reached, after which performance falls away, becoming poor at high A/F levels.

Of note, a dimensional approach to assessing fear and anxiety has also been emphasised by the NIMH's Research Domain Criteria (RDoC) framework and may help to promote investigations of the psychobiology of these symptoms across different categories of patients and non-patients [23].

3.2 Neurotransmission, Neurophysiology, and Neurophysiopathology of Apprehension/Fear

3.2.1 Genetics

Many studies have shown that anxiety- and fear-related disorders tend to run in families [24]. Specifically, the risk of these disorders is approximately four to six times higher in first-degree relatives of affected probands compared with relatives of unaffected probands [25]. Of relevance, first-degree relatives of probands with one anxiety disorder are at risk for a range of other anxiety disorders, and twin studies have documented genetic overlap among most, if not all, anxiety disorders [26]. Indeed, a growing body of evidence shows that the shared genetic component among panic disorder (PD), phobias, and generalised anxiety disorder (GAD) is substantially larger than their disorder-specific genetic components, providing further support for a dimensional approach to anxiety- and fear-related symptomatology [26, 27].

The majority of candidate genes explored in association with A/F are related to serotonin (*SLC6A4*, *HTR1A*, and *HTR2A*), γ-aminobutyric acid (GABA; *GABRB3* and *GABRA5*), stress hormones (*CRHR1*), and neuropeptides (*BDNF*, *NPSR1*, *and ACE*), but a recent comprehensive meta-analysis of the 23 most widely studied candidate variants found no robust associations [27, 28]. Indeed, like other psychiatric disorders, anxiety disorders are likely to be highly polygenic, involving thousands of genetic variants of modest effect (please see ref. [27] for an extensive review).

However, it is not difficult to conceive that several experiential factors might influence the development of high levels of A/F in any individual. A fascinating aspect of anxiety disorders is indeed the exquisite interplay of genetic and experiential factors. While there is little doubt that abnormal genes predispose individuals to pathological anxiety states, evidence clearly indicates that traumatic life events and stress are also aetiologically important [29].

3.2.2 Neurophysiology

From a neurophysiological perspective, fear and anxiety can be considered as physiological brain states, caused by external or internal stimuli, that are able to elicit specific defensive behavioural responses as well as physiological, hormonal, and autonomic reactions [30, 31]. Behavioural correlates of fear and anxiety have evolved to enable the organism to avoid or reduce harm and thus ensure its survival [30]. These behavioural responses are highly conserved across different animal species, reflecting their key relevance from an evolutionary perspective [32]. However, in humans, excessive fear and chronic anxiety represent major burdens on affected individuals and, given their high prevalence, on wider society [11, 19]. While a distinction between fear and anxiety is commonly made in both clinical and preclinical literature, the borders that define these two constructs are blurred, and the literature is sometimes controversial. Fear and anxiety have considerable overlap with respect to subjective, behavioural, physiological, and neurological characteristics; however, some key differences can be highlighted [33]. A popular distinction is that while fear is elicited by factual, acute sensory input (i.e. occurs in response to a specific object), anxiety can be evoked by potential, anticipated threats. In line with this consideration, the RDoC maps fear onto the "acute threat" domain and anxiety onto the "potential threat" domain of the negative valence system construct [16, 23].

Anxiety is less well understood than fear, and much of what constitutes this complex state remains to be elucidated. Given that excessive fear is a key component of the anxiety spectrum, it is not surprising that the search for the biological underpinnings of the anxiety domain has its roots in, and has been closely intertwined with, studies of biobehavioural fear circuits in animal models [33]. Most of our understating about neural and biobehavioural circuits underlying the fear system stems from studies using the fear conditioning paradigm (consisting of fear acquisition and extinction) [34].

This experimental paradigm consists of pairing an aversive stimulus (unconditional stimulus, e.g. an electric shock) with a neutral stimulus (e.g. a green light), resulting in the neutral stimulus becoming a signal of imminent threat (conditioned stimulus) that is now capable of eliciting a conditioned fear response. In both humans and animal models, this translates into fear acquisition [35]. Fear extinction, in turn, is induced when the conditioned stimulus is repeatedly presented without the aversive outcome, resulting in a decline of the conditioned fear response [35].

Fear acquisition involves an interplay between a number of key structures (see also Fig. 3.1). These include, but are not limited to, the amygdala, the anterior cingulate cortex (ACC), and the hippocampus [31]. In fear acquisition/learning, sensory input signals from the auditory, visual, olfactory, somatosensory, and nociceptive systems about the conditioned and unconditioned stimuli are mainly conveyed to one of the amygdala nuclei: the lateral amygdala (LA) [36]. The LA is

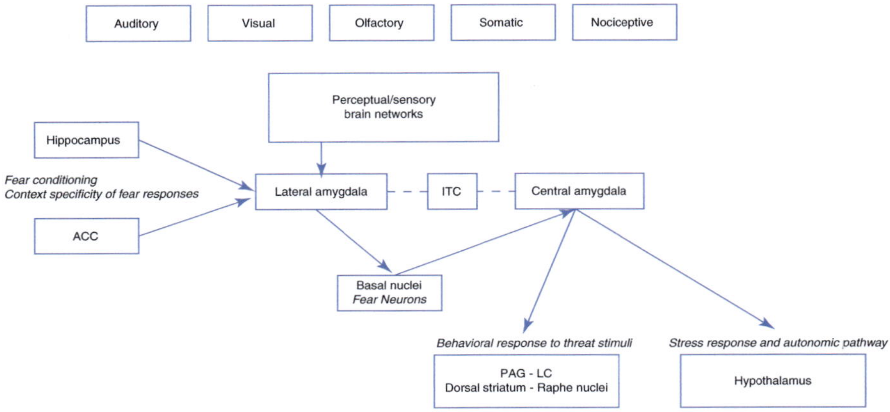

Fig. 3.1 Simplified neural circuits underlying the Apprehension/Fear dimension. The dotted lines represent inhibitory pathways. *ACC* anterior cingulate cortex, *ITC* intercalated cell masses, *PAG* periaqueductal grey, *LC* locus coeruleus

a cortex-like structure, approximately 80% of which consists of glutamatergic spiny projection neurons, which connect extensively with a number of different brain areas [30, 36].

The LA projects to the basal nuclei and the intercalated cell masses (ITC). The basal nuclei contain two subgroups of neurons: the so-called “fear neurons” and “extinction neurons” [30]. When fear is expressed, the LA activates the “fear neurons” of the basal nuclei, which in turn send excitatory projections to another relevant amygdala nucleus: the central amygdala (CEA). At the same time, the LA projections to the ITC prevent inhibition of the CEA by these nuclei. The CEA is the main output region of the amygdala, with projections to subcortical and brainstem areas [30, 31]. It coordinates the behavioural responses to threat stimuli, including freezing, opioid-mediated analgesia (through connections with the periaqueductal grey, PAG), and startle reflex potentiation (through the nucleus reticularis pontis caudalis) [37]. The CEA is also connected to monoamine systems in the brain, including the dorsal/ventral striatum (dopamine), locus coeruleus (norepinephrine), and raphe nuclei (serotonin) [2]. Finally, the CEA activates hypothalamic nuclei producing the classical peripheral stress response, with increased sympathetic arousal, hypothalamus–pituitary–adrenocortical (HPA) axis activation, and increased release of glucocorticoids (GCs) and epinephrine/norepinephrine into the bloodstream [2, 30, 31].

The hippocampus is important for context-specific fear conditioning and for encoding of contextual information during conditioning and is thus assumed to contribute to the context specificity of fear responses. Furthermore, the hippocampus is also involved in fear expression, through modulation of the activity of the dorsal anterior cingulate cortex [38].

Fear extinction involves new learning and covers three phases: acquisition, consolidation, and retrieval of extinction [2, 30, 31]. The distributed network that controls fear extinction involves many of the same brain areas that are important for fear acquisition, including the amygdala, the ACC, and the hippocampus [30, 36]. While the same structures are involved in both fear acquisition and extinction, different sets of neurons are assumed to act through different molecular mechanisms—see ref. [39] for an extensive review.

3.2.3 Neurophysiopathology

Both fear acquisition and extinction are valid means to model features of anxiety- and fear-related symptoms. Indeed, *fear acquisition* can become dysfunctional if the organism continues to display fear responses in the absence of danger, and an inability to accomplish *fear extinction* is assumed to largely contribute to the maintenance of symptoms over time [39].

A dysfunctional *fear conditioning* process may thus lead to the onset of symptoms related to the anxiety/fear dimension in humans. What triggers this dysfunction is a complex interplay between environmental and genetic factors [40]. The pathology of the anxiety/fear system has been conceptualised as learning (*fear acquisition*) under severe stress, which thus represents one of the most relevant triggers [2, 30, 31]. However, exposure to stress is a necessary—but not sufficient—condition. Indeed, stress and the stress response are important for survival and, in nature, are not pathological, but instead adaptive [41].

Genetics and epigenetic modifications of the fear system and the related neural and biobehavioural systems can lead to dysfunctions. Research progress in this area may largely contribute to the understanding of fear-related dysfunctions by providing further insight into the nature of anxiety/fear symptoms.

In terms of brain circuitry, a dysfunctional fear conditioning process and the resulting development of symptoms can be conceptualised as the consequence of an increased "bottom-up" stimulus (mainly from an exaggerated activation of the amygdala and the hippocampus in response to threating stimuli) and a decreased "top-down" control provided by the frontal lobe (mainly from the cingulate cortex) [2, 23].

It has been postulated that the inhibitory inputs from the frontal cortex to limbic regions may be disrupted in anxiety disorders, resulting in unrestrained amygdala activity and thereby an increase in A/F symptoms [30]. This view is consistent with studies showing decreased connectivity between emotion-generating areas (amygdala) and cortical regulatory regions (ACC) [42, 43]. These studies found connectivity to be inversely related to symptom severity. Further, connectivity increased after anxiety treatment [44].

Overall, these findings point to a potential deficit in the fear conditioning system caused by the hyper-activation of emotion-generating regions coupled with dysfunction in emotion-regulating regions [42, 43]. The plastic changes induced in the lateral amygdala and the dorsal hippocampus might contribute to the stabilisation of the fearful memory, a phenomenon probably associated with fear extinction process abnormalities [2, 23]. Of note, fear extinction learning has been used as a model for exposure techniques in behavioural therapy, and it could explain the results in terms of increased connectivity obtained by the previously mentioned studies [44].

3.2.4 Neurotransmission

Increased activity in emotion-processing brain regions could result from decreased inhibitory signalling by GABA or increased excitatory neurotransmission by glutamate [45]. Of note, there is strong evidence to support a role for NMDA-type glutamate receptor (NMDAR)-dependent plasticity at sensory afferents to the LA. Pharmacological blockade of NMDARs abolishes not only fear conditioning at the behavioural level but also its physiological correlates in the LA [46].

Neurotransmitters other than GABA and glutamate, such as serotonin (5-hydroxytryptamine, 5-HT), norepinephrine (NE), and dopamine (DA), seem to play a relevant role in the pathogenesis of anxiety- and fear-related symptoms [33].

The strongest human evidence suggesting a role for NE in the genesis of A/F symptoms comes from studies of the anxiogenic and anxiolytic properties of centrally acting selective noradrenergic drugs, some of which exert these actions on one of the most important NE nuclei: the locus coeruleus [47]. Clonidine and yohimbine are two well-known alpha-2 adrenergic modulators (agonist and antagonist, respectively) [48]. Alpha-2 adrenergic receptors are known to be present as autoreceptors on the cell bodies and terminals of noradrenaline neurons, where they regulate the firing rate and noradrenaline released per nerve impulse [48]. Thus, blockade of alpha-2 receptors increases locus coeruleus noradrenergic firing as well as NE release; this may explain how in patients with panic disorder, yohimbine (alpha-2 antagonist) increases anxiety and the frequency of panic attacks. The opposite effects can be observed after clonidine (alpha-2 agonist) administration in humans [47, 48].

The raphe nuclei presumably play an important role in the serotonergic aspects of the A/F dimension [49]. As mentioned above, these nuclei are extensively connected with most of the brain structures of the fear conditioning system [33, 49]. In particular, an excitatory projection from the locus coeruleus to the dorsal raphe may be important in the serotonin release observed in the ACC, amygdala, and hypothalamus in response to anxiogenic stimuli [50]. Additionally, projections from the dorsal raphe also extend to and inhibit the locus coeruleus, suggesting a possible negative feedback mechanism [30]. Of note, chronic SSRI administration suppresses locus coeruleus firing in rats [51]. To give further strength to the role of the serotonin system in the onset of A/F symptoms, a pioneer study by Hariri et al. [52], recently confirmed by two independent and large meta-analyses [53, 54], showed that increased amygdala reactivity to fear-relevant stimuli was significantly associated with the expression of the S allele (short, low-activity allele) of the serotonin transporter. Of interest, the same polymorphism and epigenetic modifications of the transcription start site of this transporter have been linked to changes in functional coupling of the amygdala with the dorsal anterior cingulate cortex and increased amygdala reactivity, respectively [55]. Other studies have provided additional evidence for the role of serotonergic system genes (HTR1A, HTR2A, HTR3A, TPH2, and MAOA) in amygdala reactivity, functional connections with the amygdala, and the onset of fear/anxiety psychopathology [16].

The dopaminergic system has also been investigated in several studies. Won and Ham [16, 56] showed an association between the COMT Met allele and greater amygdala–orbitofrontal cortex connectivity. Other dopamine-related genes have been associated with amygdala reactivity as well, including the dopamine transporter DAT1/SLC6A3 (9-repeat allele of the VNTR with higher left amygdala activation) the receptor DRD2 and the polymorphism rs1800497 in the DRD2/ANKK1 region [16, 56]. Other regulators of the activity of the A/F system include the vesicular monoamine transporter (vMAT), which packages these neurotransmitters into vesicles; the transmitter-specific serotonin transporter (SERT), and dopamine transporter (DAT); the enzyme monoamine oxidase, which degrades 5-HT, DA, and NE; and the enzyme catecholamine-*O*-methyltransferase (COMT), which degrades DA and NE [16, 56].

In the central nervous system, classic neurotransmitters often are packaged and co-released with neuropeptides, many of which are expressed in limbic regions where they can influence stress and emotion circuitry [57]. The functional implications of these limbic co-localisations have been addressed in numerous reviews [58, 59]. Neuropeptides with particularly strong links to anxiety- and fear-related psychopathology include cholecystokinin (CCK), neuropeptide Y (NPY), and substance P (Sub-P) [57–59]. CCK-B receptor agonists reportedly have an anxiogenic effect in animals and are panicogenic in patients with panic disorders [58]. NPY is synthesised in the arcuate nucleus, which receives inputs from the locus coeruleus. In rodent models, NPY has been shown to suppress the firing of the locus coeruleus and to antagonise the stress response peripherally [57]. Additionally, NPY projections to the CEA, the nucleus accumbens, PAG, and hippocampus may be involved in the NPY anxiolytic effects [57–59]. Sub-P binds to the neurokinin-1 receptor (NK-1), the activation of which, in the hypothalamus, inhibits the secretion of corticotropin-releasing hormone [57–59].

3.2.5 From Neural Circuits to Biobehavioural System Dysregulation: The Stress Response

An important aspect of A/F neurocircuitry is its overlap and interaction with the neurocircuitry that orchestrates the stress response (e.g. medial prefrontal cortex, insula, amygdala, and hippocampus: for a review, see ref. [60]).

The HPA axis is a major pathway by which stress exerts its effect on the brain and the rest of the body, and it is believed to be relevant to the development of A/F symptoms [60]. A generally consistent finding is that the HPA axis in individuals with high levels of A/F is chronically activated [2]. In the HPA axis pathway, the lateral paraventricular nucleus of the hypothalamus releases CRH, which in turn stimulates the production of ACTH by the pituitary gland; the ACTH stimulates the production of cortisol by the adrenal gland [61]. Cortisol is considered a primary stress hormone of the body, having varied effects on metabolism, the autonomic nervous system, the immune system, and brain functions [2, 60, 61]. A number of studies and review articles suggest that chronic stress, which is tightly related to the A/F dimension, is associated with mild hypercortisolemia and prolonged sympathetic nervous system activation, which in turn could favour accumulation of visceral fat, insulin resistance, and hypertension [62]. The chronic activation of the HPA axis could thus represent a potential explanation for the high incidence of cardiovascular disorders in individuals with high levels of A/F [20]. Chronic HPA axis activation also affects the immune system, including the release of humoral immune factors. These include cytokines such as interleukin-1 (IL-1) and IL-6 [39] that can in turn cause further release of CRH, which in theory serves to increase glucocorticoid effects and thereby self-limit the immune activation [39].

It has been suggested that early developmental stress exposure alters A/F circuitry via altered sensitivity and responsivity of the CRH and adrenergic systems, and recent advances in morphological work have suggested a potential mechanism for the effects of stress on fear conditioning and extinction [2, 39]. Chronic stress decreases dendritic branching in the hippocampus and ACC, but increases dendritic branching in the amygdala [39, 63]. This pattern could lead to increased conditioning and impaired extinction, and both of these processes could cause changes in anxiety-/fear-related behaviours, thus generating a vicious cycle [30].

3.3 From Psychobiology to the Clinic: The Trans-diagnostic Relevance of the Apprehension/Fear Dimension in Psychiatric Disorders

Of note, most of the above-mentioned neurocircuitry and peripheral system dysregulations (among the others: increased amygdala reactivity, reduced top-down control by the frontal lobe on the limbic system, chronic HPA axis activation, and immune system activation), which characterise individuals with high levels of A/F, are often expressed throughout the spectrum of psychiatric disorders [64–66]. This provides further evidence that A/F is a dimension that cuts across traditional diagnostic boundaries. This trans-diagnostic ability is in line with the data shown in Chap. 2, which indicate frequent comorbidity of the anxiety spectrum categorical diagnoses with other major psychiatric disorders, such as mood and psychotic disorders. Indeed, in the whole group of outpatients ($n = 1124$) and inpatients ($n = 847$)

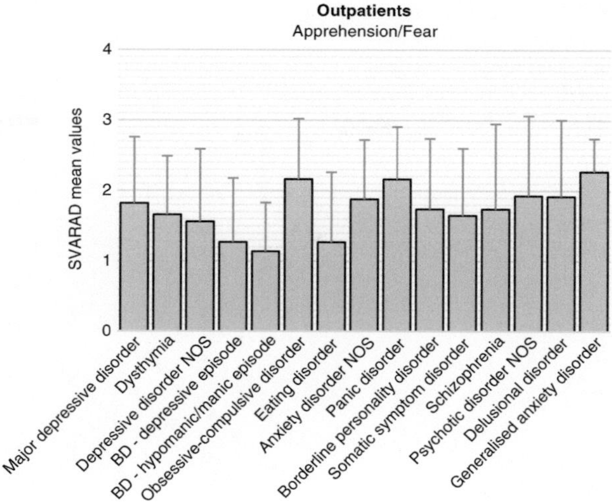

Fig. 3.2 SVARAD Apprehension/Fear dimension across outpatients' diagnostic categories: mean scores and standard deviations

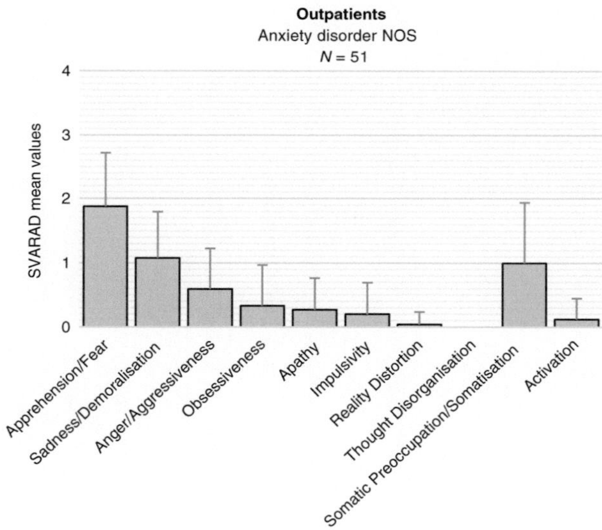

Fig. 3.3 SVARAD profile of outpatients with anxiety disorder NOS: mean scores and standard deviations. Code type: 1-2-9 ($N = 51$)

examined in the multi-parametric analysis in Chap. 2, A/F was the dimension with the highest values in all the diagnostic categories, further suggesting its trans-diagnostic relevance.

In the outpatient population, the highest scores for A/F were observed in anxiety spectrum disorders, i.e. GAD, PD, and anxiety disorders NOS—as well as in OCD. High scores of A/F were also observed in psychotic spectrum disorders, particularly in individuals with psychotic disorder NOS and delusional disorders and in patients with major depression. The lowest mean scores for A/F were observed in

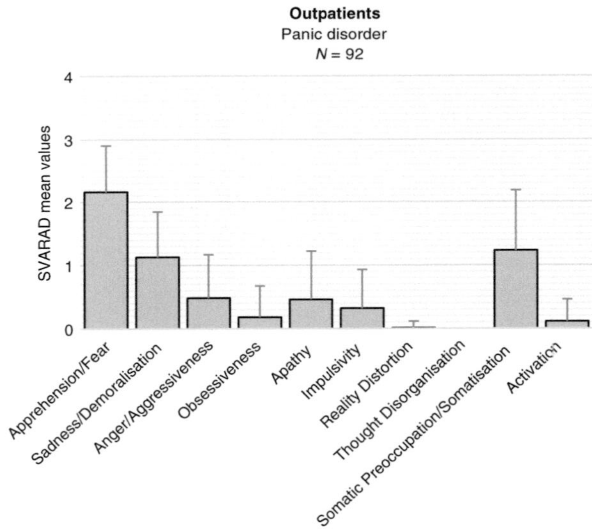

Fig. 3.4 SVARAD profile of outpatients with panic disorder: mean scores and standard deviations. Code type: 1-9-2 ($N = 92$)

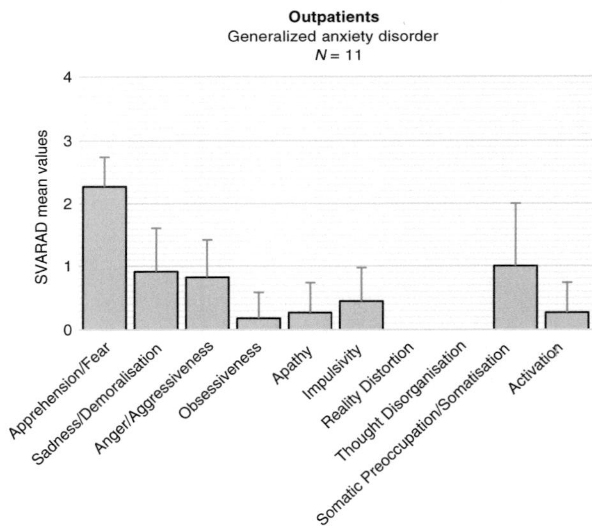

Fig. 3.5 SVARAD profile of outpatients with generalised anxiety disorder: mean scores and standard deviations. Code type: 1-9-2 ($N = 11$)

patients with BD and eating disorders (Fig. 3.2). As could be expected, within single diagnostic categories, A/F was the most relevant dimension (i.e. highest SVARAD score) among the anxiety spectrum disorders (GAD, PD, anxiety disorder NOS) (see Figs. 3.3, 3.4, and 3.5). Despite the high A/F scores in OCD, patients suffering from this disorder were also characterised by comparable levels of Sadness/ Demoralisation and higher levels of Obsessiveness (see Fig. 3.6). Similarly, the dimensional profiles of psychotic spectrum disorders, such as psychotic disorder NOS and delusional disorder, were characterised by high levels of A/F and

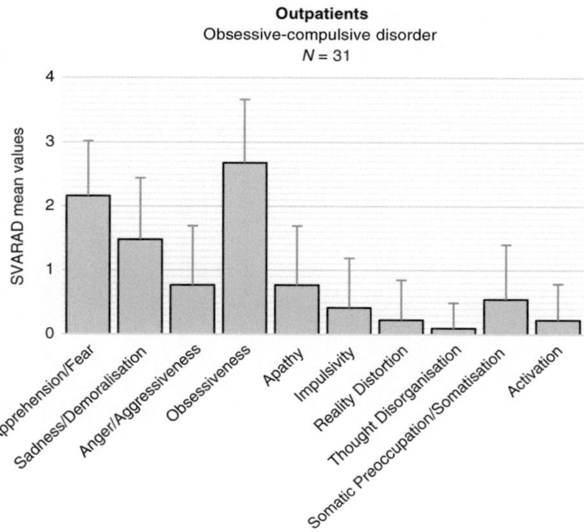

Fig. 3.6 SVARAD profile of outpatients with obsessive–compulsive disorder: mean scores and standard deviations. Code type: 4-1-2 ($N = 31$)

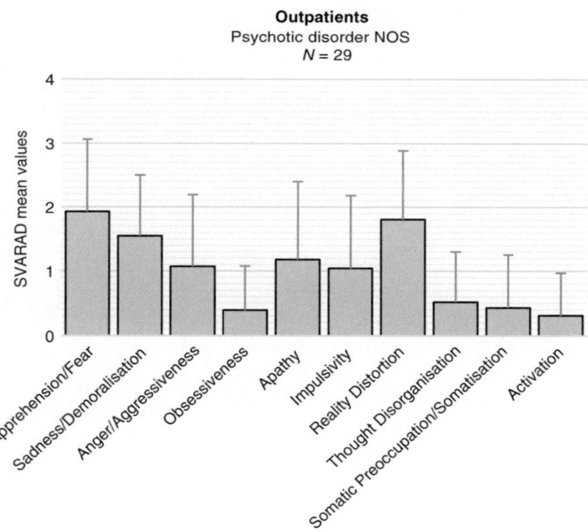

Fig. 3.7 SVARAD profile of outpatients with psychotic disorder NOS: mean scores and standard deviations. Code type: 1-7-2 ($N = 29$)

comparable or greater levels of Reality Distortion (see Figs. 3.7 and 3.8). The most relevant dimension for major depression was Sadness/Demoralisation, followed by A/F (see Fig. 3.9).

In the inpatient population, the highest scores for A/F were observed in individuals with OCD, psychotic disorder NOS, BD with mixed episodes, and major depressive disorder (see Fig. 3.10). It is worth noting that none of the inpatients described in Chap. 2 had a diagnosis of anxiety spectrum disorders such as GAD, PD, or anxiety disorders NOS. Inpatients and outpatients with OCD, psychotic disorder NOS, and major depressive disorder showed similar dimensional profiles.

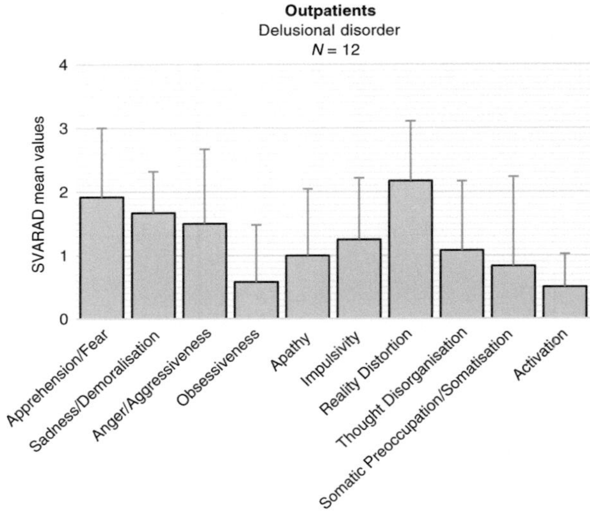

Fig. 3.8 SVARAD profile of outpatients with delusional disorder: mean scores and standard deviations. Code type: 7-1-2 ($N = 12$)

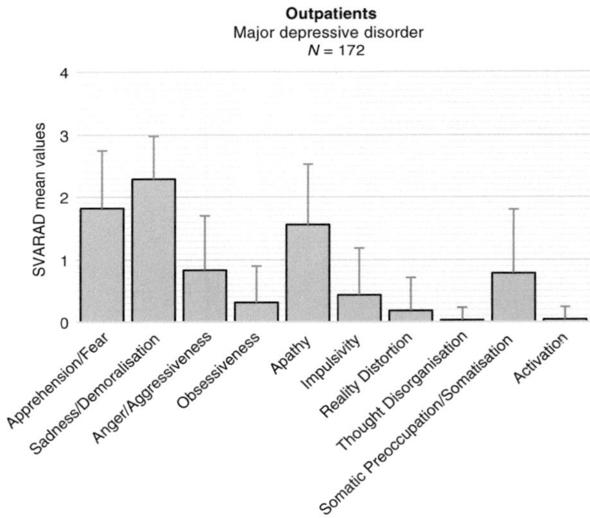

Fig. 3.9 SVARAD profile of outpatients with major depressive disorder: mean scores and standard deviations. Code type: 2-1-5 ($N = 172$)

Of note, inpatients with bipolar disorder having a mixed episode had the highest dimensional scores in A/F, followed by Sadness/Demoralisation, Impulsivity and Activation.

3.3.1 Clinical Cases

With the aim of further highlighting the relevance of the dimensional approach to anxiety and fear across different diagnostic categories and, consequentially,

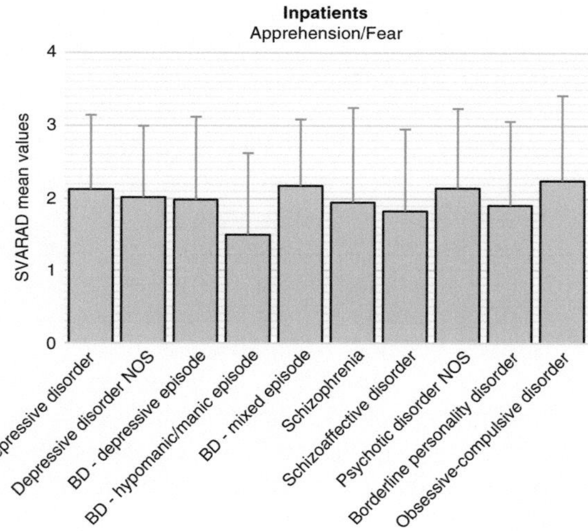

Fig. 3.10 SVARAD Apprehension/Fear dimension across inpatients' diagnostic categories: mean values and standard deviations

introducing therapeutic approaches, it is instructive to describe some exemplary clinical cases.

Case 1: A/F Within the Panic-Agoraphobic Spectrum

FM, a 37-year-old housekeeper, reported the onset of panic attacks at the age of 18. She recalled her first attack vividly: it occurred when she was in her classroom at high school. Describing the episode, she said "there was no need for me to be nervous. I was just sitting in class when my heart began to beat extremely fast, my skin began to tingle, and I felt like I was dying". Over the following years, further attacks became frequent, occurring more or less on a daily basis. In some periods, the patient experienced up to five attacks per day. With her fear of attacks, FM began to avoid crowded places. She went to church regularly, but she sat near the exit. The severity of phobic avoidance increased, and FM used to regularly ask her family members or friends to accompany her every time she needed to go out for work or shopping.

FM had not previously sought treatment since she thought that nobody could really help her. After several requests for assistance in the emergency rooms of different general hospitals, she received a diagnosis of panic disorder and was prescribed paroxetine (up to 20 mg/day). After 1 month of treatment, she was free of attacks, and within 4 months, symptoms of avoidance were also in remission. After 1 year of treatment, the paroxetine daily dose was gradually reduced to 10 mg/day, and at 4 years of follow-up, the patient had maintained good control over all anxiety symptoms. In the interim, she had divorced, since her husband proved unable to cope with a more confident and independent spouse.

Case 2: A/F in Generalised Anxiety Disorder

SC, a 21-year-old college student, presented for evaluation of "nervous" problems. He reported never having been depressed, but, for as long as he could remember, he recalled always having been anxious. He described a sharp increase in his symptom severity after he finished high school and moved away from home to attend college.

SC worried about everything: physical appearance, grades in school, friends, health conditions of his parents, and sexual inexperience. He was mildly tremulous and swallowed frequently; sweat was beaded on his brow. He knew of being constantly tense and unable to relax and was recently evaluated for stress headaches. He chewed gum to counter chronically dry mouth and often had clammy hands and a feeling of a lump in his throat. There was no apparent explanation for his chronic anxiety, but stress made his condition worse. He requested anxiolytics but agreed to also try relaxation techniques and mindfulness exercises. On the whole, these treatments brought him a sense of control over his anxiety.

Case 3: A/F Within Presentations of "Soft" Bipolar Spectrum

Sometimes in patients presenting with A/F symptoms, what appears to be an anxiety disorder may mask a hidden condition of bipolarity. This picture not infrequently presents with severe apparent symptoms of comorbid panic anxiety and is most frequent in patients with a bipolar II course. Some authors have observed that the presence of hyperthymic personality features may represent a predisposing factor. When treated with an antidepressant normally used to treat A/F symptoms, these patients may show a sharp worsening of their A/F symptoms. In some of these cases, if agitation and impulsivity increase, these patients may even develop suicidal ideation or may attempt suicide.

Patient IG, a 50-year-old building contractor, was seen in the emergency room of a general hospital, where he was taken by ambulance after a suicide attempt: taking very high doses of bromazepam and lormetazepam and quite seriously injuring his forearms with a kitchen knife.

This suicide attempt took place during a period in which the patient experienced growing symptoms of irritability, impulsivity, severe agitation, and depressive ideation associated with mood acceleration. During this period, comorbid panic attacks were quite frequent, and the clinical picture closely resembled presentations of the panic–agoraphobic spectrum, with a severe agoraphobic avoidance.

These symptoms showed a significant worsening after the patient was started on antidepressants (venlafaxine, paroxetine), just a few months before the suicide attempt. Soon after beginning treatment with these compounds, he became more irritable and impulsive and self-aggressive behaviours appeared.

The patient reported that, in the past, he had experienced mood swings and developed several periods of slight hypomania. One year before his suicide attempt, the patient had experienced serious economic problems for which he had to sell off his company.

3.4 Treatment of Apprehension/Fear

3.4.1 Psychotherapy

The systems underlying the A/F dimension can be targeted with psychotherapy interventions (see ref. [64] for an extensive review). Cognitive-behavioural therapy for A/F has been consistently associated with increased prefrontal cortex activity and increased ACC–amygdala functional connectivity, consistent with the hypothesised top-down cortical substrates suggested in Sect. 3.2.3. Minor evidence also supports psychodynamic psychotherapy. Causative factors (predisposing, precipitating, and perpetuating) should be identified, and wherever possible, attempts should be made to tackle these. Helpful reading materials, e.g. information brochures for each anxiety condition, with contact details for agencies catering to the counselling and support of persons with psychiatric problems (including anxiety disorders) are recommended. Psychological treatments play an important role in the management of anxiety disorders; however, patient preference and motivation determine choice of treatment. General practitioners and nurses can be trained to deliver a range of specific anxiety management strategies, including breathing control, relaxation, and problem-solving techniques. However, extensive training is essential before specific interventions, such as cognitive-behavioural therapy, can be done safely and effectively. The effectiveness of therapy depends on a good therapeutic relationship, with a fundamental agreement on the goals and tasks of therapy and commitment to the working relationship between therapist and patient. The duration, frequency, and nature of treatment should be collaboratively agreed upon at the outset. The patients' social, cultural, and religious/spiritual views and beliefs should be respected by the therapist or treating clinician. Cognitive-behavioural therapy is a pragmatic combination of concepts and techniques from cognitive-behavioural therapies. Cognitive techniques (e.g. identification and modification of negative automatic thoughts and dysfunctional assumptions and schemas/core beliefs) in combination with behavioural techniques (e.g. exposure to feared situations/objects) are used with the aim of achieving symptom relief and relapse prevention. Phobias and obsessional fears tend to persist when there is avoidance of the feared situation. In exposure therapy, the patient is gradually exposed to a graded set of feared situations/objects/thoughts until fears spontaneously reduce (termed "habituation"). Exposure must be of sufficient duration for habituation to occur. Repeated exposure brings about further reduction of anxiety and a concomitant increase in a sense of mastery over the fear.

With regard to anxiety syndrome within the panic spectrum, the goal of psychotherapy treatment is to eliminate panic attacks, anticipatory anxiety, and avoidance. Psychoeducation for patients with anxiety disorders involves teaching patients about the disorder and discussing treatment options, modalities of treatment, and coping strategies. The support of family members, friends, support groups, and community organisations can also benefit the patient. Psychoeducation has been shown to improve quality of life, reduce symptoms, and improve treatment outcomes. Cognitive-behavioural therapy is the only type of psychotherapy shown to be efficacious in the treatment of panic disorder, with or without agoraphobia [67]. The treatment components of cognitive-behavioural therapy may include psychoeducation, in vivo exposure to feared situations, interoceptive exposure, cognitive restructuring, continuous panic monitoring, and breathing retraining. Pharmacotherapy and psychotherapy can be used in combination for treatment of panic disorder with or without agoraphobia. Although monotherapy with SSRIs is effective, meta-analytic studies have demonstrated the superiority of combined cognitive-behavioural therapy with pharmacological treatment over monotherapy [68].

Regarding GAD, cognitive-behavioural therapy administered by experienced therapists has shown good evidence of efficacy in generalised anxiety disorder [69]. However, combination therapy is not demonstrably superior to either cognitive-behavioural therapy or pharmacotherapy alone [70] and may add significantly to the cost of treatment for the patient. While combination therapy (addition of medication to cognitive-behavioural therapy) will enhance short-term outcomes, there is no evidence to determine, at present, whether or not combination therapy will improve long-term outcomes.

When treating specific phobias, the goals of treatment are the mastery of fear and the recovery of function. Components of cognitive-behavioural therapy for a specific phobia may include systematic desensitisation, imaginal exposure, and in vivo exposure. Medications alone are of little benefit in specific phobia, except in cases where there has been substantial reductions in quality of life. As much as 70–85% of specific phobias could be effectively treated by exposure therapy [64, 71].

When treating patients with social anxiety, exposure to feared situations is a crucial component of cognitive-behavioural therapy. Group cognitive-behavioural therapy approaches are also useful and often include elements of social skills training. Cognitive-behavioural therapy interventions include in vivo exposure, cognitive restructuring, relaxation training, and self-control desensitisation, of which exposure-based interventions are the most efficacious for social anxiety [64, 71].

Evidence suggests that cognitive-behavioural therapy is an effective treatment for post-traumatic stress disorder [64, 72]. The components of cognitive-behavioural therapy include prolonged exposure to memories of the traumatic event.

3.4.2 Pharmacotherapy

Over the last two decades, pharmacological treatments for anxiety disorders have become more effective and tolerable (see ref. [64] for an extensive review). At the same time, research has yielded a vastly improved understanding of the neurobiological and physiological mechanisms involved in chronic anxiety and stress responses, suggesting new approaches to the treatment of anxiety disorders. Despite these impressive changes, however, between one-third and one-half of patients on a modern antidepressant do not achieve sustained remission from anxiety [73]. Unfortunately, although patients often use antidepressant medications for years, high-quality data on the drugs' long-term efficacy are limited. The problem is compounded by the growing number of different drug classes, which has prompted clinicians to combine drugs and change dosing regimens without good data on optimal treatment combinations.

3.4.2.1 Selective Serotonin Reuptake Inhibitors (SSRIs) and Serotonin–Norepinephrine Reuptake Inhibitors (SNRIs)

The widely studied SSRIs, and to a growing degree, the SNRIs, are considered the first-line pharmacological treatments for anxiety disorders [74]. Specific phobia is the exception. In specific phobia, these medications have rarely been studied or used clinically because exposure therapy is considered the first-line treatment. The few studies comparing SNRIs to SSRIs show similar responses. SSRIs and SNRIs work by blocking the reuptake of serotonin or norepinephrine, respectively, which increases synaptic levels of 5-HT or norepinephrine in the synapse. This starts a cascade of downstream effects on other neurotransmitters, second messengers, and immediate early genes, ultimately producing long-term neurochemical changes in the brain.

It is thought that anxious patients are more sensitive to jitteriness with these agents, though this has not been conclusively studied. These effects can be minimised by starting at a low dose and increasing the dose gradually over 2–4 weeks. SSRIs have been associated with increased suicidal ideation, prompting the US Food and Drug Administration's (FDA) "black box" warning for individuals 24 years old or younger. The evidence behind this warning has been widely criticised [75], and many experts consider it appropriate to prescribe SSRIs to children with severe functional impairment when followed by careful monitoring.

In addition to SSRI-like side effects, venlafaxine is associated with elevations in blood pressure, making this a safety issue with older adults and those with cardiovascular issues. Data from patients with depression, and some uncontrolled data with anxiety, suggest that about 20% of patients may need 10–12 weeks or longer before responding. Thus, increasing the dose to the highest level tolerated is always recommended for any patient with an incomplete response (i.e. not having achieved remission). Psychological factors, including negative beliefs about perceived harmful effects, stigma, and lack of "buy in" to the treatment rationale, are negatively associated with adherence and desired outcome [76]. These issues can be addressed through careful psychoeducation and monitoring.

There are serious questions about how much, and in whom, the placebo effect contributes to antidepressant response. In a recent meta-analysis, Fournier et al. [77] found that for patients with mild or moderate depression symptoms, drug response (compared with placebo) may be minimal or nonexistent; however, for patients with very severe depression, the benefit of antidepressants over placebo is substantial. However, a recent study with a much larger and more complete database suggests that initial severity of depression is unrelated to antidepressant response [78]. The relationship of response to initial severity should be systematically examined in the anxiety disorders as well.

3.4.2.2 Other Antidepressants

Extensive studies of TCAs show that they have similar efficacy to SSRIs for panic disorder and generalised anxiety disorder [79]. TCAs are lethal in overdose and, compared with SSRIs, have a markedly broader, more problematic, and less tolerable side effect profile, including dry mouth, blurred vision, constipation, urinary retention, cardiac arrhythmia, tachycardia, sedation, postural hypotension, dizziness, and headache. Nonetheless, TCAs may work when first-line agents do not.

Monoamine oxidase inhibitors (MAOIs) are effective for both panic symptoms and for social anxiety and are thought by some experts to be excellent options for severe, treatment-resistant anxiety disorders [80]. However, they have the worst side effect profile and greatest safety burden of all antidepressants. Patients on an MAOI can experience dangerous hypertensive reactions if they consume foods that contain tyramine (e.g. cheese, beer, and wine) or use certain drugs (e.g. meperidine, decongestants, or energy drinks containing ephedrine or phenylpropylamine). They may also gain weight, lose sleep, and feel sedated during the day while taking MAOIs. Thus, clinicians do not routinely prescribe MAOIs to their patients with anxiety disorders, although they are probably not considered frequently enough in treatment-resistant patients.

Few double-blind, placebo-controlled RCTs have examined the efficacy of other antidepressants for anxiety disorders. Mirtazapine may be efficacious in SAD [81] and post-traumatic stress disorder [82]; nefazodone in PTSD but not GAD [83]; and trazodone in GAD [84]. Finally, bupropion has not demonstrated efficacy in PTSD or PD [85], although it is often used as an adjunctive antidepressant across the anxiety disorders.

3.4.2.3 Benzodiazepines

Benzodiazepines bind to a specific receptor site on the gamma-aminobutyric acid–A receptor (GABA–A) complex and facilitate GABA inhibitory effects by acting on a chloride ion channel. They were initially considered first-line treatments for anxiety because of their tolerability and equal efficacy to TCAs, but became second-line options when it became clear that SSRIs were both more tolerable and efficacious. Currently, benzodiazepines are primarily used for individuals who have had suboptimal responses to antidepressants [86].

Benzodiazepines are also used for their potent, short-term effects (e.g. flying on an airplane) or to help reduce anxiety during the initial weeks of an antidepressant when anxiolytic effects have yet to occur. These uses are appealing to the patient but not always desirable, as they can reinforce pill taking, serve as a safety signal that undermines self-efficacy [87], and become incorporated into the conditioned fear response. These concerns are exacerbated when benzodiazepines are taken on an as-needed basis. As-needed use links pill taking to rapid reduction in anxiety, powerfully reinforcing avoidance in anxiety-provoking situations and encouraging longer-term reliance on the drug.

Chronic benzodiazepine use is associated with physiological dependence, short-term cognitive and psychomotor impairment, and rebound anxiety upon discontinuation. Patients with a history of substance abuse are at increased risk of abusing benzodiazepines. Where clinically indicated, benzodiazepines can be gradually tapered and eventually discontinued over a period of several months while starting another medication or CBT [88].

3.4.2.4 Alpha–Delta Calcium Channel Anticonvulsants

Anticonvulsants of the alpha–delta calcium channel class, including both gabapentin and the newer agent pregabalin, widely reduce neuronal excitability and resemble the benzodiazepines in their ability to alter the balance between inhibitory and excitatory neuronal activity. Also, similar to benzodiazepines, these drugs have a rapid onset of action and are superior to placebo for GAD [89] and SAD [90]. Meta-analytic evidence suggests that pregabalin may even reduce depressive symptoms that co-occur with GAD [91]. These drugs have fewer problems with abuse, tolerance, and withdrawal than benzodiazepines and, in fact, have been used as treatments for both alcohol [92] and stimulant dependence [93].

Gabapentin has been examined as a therapy for treating social phobia, panic and somatoform disorders, anxiety in breast cancer survivors, and surgery-associated anxiety, with mixed results. Gabapentin was superior to placebo in the treatment of symptoms associated with social phobia, according to both patient- and clinician-rated scales, and proved superior to placebo in reducing hot flashes and anxiety in breast cancer patients who had completed chemotherapy cycles [94]. Several studies report gabapentin as effective in reducing perisurgical anxiety in otherwise psychologically healthy patients. The available data suggest that gabapentin is a potentially effective adjuvant agent in the treatment of PTSD. However, monotherapy gabapentin appears ineffective for the prevention of PTSD after a traumatic event has occurred.

3.5 The Somatic Preoccupation/Somatisation Dimension: General Considerations

The overarching category of somatic symptom disorders includes conditions that share the common feature of physical symptoms that induce undue discomfort, distress, or dysfunction. Abdominal pain, bloating, dizziness, chest pain, pelvic pain, intolerance of food, palpitations, and joint pains are common symptoms and typical reasons for doctor visits [95]. Some of these symptoms have base rates of more than 30% in the general population [96]. Although the probability of remittance is substantial for individual symptoms, many affected people have multiple symptoms that tend to be persistent. A diversity of diagnoses and labels have been suggested for these complaints, e.g. the complaints have been associated with unexplained physical symptoms, subjective health complaints, fibromyalgia, chronic fatigue syndrome, and multiple chemical sensitivity syndrome [97]. Patients affected by these disorders are associated with high utilisation of health-care systems and high cost.

Hypochondria, also known as hypochondriasis, is an extreme depression of mind or spirits often centred on imaginary physical ailments. It is also sometimes called health phobia and is referred to as an excessive preoccupation or worry about having a serious illness. Its cause is still unknown and persists even after a physician has assured the patient that he or she is healthy. The term hypochondria comes from the Greek hypo (below) and chondros (cartilage of the breast bone) and is thought to have been originally coined by Hippocrates. It was thought by many Greek physicians of antiquity that many ailments were caused by the movement of the spleen, an organ located near the hypochondrium (the upper region of the abdomen just below the ribs on either side of the epigastrium). Later use in the nineteenth century employed the term to mean, "illness without a specific cause", and it is thought that around that time period the term evolved to be the male counterpart to female hysteria. In modern usage, the term hypochondriac is often used as a pejorative label for individuals who hold the belief that they have a serious illness despite repeated reassurance from physicians that they are perfectly healthy. Hypochondria is sometimes also confused with malingering, an intentional falsification of illness.

Psychiatric nosology has never readily accommodated patients with physical symptoms that lack an organic basis and in whom psychological factors are thought to be aetiologically relevant. The terms "hysteria", "hypochondriasis", and "functional disorder" have been used more or less interchangeably, since the end of the eighteenth century, to describe such disorders. But any historical account of these disorders has to consider the term "neurosis", introduced in 1769 by Cullen. He insisted that the neuroses never resulted from "local" but from "general" alterations of the nervous system. The first description of medically unexplained symptoms in contemporary nosology was given by Paul Briquet in 1859. Patients with Briquet's syndrome feel that they have been sickly most of their lives and complain of a multitude of symptoms referable to numerous different organ systems. This conviction of illness persists despite repeatedly negative and unrevealing consultations, hospitalisations, and diagnostic procedures. Most patients gradually fall ill in their teenage years. Briquet's syndrome, currently known as somatisation disorder, is rare in males. In the mid-nineteenth century, along with the development of a technologically and anatomically oriented medicine, the concept of neurosis came under attack. These advances led to a reductionist view, based on localisation and a reduction to the anatomical level, that was in conflict with the concept of neurosis. During the latter part of the nineteenth century, neurasthenia was regarded as the quintessential "functional disorder". Neurasthenia provided the most respectable label for distressing, but not life-threatening, complaints. It was generally preferred by clinicians to its nearest alternatives: hypochondria, hysteria, and insanity. However, towards the end of the nineteenth century, the organic cause of neurasthenia became difficult to sustain. Together with other influences, such as the rise of neurology as a medical specialty and Freud's attempt to detach "anxiety attacks" from the main group of neurasthenic disorders, interest in neurasthenia began to wane. In the opinion of clinicians, neurasthenia was no longer considered as an organically determined illness managed by a neurologist; instead, it came to be viewed as a psychiatric disorder connoting vulnerability and constitutional deficiency. The word "functional" also became equated with "psychologically determined", and this use of the word has continued to the present day.

More recently the term "somatisation" has been introduced to describe patients with somatic complaints that do not have an organic basis. This term, however, is generic and subsumes a wide range of clinical phenomena. It is now used to describe the variety of processes that lead patients to seek medical help for bodily symptoms that are misattributed to organic disease. It may be acute, subacute, or chronic, and may be applied transnosographically, since it is more optimally seen as a process rather than as a disease entity. According to Lipowski, somatisation does not necessarily imply that the patient does not have a concurrent physical illness.

In the DSM-V, the disorders included under the "Somatic Symptom Disorders" heading are somatic symptom disorder, illness anxiety disorder, and conversion disorder. However, beyond those patients whose symptoms satisfy diagnostic criteria for these disorders, in the "real-world" patients with psychiatric diagnoses belonging to other areas, who present several "unexplained medical symptoms", are extremely frequent (trans-nosology of somatisation) [98].

In this section, we will refer also to those patients whose physical symptoms are not strictly confined to diagnoses of the "somatoform" area. These presentations account for 15–30% of primary health-care consultations and up to 20% of internal and neurological inpatient populations [99]. Two terms commonly used in any discussion of somatoform disorders include unexplained or "functional" somatic symptoms and hypochondriasis (illness anxiety). These terms differ in crucial ways. The former is a term used to describe somatic symptoms not caused by physical disease or tissue damage. The latter is a term that indicates an unrealistic fear or belief that one has a disease, most often based on the perception of an unexplained somatic symptom. In the case of illness anxiety disorder, the disorders carry the additional component of intrusive unpleasant thoughts about disease, compulsions to check for reassurance, and an accompanying negative appraisal of bodily symptoms that results in fear or avoidance. Because the terms illness anxiety disorder and somatic symptom disorder are often used interchangeably by primary care clinicians, it is worth emphasising that in illness anxiety disorder, the fear of a serious illness preoccupies the patient, and the compulsive checking serves to temporarily reduce the anxiety, creating a mental state and behavioural response that is quite similar to obsessive–compulsive disorder. In somatic symptom disorder, on the other hand, the primary concern is not catastrophic, life-threatening illness but concern about multiple unexplained somatic symptoms.

A somatoform disorder that presents with several different features is conversion disorder [100]. Conversion disorder, unlike the other somatoform disorders, requires a stressor to precede the onset of the loss of function. Given the often-cited symbolic significance of the part of the nervous system that is affected and given the lack of conscious awareness by the patient of the relationship between the stressor and the area of somatic dysfunction, it is clear that patients with conversion symptoms have more of a dissociative process at work rather than a primarily somatising one or obsessional thinking.

According to other authors, DSM-V somatic symptom disorders seem to neglect important clinical phenomena, such as illness denial, resulting in a narrow view of patients' functioning. In this perspective, some innovative concepts in this field may be found in the Diagnostic Criteria for Psychosomatic Research (DCPR), which were introduced in 1995 by an international group of investigators to expand the traditional domains of the disease model. Data from recent studies proved that the additional information provided by the DCPR was able to enhance the decision-making process. The DCPR are a set of 12 "psychosomatic syndromes" that provide operational tools for psychosocial variables with prognostic and therapeutic implications in clinical settings. Eight syndromes concern the main manifestations of abnormal illness behaviour: somatisation, hypochondriacal fears and beliefs, and illness denial. The other four syndromes (alexithymia, type A behaviour, demoralisation, and irritable mood) refer to the domain of psychological factors affecting medical conditions. The DCPR were found to be more sensitive than the DSM-IV in identifying subthreshold psychological distress and characterising patients' psychological response to medical illness [101].

3.6 Neurotransmission, Neurophysiology, and Neurophysiopathology of Somatic Preoccupation/Somatisation

As extensively reviewed by Rief and colleagues [102], most models of somatoform symptoms emphasise the interaction of cognitive and perceptual processes with behavioural, affective, and biological changes. Although there is evidence that all of these features contribute to the perception of physical complaints, somatoform disorders are frequently misunderstood as mere cognitive-attributional phenomena [102].

3.6.1 Autonomic Physiological Arousal

Somatoform symptoms could in theory result from heightened physiological activity. Increased physiological activation increases the likelihood of perception and misattribution of bodily signals [103]. Although it is evident that perception is substantially influenced by psychological factors, physiological activation of variables such as heart rate or others could still play an independent role. Empirical investigation of this basic hypothesis is less frequent than expected. In Pennebaker's model, the perception of physical symptoms is determined by the intensity of the interoceptive signal itself divided by the intensity of external stimulation ("distraction") [104]. This model would suggest a direct relationship between the intensity of physiological signals and the severity of somatoform symptoms. Physiological hyperreactivity would therefore be a risk factor for the development of physical symptoms. In a study by Rief and Auer [105], the psychophysiological reactivity of patients with multiple somatoform symptoms was assessed during relaxation and mental distress. The span of apprehension test, which requires continuous attentional processing, was used as a mental stressor. This is a choice reaction time task with visual stimuli including differing numbers of distracting elements. For most physiological signals included in this study (such as muscular reactivity, electrodermal responses, and peripheral circulation), no significant differences between healthy controls and patients with somatisation syndrome were found. In healthy controls, the change from attention tasks to rest periods was associated with a substantial decrease in heart rate ("recovery response"). This reduced physiological activity after mentally distressing tasks was not found in patients with somatisation syndrome. This effect was not determined by depression or anxiety.

3.6.2 The Endocrine System

The endocrine system, in particular the hypothalamic–pituitary–adrenal axis (HPA), is activated by stress and also influences pain perception. Stress and pain perceptions are both relevant to somatoform disorders. Even though among the hormones of the HPA axis, cortisol has been investigated the most, the results of these studies are equivocal. Some authors emphasise the comparability with post-traumatic stress disorders, finding that distressed patients with "unexplained" physical symptoms showed a tendency for hypocortisolism [106]. Other studies have found normal or even increased concentrations of free cortisol [107, 108]. Some authors argue that the activity of the HPA axis changes depending on the timeline of the stressors, with different HPA axis responses for acute stress vs. long-lasting, chronic stress. These variations might interact with changes in symptoms and pain perception, with hypoalgesia during acute stress, but hyperalgesia following chronic states of distress. In another study, Gaab and others confirmed normal cortisol concentrations in somatoform-associated disorders (chronic fatigue syndrome), but found evidence for enhanced glucocorticoid sensitivity in response to in vitro dexamethasone stimulation [109].

3.6.3 Somatoform Symptoms and Immunology

Immune stimulation seems to activate both analgesia and hyperalgesia circuitry [110]. Some immune parameters seem to be associated with the subjective feeling of being ill. Lekander and others [111] demonstrated that there is a correlation between self-rated health and levels of circulating cytokines. Activation of the immune system seems to induce illness behaviour patterns that are similar to those seen in depression and somatisation. Dantzer and his group investigated the effect of injecting the proinflammatory cytokine IL-1 into the brain of rats, and they could show that this induces sickness behaviours such as social withdrawal, reduction of physical activity, and others [112]. These results suggest that in some cases, immune changes can induce behaviour changes that are relevant for somatisation syndrome. However, it remains unclear whether this causality can also be bidirectional and whether it contributes especially to the development and maintenance of somatoform symptoms in humans. In depressed people, higher concentrations of parameters of the proinflammatory system have been described. Increased concentrations of soluble CD8 T-lymphocytes in depressives have been described, but the concentrations in patients with somatisation syndrome were decreased [113]. Also, the concentrations of interleukin-6, one of the major cytokines confirmed to play a role, reduced proinflammatory capacity in patients with somatoform disorders.

3.6.4 Neurotransmission

Serotonin plays a major role in various pain conditions, such as migraine [114]. Serotonin-associated disorders such as depression are typically associated with altered pain perception thresholds [115]. However, this is not the only rationale for investigating amino acids in somatoform disorders. Physical weakness, bodily exhaustion, and fatigue are not only possibly triggered by the central nervous system but can also have peripheral sources, such as energy metabolism in the muscles [116]. The concentration of branched-chain amino acids (such as valine, leucine, and isoleucine) was found to be different among patients with somatisation syndrome, depressives, and controls, with reduced concentrations in both clinical groups. These reductions were, however, more pronounced in somatisation than in depression [117]. These amino acids compete with other amino acids (such as tryptophan) at the blood–brain barrier. In addition, however, they are also relevant for energy metabolism in the muscles. Therefore, this might be a correlate of the subjective feeling of weakness which is a typical symptom not only of somatisation syndrome but also of somatoform-associated disorders such as chronic fatigue syndrome. Schwarz [116] and others found that low levels of 5-HIAA and tryptophan were related to higher pain scores in fibromyalgia patients, a result that might be relevant to somatisation syndrome. Moreover, there was a tendency towards an association of higher pain scores with higher serum concentrations of the neuropeptide substance P, suggesting an antagonistic relationship between substance P and the serotonergic system in nociception.

3.6.5 Brain Circuitries

The conscious perception of symptoms takes place in the brain. Evoked potentials reflect both attention and filtering processes [118]. Modern brain imaging techniques have been widely used in pain research, but few studies involve somatoform disorder patients. In pain research, the existence of a "pain matrix", including structures of the spinal cord, brainstem, hypothalamus, amygdala–hippocampus, prefrontal and cingulate cortex, as well as the thalamus and somatosensory cortices, is widely accepted [119]. It can be expected that at least some of these areas are also involved in the perception of somatoform symptoms.

Modern brain imaging techniques are only beginning to be applied to somatoform symptoms. It can be expected that somatosensory areas are involved, but also, both prefrontal and right parietal regions are thought to be components of a distributed neural network that integrates processes of attention and awareness [120]. Hakala and colleagues compared PET results from ten women with multiple somatoform symptoms to healthy controls [121]. They found lower glucose metabolism rates in caudate nuclei, left putamen, and right precentral gyrus. The patients also showed bilateral enlargement of caudate nuclei volumes. Abnormalities of the caudate nuclei have also been found for body dysmorphic disorder [122]. However, the specificity of these findings remains unclear. Brain imaging techniques in somatoform disorders have also been used in patients with conversion symptoms. Altered somatosensory-evoked responses in specific forebrain areas have been described [123], as well as decreased regional blood flow in the thalamus and basal ganglia contralateral to the sensorimotor deficit [124]. Experimental fMRI studies provide perhaps one of the most exciting insights into brain processes involved in the maintenance of chronic complaints. The effect of distraction on pain perception was demonstrated by Bantick et al., who found that distraction leads to reduced activity in pain-associated centres [125], again supporting a signal-filter model as presented below.

3.6.6 A Signal-Filter Model of Somatoform Symptoms

Somatoform disorders can be understood as disorders in the perception of bodily signals. Therefore, as suggested by Rief et al. [102], all biological approaches have to be discussed in light of their possible influence on the perception of bodily signals.

Most body parts send sensory signals to the brain. Due to neural filtering processes, most of these signals do not come to consciousness in healthy people. This is also the basis of the gate control theory in pain research. In somatoform disorders, physical sensations are perceived and interfere with planned behaviour and intentional thinking. Consequently, as Rief and colleagues suggest [102]: "reasons for these misperceptions can be either amplified sensory signals (e.g. strong sensory input), reduced filtering capacities, or further factors influencing the strength of the signal or the capacity of sensory filters" (e.g. selective attention because of health anxiety or immunological changes during infections: see Fig. 3.11).

In the model in Fig. 3.11, possible psychobiological and psychological influences in somatoform disorders are grouped to signal amplifying versus filtering reduction effects. The general cognitive activation theory of stress [126] can be combined easily with this signal-filter model of somatoform symptoms. The primary stress response leads to an activation, which increases physiological signals. In most people, this does not lead to prompt symptom perception, as most distressing situations offer substantial distraction. Only when the situational distraction ends and the physiological activation continues does the risk for the perception of bodily signals increase. This is especially the case in chronic states of distress.

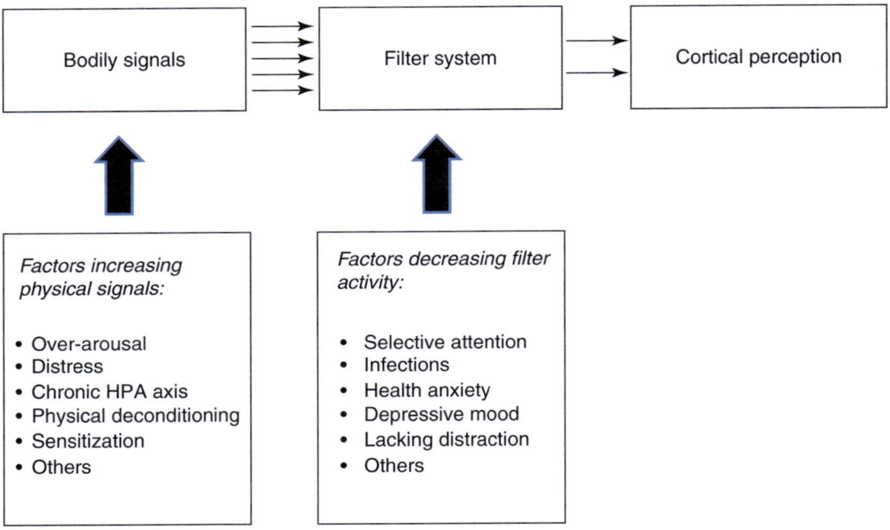

Fig. 3.11 The filter model for the Somatic Preoccupation/Somatisation dimension

As Ursin and Eriksen point out, only sustained high arousal levels constitute a potential health risk. In somatoform disorders, sensitisation might also play a role [127]. Sensitisation describes the fact that the same signals can lead to increasingly amplified perceptions. Although the bodily signal may continue to be of minor amplitude, it might be perceived as more and more intense. The repeated perception of physical signals in combination with uncertainty about the origin of the sensations can hinder the habituation that would ordinarily be expected. The cognitive component of this model has already been well described in the somatosensory amplification model [128], but sensitisation also refers to a neuronal process.

In sum, somatoform symptoms have biological components that have an important role in creating a vicious circle together with cognitive, behavioural, and emotional features. However, most of the studies cited above have specific shortcomings, e.g. most are cross-sectional in nature. Therefore, these results do not allow any conclusions regarding the sequence and timeline of the single components during development and persistence.

3.7 From Psychobiology to Clinic: The Trans-diagnostic Relevance of the Somatic Preoccupation/Somatisation Dimension in Psychiatric Disorders

The data presented in Chap. 2 further provide evidence for the existence of a Somatic Preoccupation/Somatisation (S/S) dimension cutting across traditional diagnostic boundaries.

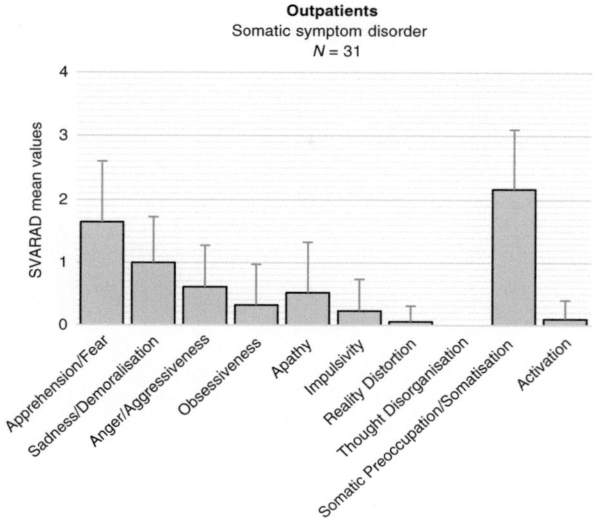

Fig. 3.12 SVARAD profile of outpatients with somatic symptom disorder: mean scores and standard deviations. Code type: 9-1-2 ($N = 31$)

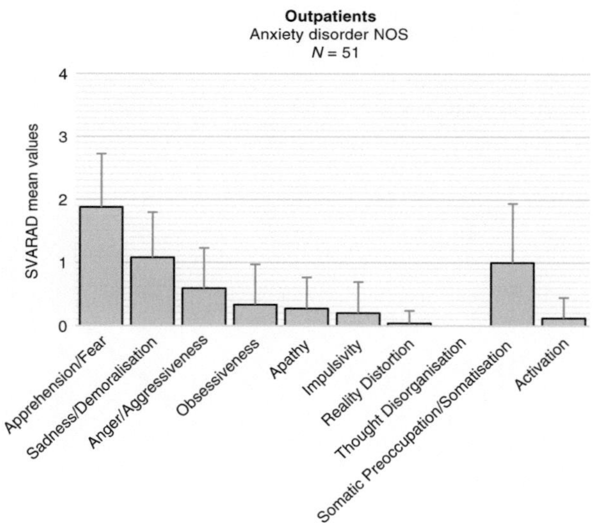

Fig. 3.13 SVARAD profile of outpatients with anxiety disorder NOS: mean scores and standard deviations. Code type: 1-2-9 ($N = 51$)

In the outpatient population ($n = 1124$), the highest S/S scores were observed in somatic symptom disorder (Fig. 3.12). High scores for S/S were also observed in anxiety spectrum disorders, particularly in individuals with anxiety disorder NOS and panic disorder, as well as in eating disorders (Figs. 3.13, 3.14, and 3.15). Not surprisingly, within single diagnostic categories, S/S was the most relevant dimension (i.e. the one having the highest scores) among those investigated using the SVARAD for somatic symptom disorders (Fig. 3.12). Despite the high values of S/S in panic disorder and anxiety NOS, patients suffering from these disorders were characterised by greater levels of A/F and comparable levels of Sadness/

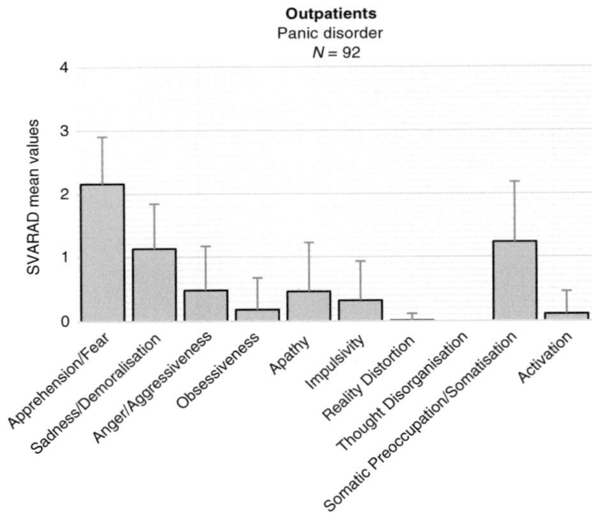

Fig. 3.14 SVARAD profile of outpatients with panic disorder: mean scores and standard deviations. Code type: 1-9-2 ($N = 92$)

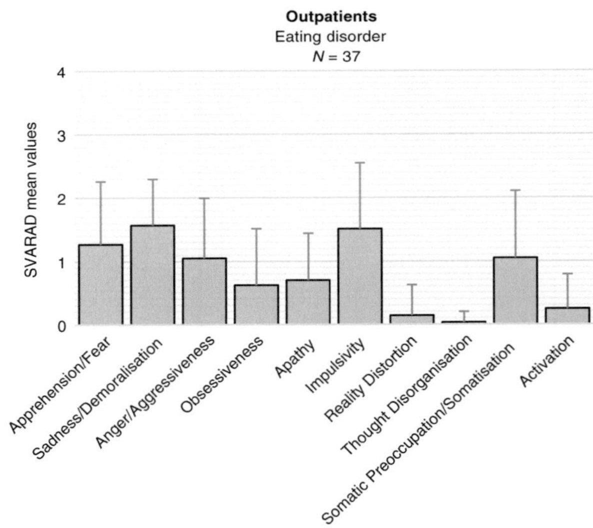

Fig. 3.15 SVARAD profile of outpatients with eating disorders: mean scores and standard deviations. Code type: 2-6-1 ($N = 37$)

Demoralisation (Figs. 3.13 and 3.14). Similarly, the dimensional profiles of eating disorders were characterised by high levels of S/S, but greater levels of A/F, Sadness/Demoralisation, and Impulsivity (Fig. 3.15).

In the inpatient population ($n = 847$), the highest S/S scores were observed in individuals with schizophrenia and schizoaffective disorders, unipolar depression (major depressive disorder and depressive disorder NOS), and OCD. Despite this, other dimensions better characterised these groups of disorders, such as Reality Distortion and A/F (for schizophrenia and schizoaffective disorder) and Obsessiveness, A/F, and Sadness/Demoralisation (for OCD and unipolar depression).

3.7.1 Clinical Cases

Case 1: Somatisation

CM, a 26-year-old homemaker, complained of symptoms of weakness and malaise of a 1-year duration. She reported abdominal pain, nausea and vomiting, headache, a burning pain in her eyes, and low back pain. She recalled having experienced sharp rectum pain when walking and mucus in her stools, a few months earlier. She was given a diagnosis of ulcerative colitis and was started on sulfasalazine therapy. Additional symptoms that had been present for short periods before the visit were urinary urgency, cough incontinence, malodourous stools, and tingling in the hands and feet.

After 20 years, at the age of 46, CM was seen for a second time in the same clinic. Her multiple somatic complaints were quite similar to those described earlier, and she reported that she had never been free of them. In addition, she was concerned that her skin was becoming darker and that her scalp hair was falling out.

Six years later, she was admitted to a psychiatric service. During the intervening years, she had received a total hysterectomy and oophorectomy, but apart from menstruation-related symptoms, she continued to have the same unrelenting physical complaints.

Case 2: Illness Anxiety

MV, a 76-year-old retired Air Force officer, was referred to a general hospital psychiatric outpatient clinic for a 9-month preoccupation with having colon cancer. He was in a general good health condition but reported having been diagnosed with coronary artery disease and diabetes (under control with oral hypoglycemics). He had no history of mental disorders. During his visit, he explained his concern of having colon cancer, a disorder that a brother and a sister had developed. As evidence of having developed a tumour, he reported diffuse abdominal pain and cited an abnormal barium enema examination 1 year earlier (that examination had revealed diverticulosis). Since he became preoccupied with having cancer, he had seen 13 physicians, all of whom had failed to reassure him that he did not have cancer. Despite his complaint, the patient denied depression. He reported, however, sleeping less than usual but attributed this problem to his abdominal discomfort. He was pleasant and cooperated well with the ward team but remained firmly convinced of having no need for any psychiatric help and remained convinced of the possibility of having cancer, despite the reassurances he was given. He refused any type of psychiatric treatment and accepted only a benzodiazepine prescription in order to improve the quality and duration of sleep.

3.8 Treatment of Somatic Preoccupation/Somatisation

3.8.1 Psychotherapy

Randomised trials have demonstrated the value of physician education in the management of the patient with somatisation [129]. Cognitive-behavioural psychotherapy strategies may be specifically helpful in reducing distress and high medical use. Psychosocial interventions directed by physicians form the basis for successful treatment. A strong relationship between the patient and the primary care physician can assist in long-term management (see ref. [102] for an extensive review). Psychoeducation can be helpful through letting the patient know that physical symptoms may be exacerbated by anxiety or other emotional problems. However, it is important to be careful because patients are likely to resist suggestions that their condition is due to emotional rather than physical problems.

The primary care physician should inform the patient that the symptoms do not appear to be due to a life-threatening, disabling medical condition and should schedule regular visits for reassessment and reinforcement of the lack of a severe physical condition underlying the ongoing symptoms. The patient may also be told that some patients with similar symptoms have had spontaneous improvement, implying that spontaneous improvement may occur. However, the physician should accept the patient's physical symptoms and not pursue a goal of symptom resolution. Indeed, regular, non-invasive medical assessment reduces anxiety and limits health-care-seeking behaviour; this may be facilitated by regularly scheduled visits with the patient's primary care physician. Patients should be encouraged to remain active and limit the effect of target symptoms on the quality of life and daily functioning. Family members should not become preoccupied with the patient's physical symptoms or medical care. Family members should direct the patient to report symptoms to his or her primary care physician.

In somatic symptom disorder, patients may resist suggestions for individual or group psychotherapy because they view their illness as a medical problem. Patients who accept psychotherapy may be able to reduce health-care utilisation. Psychosocial interventions that focus on maintaining social and occupational function despite chronic medical symptoms may be helpful [130]. Limited studies about specific types of psychotherapy exist for conversion disorder. Behaviour therapy or hypnosis may be effective. Symptoms often resolve spontaneously.

Cognitive-behavioural therapy is the most consistently supported treatment for the full spectrum of somatoform disorders [102]. A 2007 review [131] of randomised controlled trials concluded that cognitive-behavioural therapy alleviated symptoms and improved the ability to function better than a control situation or another type of therapy. Cognitive-behavioural therapy helps patients find ways to reframe and gain control of their situation and thereby break what can become a self-fulfilling cycle of pain and despair. Specific techniques used include relaxation training, problem-solving, visualisation, biofeedback, exercise, and breathing techniques. Such a multipronged approach may be exactly what is necessary in somatoform disorders. As reviewed by Rief et al. [102], the rationale to treat somatisation disorders with cognitive-behavioural techniques is based on the following determinants:

(a) Catastrophisation: anxious thoughts about the symptoms occur, and the dangerous or humiliating nature of the symptoms is overrated. For this reason, the patient gets more and more preoccupied with his symptoms; he continuously monitors himself and neglects other areas of life (e.g. amusement, leisure time activities, and certain work areas) or subordinates them to the illness. On account of this, the patient gets more and more isolated, and his or her social competence decreases. On account of the increased attention and the anxieties, the threshold of pain and the threshold of stimulus connected to body perception lowers. In turn, in response to the lowered stimulus threshold, pain and other symptoms strengthen subjectively, and this further increases the anxieties and catastrophic thinking, leading to a self-destructive cycle of catastrophisation and self-monitoring leading to further lowering of the threshold of stimulus and strengthening of symptoms.
(b) Reassurance-seeking behaviour: the patients are anxious that their complaints are the signs of some serious disease, and they see doctors partly because they seek reassurance. However, since the complaints exist despite the examination results, the value of the reassurances continuously decreases, and a distrust of the health system ("they cannot diagnose my problem…") may evolve in the end.

In illness anxiety, physicians should attempt to answer questions and reduce the patient's fear of a specific illness. Group psychotherapy may provide social support and reduce anxiety. Cognitive therapy strategies may help by focusing on distorted disease-related cognitions [132], while individual insight-oriented psychotherapies have not been proven effective. Other studies [133] have shown that cognitive-behavioural therapy reduces depressive symptoms in people with somatic diseases. In particular, this type of therapy is especially effective for patients who fit the criteria for a depressive disorder. Cognitive-behavioural therapy was superior to control conditions, with even greater effects in groups restricted to participants with depressive disorder [134]. Psychological cognitive-behavioural therapy interventions to address health management and service use were found to be feasible, cost-effective, and well-accepted in long-term frequent attenders (FAs) in primary care. On the whole, by the end of the treatment, 462 FAs cut their contacts with the health service in half [135].

3.8.2 Psychopharmacotherapy

To the extent that the two terms of functional somatic symptoms and hypochondriasis indicate different phenomena and perhaps different pathophysiology, the treatment response to one type of somatoform disorder (e.g. hypochondriasis) may have only limited bearing on the treatment responsiveness of another type of somatoform disorder (e.g. somatic symptom disorder) (please see ref. [136] for an extensive review).

The overarching category of somatic symptom disorders includes conditions that share the common feature of physical symptoms that induce undue discomfort, distress, or dysfunction. In the case of illness anxiety disorder, the disorders carry the additional component of intrusive unpleasant thoughts about disease, compulsions to check for reassurance, and an accompanying negative appraisal of bodily symptoms that results in fear or avoidance.

In these disorders, the meaning and implications of the symptoms are more distressing than the symptoms themselves. In the case of somatisation disorder and pain disorder, the symptoms themselves are the primary focus of discomfort and distress. Because the terms illness anxiety disorder and somatic symptom disorder are often used interchangeably by primary care clinicians, it is worth emphasising that in illness anxiety disorder, the fear of a serious illness preoccupies the patient and the compulsive checking serves to temporarily reduce the anxiety, creating a mental state and behavioural response that is quite similar to obsessive–compulsive disorder. In somatic symptom disorder, on the other hand, the primary concern is not catastrophic, life-threatening illness but concern about multiple unexplained somatic symptoms.

Illness anxiety disorder therefore might be considered to fall primarily within an "obsessional/cognitive cluster", whereas somatisation would fall primarily within a "somatic/sensory cluster". A somatoform disorder that may not fit well into either of these clusters is conversion disorder. Conversion disorder, unlike the other somatoform disorders, requires a stressor to precede the onset of the loss of function. Given the often-cited symbolic significance of the part of the nervous system that is affected, and given the lack of conscious awareness by the patient of the relationship between the stressor and the area of somatic dysfunction, it is clear that patients with conversion symptoms have more of a dissociative process at work than a primarily obsessional or somatising one [136].

The majority of research on the pharmacotherapy of somatoform disorders has been conducted on the obsessional cluster of somatoform disorders (hypochondriasis). To the extent that illness anxiety falls within the domain of "obsessive–compulsive spectrum" disorders [137], it should not be surprising that patients with these disorders would have a preferential pharmacologic response to agents also found to be helpful for the obsessive–compulsive disorders.

It is well known that hypochondriasis may emerge as a secondary feature of other primary psychiatric disorders, such as panic disorder or "masked" major depression, and that treatment of the underlying disorder will lead to a resolution of the hypochondriacal preoccupation. Kellner et al. [138] demonstrated that about one-third of patients with melancholic depression had scores on a hypochondriasis scale that reached a threshold identified as being characteristic of patients with hypochondriasis. After these patients were treated with amitriptyline, the hypochondriacal features resolved along with the depression. Similarly, Noyes et al. [139] reported that hypochondriasis scores among patients with panic disorder declined in parallel with the resolution of the panic attacks as a result of pharmacotherapy.

As SRIs became available and more widely used, case reports and clinical case series suggested that these agents might be helpful for hypochondriasis:

clomipramine [140, 141], fluvoxamine [142], fluoxetine [137, 143], and citalopram [144]. In one case report [142], a patient who showed no benefit to 80 mg/day of fluoxetine for 12 weeks subsequently responded very well to 300 mg/day of fluvoxamine. Therefore, patients with hypochondriasis who fail to respond to one SRI may experience benefit from an alternative SRI.

Uncontrolled open-label series have suggested efficacy associated with fluoxetine [145], fluvoxamine [146], paroxetine [147], nefazodone [148], and imipramine [149]. The fluoxetine trial lasted 12 weeks and used a flexible dosing regimen, such that patients started on 20 mg/day and had dose increases as needed to 80 mg/day. In this trial, 10 of 14 (70%) of the patients who completed the study were responders, with 4 of the 14 rated as being nearly symptom-free. Of interest, that trial demonstrated that patients without other axis I comorbidity (six of seven patients) were as likely or more likely to benefit than patients with axis I comorbidity (four of seven patients). Also, as measured by the Whiteley Index, although disease conviction and disease fear improved significantly, bodily preoccupation did not improve. The fluvoxamine trial consisted of 2 weeks of placebo followed by 10 weeks of fluvoxamine, starting at 50 mg/day and increasing weekly by 50 mg to the target dose of 300 mg/day. The responder rate to fluvoxamine of 72.7% among the 11 patients who completed at least 6 weeks was comparable to the rate reported in the fluoxetine study. Unlike the fluoxetine study, there was significant improvement in bodily preoccupation, as well as disease phobia and conviction. The 12-week paroxetine trial began with 11 patients and used a flexible dosing schedule to a target maximum of 60 mg/day. Of the nine patients who completed the trial, eight were rated as improved in hypochondriasis, five of whom were considered virtually symptom-free. The mean dose for the patients who completed the trial was 31 mg/day (S.D. 17.9 mg). In the nefazodone open-label trial, 11 patients entered the study, and nine completed the full 8 weeks of treatment (mean dose, 432 mg/day). Of these, five were rated as "much improved", with significant improvement noted in a variety of areas on the Kellner Illness Attitudes Scale (illness worry, concern about pain, hypochondriacal beliefs, and body preoccupation). The imipramine trial among hypochondriacal patients without major depression lasted for 8 weeks, with a dose schedule that increased to 150 mg/day. In this study, eight of ten patients completed at least 4 weeks, and each of these eight was at least moderately improved, although only one of the eight patients was considered symptom-free at the end of the study. In reviewing the above open-label trials, it appears that the percentage of patients considered virtually symptom-free was greater among patients given serotonin reuptake inhibitors (range, 28.6–55.6%) than among patients given the tricyclic imipramine (12.5%).

There is one published report of a placebo-controlled trial of serotonin-reuptake inhibitors in hypochondriasis. This preliminary report [142] was a mid-study analysis of the first 25 patients to enter the trial. In this 10-week study, the dose started at 20 mg/day and increased as needed to 80 mg/day. A 2-week placebo run-in before randomisation served to identify and exclude immediate placebo responders. Sixteen patients completed a minimum treatment of at least 6 weeks. Of these 16 patients, eight of ten (80%) randomised to fluoxetine were responders vs. three of six randomised to placebo (50%). In this study, five of ten patients given fluoxetine were virtually symptom-free by the end of the trial vs. only one of six placebo-treated patients. This preliminary report supported the effectiveness of fluoxetine for hypochondriasis. A presentation [150] on the final analysis of this placebo-controlled trial indicated that patients given fluoxetine did indeed experience significantly greater improvement in hypochondriasis than did patients given placebo [142].

Over the last several decades, multiple terms have been employed to cluster together unexplained physical symptoms. These include Briquet's syndrome, somatisation disorder, and abridged somatisation. Although the number of symptoms included in each syndromic entity varies considerably, the essential unifying features include medically unexplained symptoms that are associated with considerable disability, psychopathology, and high levels of health-care utilisation [151]. A literature search did not reveal any published controlled studies evaluating the efficacy of pharmacotherapy for either the full or abridged somatisation disorder diagnosis. One open-label study, however, included patients with a variety of somatoform disorders [152]. In this uncontrolled trial of fluvoxamine, 29 patients were treated—18 of whom had one somatoform disorder and 11 of whom had two somatoform disorders. Among the 23 patients who completed at least 2 weeks of treatment with fluvoxamine, 61% were at least moderately improved on global psychiatric and functional status measures. Moderate or greater improvement was noted among four of seven patients with somatisation disorder, three of six patients with pain disorder, four of seven patients with hypochondriasis, six of nine with undifferentiated somatoform disorder, and one of two with conversion disorder; some subjects had multiple disorders. Patients with comorbid major depression were just as likely to respond as those who did not have this comorbid disorder. Significant improvement in somatisation was noted in the Brief Symptom Inventory subscale index when comparing pre- and post-treatment scores. Significant improvement was also noted in anxiety, depression, and insomnia. Pain, however, proved less responsive to fluvoxamine.

References

1. Crocq M-AA. history of anxiety: from Hippocrates to DSM. Dialogues Clin Neurosci. 2015;17:319–25.
2. Kessler RC, Berglund P, Demler O, Jin R, Merikangas KR, Walters EE. Lifetime prevalence and age-of-onset distributions of DSM-IV disorders in the national comorbidity survey replication. Arch Gen Psychiatry. 2005;62:593.
3. Steimer T. The biology of fear- and anxiety-related behaviors. Dialogues Clin Neurosci. 2002;4:231–49.
4. Aldao A, Jazaieri H, Goldin PR, Gross JJ. Adaptive and maladaptive emotion regulation strategies: interactive effects during CBT for social anxiety disorder. J Anxiety Disord. 2014;28:382–9.
5. Narrow WE, Kuhl EA. Dimensional approaches to psychiatric diagnosis in DSM-5. J Ment Health Policy Econ. 2011;14:197–200.
6. Parker G, Rosen A, Trauer T, Hadzi-Pavlovic D. Disability associated with mood states and comparator conditions: application of the Life Skills Profile measure of disability. Bipolar Disord. 2007;9:11–5.

7. Kawa S, Giordano J. A brief historicity of the Diagnostic and Statistical Manual of Mental Disorders: issues and implications for the future of psychiatric canon and practice. Philos Ethics Humanit Med. 2012;7:2.
8. Shorter E. The history of nosology and the rise of the Diagnostic and Statistical Manual of Mental Disorders. Dialogues Clin Neurosci. 2015;17:59–67.
9. Klein DF. Delineation of two drug-responsive anxiety syndromes. Psychopharmacologia. 1964;5:397–408.
10. Regier DA, Kuhl EA, Kupfer DJ. The DSM-5: classification and criteria changes. World Psychiatry. 2013;12:92–8.
11. Kogan CS, Stein DJ, Maj M, First MB, Emmelkamp PMG, Reed GM. The classification of anxiety and fear-related disorders in the ICD-11. Depress Anxiety. 2016;33:1141–54.
12. Insel T, Cuthbert B, Garvey M, Heinssen R, Pine DS, Quinn K, et al. Research domain criteria (RDoC): toward a new classification framework for research on mental disorders. Am J Psychiatry. 2010;167:748–51.
13. Insel TR. The NIMH Research Domain Criteria (RDoC) project: precision medicine for psychiatry. Am J Psychiatry. 2014;171:395–7.
14. Casey BJ, Craddock N, Cuthbert BN, Hyman SE, Lee FS, Ressler KJ. DSM-5 and RDoC: progress in psychiatry research? Nat Rev Neurosci. 2013;14:810–4.
15. Lang PJ, McTeague LM. The anxiety disorder spectrum: fear imagery, physiological reactivity, and differential diagnosis. Anxiety Stress Coping. 2009;22:5–25.
16. Savage JE, Sawyers C, Roberson-Nay R, Hettema JM. The genetics of anxiety-related negative valence system traits. Am J Med Genet Part B Neuropsychiatr Genet. 2017;174:156–77.
17. Biondi M, Pasquini M. Dimensional psychopharmacology in somatising patients. Adv Psychosom Med. 2015;2:24–35.
18. Paulus MP. The breathing conundrum-interoceptive sensitivity and anxiety. Depress Anxiety. 2013;30:315–20.
19. Nash-Wright J. Dealing with anxiety disorders in the workplace. Prof Case Manag. 2011;16:55–9.
20. Emdin CA, Odutayo A, Wong CX, Tran J, Hsiao AJ, Hunn BHM. Meta-analysis of anxiety as a risk factor for cardiovascular disease. Am J Cardiol. 2016;118(4):511–9. Available from: http://www.ncbi.nlm.nih.gov/pubmed/27324160.
21. Calabrese EJ. Stress biology and hormesis: the Yerkes–Dodson Law in psychology—a special case of the hormesis dose response. Crit Rev Toxicol. 2008;38:453–62.
22. Selye H. Stress in health and disease: Elsevier Science; 1976. p. 1301
23. Zoellner LA, Foa EB. Applying Research Domain Criteria (RDoC) to the study of fear and anxiety: a critical comment. Psychophysiology. 2016;53:332–5.
24. Lenze EJ, Wetherell JLA. lifespan view of anxiety disorders. Dialogues Clin Neurosci. 2011;13:381–99.
25. Hettema JM, Neale MC, Kendler KS. A review and meta-analysis of the genetic epidemiology of anxiety disorders. Am J Psychiatry. 2001;158:1568–78.
26. Smoller JW. Disorders and borders: psychiatric genetics and nosology. Am J Med Genet Part B Neuropsychiatr Genet. 2013;162:559–78.
27. Smoller JW. The genetics of stress-related disorders: PTSD, depression, and anxiety disorders. Neuropsychopharmacology. 2016;41:297–319.
28. Howe AS, Buttenschøn HN, Bani-Fatemi A, Maron E, Otowa T, Erhardt A, et al. Candidate genes in panic disorder: meta-analyses of 23 common variants in major anxiogenic pathways. Mol Psychiatry. 2016;21:665–79.
29. Martin EI, Ressler KJ, Binder E, Nemeroff CB. The neurobiology of anxiety disorders: brain imaging, genetics, and psychoneuroendocrinology. Psychiatr Clin North Am. 2009;32:549–75.
30. Tovote P, Fadok JP, Lüthi A. Neuronal circuits for fear and anxiety. Nat Rev Neurosci. 2015;16:317–31.
31. Shin LM, Liberzon I. The neurocircuitry of fear, stress, and anxiety disorders. Neuropsychopharmacology. 2010;35:169–91.

32. Price JS. Evolutionary aspects of anxiety disorders. Dialogues Clin Neurosci. 2003;5:223–36.
33. Perusini JN, Fanselow MS. Neurobehavioral perspectives on the distinction between fear and anxiety. Learn Mem. 2015;22:417–25.
34. Fanselow MS, Lester LS, Helmstetter FJ. Changes in feeding and foraging patterns as an antipredator defensive strategy: a laboratory simulation using aversive stimulation in a closed economy. J Exp Anal Behav. 1988;50:361–74.
35. Hoffman AN, Lorson NG, Sanabria F, Foster Olive M, Conrad CD. Chronic stress disrupts fear extinction and enhances amygdala and hippocampal Fos expression in an animal model of post-traumatic stress disorder. Neurobiol Learn Mem. 2014;112:139–47.
36. Rainnie DG, Mania I, Mascagni F, McDonald AJ. Physiological and morphological characterization of parvalbumin-containing interneurons of the rat basolateral amygdala. J Comp Neurol. 2006;498:142–61.
37. LeDoux JE. Coming to terms with fear. Proc Natl Acad Sci U S A. 2014;111:2871–8.
38. Anagnostaras SG, Gale GD, Fanselow MS. Hippocampus and contextual fear conditioning: recent controversies and advances. Hippocampus. 2001;11:8–17.
39. Stockhorst U, Antov MI. Modulation of fear extinction by stress, stress hormones and estradiol: a review. Front Behav Neurosci. 2015;9:359.
40. Charney DS. Psychobiological mechanisms of resilience and vulnerability: implications for successful adaptation to extreme stress. Am J Psychiatry. 2004;161:195–216.
41. McEwen BS, Gianaros PJ. Central role of the brain in stress and adaptation: links to socioeconomic status, health, and disease. Ann N Y Acad Sci. 2010;1186:190–222.
42. Hakamata Y, Komi S, Moriguchi Y, Izawa S, Motomura Y, Sato E, et al. Amygdala-centred functional connectivity affects daily cortisol concentrations: a putative link with anxiety. Sci Rep. 2017;7:8313.
43. Banks SJ, Eddy KT, Angstadt M, Nathan PJ, Phan KL. Amygdala-frontal connectivity during emotion regulation. Soc Cogn Affect Neurosci. 2007;2:303–12.
44. Klumpp H, Keutmann MK, Fitzgerald DA, Shankman SA, Phan KL. Resting state amygdala-prefrontal connectivity predicts symptom change after cognitive behavioral therapy in generalized social anxiety disorder. Biol Mood Anxiety Disord. 2014;4:14.
45. Leweke FM, Giuffrida A, Koethe D, Schreiber D, Nolden BM, Kranaster L, et al. Anandamide levels in cerebrospinal fluid of first-episode schizophrenic patients: impact of cannabis use. Schizophr Res. 2007;94:29–36.
46. Weisskopf MG, Bauer EP, LeDoux JE, Bauer E, Sweatt J, LeDoux J. L-type voltage-gated calcium channels mediate NMDA-independent associative long-term potentiation at thalamic input synapses to the amygdala. J Neurosci. 1999;19:10512–9.
47. Dell'Osso B, Buoli M, Baldwin DS, Altamura AC. Serotonin norepinephrine reuptake inhibitors (SNRIs) in anxiety disorders: a comprehensive review of their clinical efficacy. Hum Psychopharmacol Clin Exp. 2010;25:17–29.
48. Verwaerde P, Tran MA, Montastruc JL, Senard JM, Portolan G. Effects of yohimbine, an alpha 2-adrenoceptor antagonist, on experimental neurogenic orthostatic hypotension. Fundam Clin Pharmacol. 1997;11:567–75.
49. Andrade TG, Zangrossi H, Graeff FG. The median raphe nucleus in anxiety revisited. J Psychopharmacol. 2013;27:1107–15.
50. Davis M, Whalen PJ. The amygdala: vigilance and emotion. Mol Psychiatry. 2001;6:13–34.
51. Darling RD, Alzghoul L, Zhang J, Khatri N, Paul IA, Simpson KL, et al. Perinatal citalopram exposure selectively increases locus ceruleus circuit function in male rats. J Neurosci. 2011;31:16709–15.
52. Hariri AR, Mattay VS, Tessitore A, Kolachana B, Fera F, Goldman D, et al. Serotonin transporter genetic variation and the response of the human amygdala. Science. 2002;297(5580):400–3.
53. Munafò MR, Brown SM, Hariri AR. Serotonin transporter (5-HTTLPR) genotype and amygdala activation: a meta-analysis. Biol Psychiatry. 2008;63:852–7.
54. Miller R, Wankerl M, Stalder T, Kirschbaum C, Alexander N. The serotonin transporter gene-linked polymorphic region (5-HTTLPR) and cortisol stress reactivity: a meta-analysis. Mol Psychiatry. 2013;18:1018–24.

55. Fang Z, Zhu S, Gillihan SJ, Korczykowski M, Detre JA, Rao H. Serotonin transporter genotype modulates functional connectivity between amygdala and PCC/PCu during mood recovery. Front Hum Neurosci. 2013;7:704.
56. Won E, Ham B-J. Imaging genetics studies on monoaminergic genes in major depressive disorder. Prog Neuro-Psychopharmacol Biol Psychiatry. 2016;64:311–9.
57. Enman NM, Sabban EL, McGonigle P, Van Bockstaele EJ. Targeting the neuropeptide y system in stress-related psychiatric disorders. Neurobiol Stress. 2015;1:33–43.
58. Heilig M. The NPY system in stress, anxiety and depression. Neuropeptides. 2004;38: 213–24.
59. Eiden LE. Neuropeptide-catecholamine interactions in stress. Adv Pharmacol. 2013;68: 399–404.
60. Heim C, Nemeroff CB. The role of childhood trauma in the neurobiology of mood and anxiety disorders: preclinical and clinical studies. Biol Psychiatry. 2001;49:1023–39.
61. Smith SM, Vale WW. The role of the hypothalamic-pituitary-adrenal axis in neuroendocrine responses to stress. Dialogues Clin Neurosci. 2006;8:383–95.
62. Peeke PM, Chrousos GP. Hypercortisolism and obesity. Ann N Y Acad Sci. 1995;771:665–76.
63. Willette AA, Kapogiannis D. Does the brain shrink as the waist expands? Ageing Res Rev. 2015;20:86–97.
64. Lim L, Chan HN, Chew PH, Chua SM, Ho C, Kwek SK, Lee TS, Loh P, Lum A, Tan YH, Wan YM, Woo M, Yap HL. Ministry of healthclinical practice guidelines: anxiety disorders. Singapore Med J. 2015;56:310–5.
65. Mahadevan J, Sundaresh A, Rajkumar RP, Muthuramalingam A, Menon V, Negi VS, et al. An exploratory study of immune markers in acute and transient psychosis. Asian J Psychiatr. 2017;25:219–23.
66. Ryan MCM, Sharifi N, Condren R, Thakore JH. Evidence of basal pituitary-adrenal overactivity in first episode, drug naïve patients with schizophrenia. Psychoneuroendocrinology. 2004;29:1065–70.
67. Stewart RE, Chambless DL. Cognitive-behavioral therapy for adult anxiety disorders in clinical practice: a meta-analysis of effectiveness studies. J Consult Clin Psychol. 2009;77:595–606.
68. Bandelow B, Seidler-Brandler U, Becker A. Meta-analysis of randomized controlled comparisons of psychopharmacological and psychological treatments for anxiety disorders. World J Biol Psychiatry. 2007;8:175–87.
69. Butler AC, Chapman JE, Forman EM, Beck AT. The empirical status of cognitive-behavioral therapy: a review of meta-analyses. Clin Psychol Rev. 2006;26:17–31.
70. Hoffmann SG, Sawyer AT, Korte KJ, Smits JA. Is it beneficial to add pharmacotherapy to cognitive-behavioral therapy when treating anxiety disorders? A meta-analytic review. Int J Cogn Ther. 2009;2:160–75.
71. Borgeat F, Stankovic M, Khazaal Y, Rouget BW, Baumann MC, Riquier F, et al. Does the form or the amount of exposure make a difference in the cognitive-behavioral therapy treatment of social phobia? J Nerv Ment Dis. 2009;197:507–13.
72. Bisson J, Andrew M. Psychological treatment of post-traumatic stress disorder (PTSD). Cochrane Database Syst Rev. 2007:CD003388.
73. Pollack MH, Otto MW, Roy-Byrne PP, Coplan JD, Rothbaum BO, Simon NM, et al. Novel treatment approaches for refractory anxiety disorders. Focus. 2008;6:486–95.
74. Ravindran LN, Stein MB. The pharmacologic treatment of anxiety disorders. J Clin Psychiatry. 2010;71:839–54.
75. Kaizar EE, Greenhouse JB, Seltman H, Kelleher K. Do antidepressants cause suicidality in children? A Bayesian meta-analysis. Clin Trials. 2006;3:73–98.
76. Warden D, Rush AJ, Carmody TJ, Kashner TM, Biggs MM, Crismon ML, et al. Predictors of attrition during one year of depression treatment: a roadmap to personalized intervention. J Psychiatr Pract. 2009;15:113–24.
77. Fournier JC, DeRubeis RJ, Hollon SD, Dimidjian S, Amsterdam JD, Shelton RC, et al. Antidepressant drug effects and depression severity. JAMA. 2010;303:47–53.

78. Gibbons RD, Hur K, Brown CH, Davis JM, Mann JJ. Benefits from antidepressants: synthesis of 6- week patient-level outcomes from double-blind placebo-controlled randomized trials of fluoxetine and venlafaxine. Arch Gen Psychiatry. 2012;21:572–9.
79. Schmitt R, Gazalle FK, Lima MS, de Cunha Ã, Souza J, Kapczinski F. The efficacy of antidepressants for generalized anxiety disorder: a systematic review and meta-analysis. Rev Bras Psiquiatr. 2005;27:18–24.
80. Bakish D, Hooper CL, West DL, Miller C, Blanchard A, Bashir F. Moclobemide and specific serotonin re-uptake inhibitor combination treatment of resistant anxiety and depressive disorders. Hum Psychopharmacol Clin Exp. 1995;10:105–9.
81. Schutters SIJ, Van Megen HJGM, Van Veen JF, Denys DAJP, Westenberg HGM. Mirtazapine in generalized social anxiety disorder: a randomized, double-blind, placebo-controlled study. Int Clin Psychopharmacol. 2010;25:302–4.
82. Davidson JRT, Weisler RH, Butterfield MI, Casat CD, Connor KM, Barnett S, et al. Mirtazapine vs. placebo in posttraumatic stress disorder: a pilot trial. Biol Psychiatry. 2003;53:188–91.
83. Van Ameringen M, Mancini M, Oakman C, Walker J, Kjernisted J, Chokka P, et al. Nefazodone in the treatment of generalized social phobia: a randomized, placebo-controlled trial. J Clin Psychiatry. 2007;68:288–95.
84. Rickels K, Downing R, Schweizer E, Hassman H. Antidepressants for the treatment of generalized anxiety disorder: a placebo-controlled comparison of imipramine, trazodone, and diazepam. Arch Gen Psychiatry. 1993;50:884–95.
85. Becker ME, Hertzberg MA, Moore SD, Dennis MF, Bukenya DS, Beckham JC. A placebo-controlled trial of bupropion SR in the treatment of chronic posttraumatic stress disorder. J Clin Psychopharmacol. 2007;27:193–7.
86. Simon NM, Otto MW, Worthington JJ, Hoge EA, Thompson EH, LeBeau RT, et al. Next-step strategies for panic disorder refractory to initial pharmacotherapy. J Clin Psychiatry. 2009;70:1563–70.
87. Westra HA, Stewart SH, Conrad BE. Naturalistic manner of benzodiazepine use and cognitive behavioral therapy outcome in panic disorder with agoraphobia. J Anxiety Disord. 2002;16:233–46.
88. Otto MW, McHugh RK, Simon NM, Farach FJ, Worthington JJ, Pollack MH. Efficacy of CBT for benzodiazepine discontinuation in patients with panic disorder: further evaluation. Behav Res Ther. 2010;48:720–7.
89. Feltner DE, Wittchen H-U, Kavoussi R, Brock J, Baldinetti F, Pande AC. Long-term efficacy of pregabalin in generalized anxiety disorder. Int Clin Psychopharmacol. 2008;23:18–28.
90. Feltner DE, Liu-Dumaw M, Schweizer E, Bielski R. Efficacy of pregabalin in generalized social anxiety disorder. Int Clin Psychopharmacol. 2011;26:213–20.
91. Stein DJ, Baldwin DS, Baldinetti F, Mandel F. Efficacy of pregabalin in depressive symptoms associated with generalized anxiety disorder: a pooled analysis of 6 studies. Eur Neuropsychopharmacol. 2008;18:422–30.
92. Furieri FA, Nakamura-Palacios EM. Gabapentin reduces alcohol consumption and craving: a randomized, double-blind, placebo-controlled trial. J Clin Psychiatry. 2007;68:1691–700.
93. Urschel HC, Hanselka LL, Baron M. A controlled trial of flumazenil and gabapentin for initial treatment of methylam-phetamine dependence. J Psychopharmacol. 2011;25:254–62.
94. Houghton KT, Forrest A, Awad A, Atkinson LZ, Stockton S, Harrison PJ, Geddes JR, Cipriani A. Biological rationale and potential clinical use of gabapentin and pregabalin in bipolar disorder, insomnia and anxiety: protocol for a systematic review and meta-analysis. BMJ Open. 2017;7:1–8.
95. Katon WJ, Walker EA. Medically unexplained symptoms in primary care. J Clin Psychiatry. 1998;59(Suppl 20):15–21.
96. van Eck van der Sluijs J, Ten Have M, Rijnders C, van Marwijk H, de Graaf R, van der Feltz-Cornelis C. Medically unexplained and explained physical symptoms in the general population: association with prevalent and incident mental disorders. PLoS One. 2015;10:4.

97. Rosendal M, Olde Hartman TC, Aamland A, van der Horst H, Lucassen P, Budtz-Lilly A, et al. "Medically unexplained" symptoms and symptom disorders in primary care: prognosis-based recognition and classification. BMC Fam Pract. 2017;18:18.
98. Dowrick C. Medically unexplained symptoms in primary care: how can doctors help, not hinder? Ment Health Fam Med. 2010;7:191–2.
99. Barsky AJ, Peekna HM, Borus JF. Somatic symptom reporting in women and men. J Gen Intern Med. 2001;16:266–75.
100. Harvey SB, Stanton BR, David AS. Conversion disorder: towards a neurobiological understanding. Neuropsychiatr Dis Treat. 2006;2:13–20.
101. Sirri L, Fava GA. Diagnostic criteria for psychosomatic research and somatic symptom disorders. Int Rev Psychiatry. 2013;25:19–30.
102. Rief W, Barsky AJ. Psychobiological perspectives on somatoform disorders. Psychoneuroendocrinology. 2005;30:996–1002.
103. Liu L, Cohen S, Schulz MS, Waldinger RJ. Sources of somatization: exploring the roles of insecurity in relationships and styles of anger experience and expression. Soc Sci Med. 2011;73:1436–43.
104. Pennebaker JW. Putting stress into words: health, linguistic, and therapeutic implications. Behav Res Ther. 1993;31:539–48.
105. Rief W, Auer C. Is somatization a habituation disorder? Physiological reactivity in somatization syndrome. Psychiatry Res. 2001;101:63–74.
106. Heim C, Ehlert U, Hellhammer DH. The potential role of hypocortisolism in the pathophysiology of stress-related bodily disorders. Psychoneuroendocrinology. 2000;25:1–35.
107. Rief W, Shaw R, Fichter MM. Elevated levels of psychophysiological arousal and cortisol in patients with somatization syndrome. Psychosom Med. 1998;60:198–203.
108. Rief W, Auer C. Cortisol and somatization. Biol Psychol. 2000;53:13–23.
109. Gaab J, Rohleder N, Nater UM, Ehlert U. Psychological determinants of the cortisol stress response: the role of anticipatory cognitive appraisal. Psychoneuroendocrinology. 2005;30:599–610.
110. Watkins LR, Maier SF. The pain of being sick: implications of immune-to-brain communication for understanding pain. Annu Rev Psychol. 2000;51:29–57.
111. Lekander M, Elofsson S, Neve IM, Hansson LO, Unden AL. Self-rated health is related to levels of circulating cytokines. Psychosom Med. 2004;66:559–63.
112. Dantzer R, Bluthé RM, Layé S, Bret-Dibat JL, Parnet P, Kelley KW. Cytokines and sickness behavior. Ann N Y Acad Sci. 1998;840:586–90.
113. Rief W, Pilger F, Ihle D, Bosmans E, Egyed B, Maes M. Immunological differences between patients with major depression and somatization syndrome. Psychiatry Res. 2001; 105:165–74.
114. Hamel E, Currents H. Serotonin and migraine: biology and clinical implications. Cephalalgia. 2007;27:1293–300.
115. Marazziti D, Castrogiovanni P, Rossi A, Rosa C, Ghione S, Di Muro A, et al. Pain threshold is reduced in depression. Int J Neuropsychopharmacol. 1998;1:1.
116. Schwarz MJ, Späth M, Müller-Bardorff H, Pongratz DE, Bondy B, Ackenheil M. Relationship of substance P, 5-hydroxyindole acetic acid and tryptophan in serum of fibromyalgia patients. Neurosci Lett. 1999;259:196–8.
117. Rief W, Pilger F, Ihle D, Verkerk R, Scharpe S, Maes M. Psychobiological aspects of somatoform disorders: contributions of monoaminergic transmitter systems. Neuropsychobiology. 2004;49:24–9.
118. Mobascher A, Brinkmeyer J, Warbrick T, Musso F, Wittsack HJ, Saleh A, et al. Laser-evoked potential P2 single-trial amplitudes covary with the fMRI BOLD response in the medial pain system and interconnected subcortical structures. Neuroimage. 2009; 45:917–26.
119. Jones AKP, Kulkarni B, Derbyshire SWG. Pain mechanisms and their disorders. Br Med Bull. 2003;65:83–93.
120. Boeckle M, Schrimpf M, Liegl G, Pieh C. Neural correlates of somatoform disorders from a meta-analytic perspective on neuroimaging studies. NeuroImage Clin. 2016;11:606–13.

121. Hakala M, Karlsson H, Ruotsalainen U, Koponen S, Bergman J, Stenman H. Severe somatization in women is associated with altered cerebral glucose metabolism. Psychol Med. 2002;3:1379–85.
122. Rauch SL, Phillips KA, Segal E, Makris N, Shin LM, Whalen PJ. A preliminary morphometric magnetic resonance imaging study of regional brain volumes in body dysmorphic disorder. Psychiatry Res. 2003;122:13–9.
123. Mailis-Gagnon A, Giannoylis I, Downar J, Kwan CL, Mikulis DJ, Crawley AP. Altered central somatosensory processing in chronic pain patients with "hysterical" anesthesia. Neurology. 2003;60:1501–7.
124. Vuilleumier P, Chicherio C, Assal F, Schwartz S, Slosman D, Landis T. Functional neuroanatomical correlates of hysterical sensorimotor loss. Brain. 2001;124:1077–90.
125. Bantick SJ, Wise RG, Ploghaus A, Clare S, Smith SM, Tracey I. Imaging how attention modulates pain in humans using functional MRI. Brain. 2002;125:310–9.
126. Eriksen H, Ursin H. Subjective health complaints, sensitization, and sustained cognitive activation (stress). J Psychosom Res. 2004;56:445–8.
127. Ursin H. Sensitization, somatization, and subjective health complaints. Int J Behav Med. 1997;4:105–16.
128. Barsky AJ. Amplification, somatization, and the somatoform disorders. Psychosomatics. 1992;33:28–34.
129. Martin A, Rauh E, Fichter M, Rief W. A one-session treatment for patients suffering from medically unexplained symptoms in primary care: a randomized clinical trial. Psychosomatics. 2007;48:294–303.
130. Smith GR Jr, Monson RA, Ray DC. Psychiatric consultation in somatization disorder. A randomized controlled study. N Engl J Med. 1986;314:1407–13.
131. Kroenke K. Efficacy of treatment for somatoform disorders: a review of randomized controlled trials. Psychosom Med. 2007;69:881–8.
132. Barsky AJ, Ahern DK. Cognitive behavior therapy for hypochondriasis: a randomized controlled trial. JAMA. 2004;291:1464–70.
133. Bleichhardt G, Timmer B, Rief W. Cognitive-behavioural therapy for patients with multiple somatoform symptoms--a randomised controlled trial in tertiary care. J Psychosom Res. 2004;56:449–54.
134. Beltman MW, Voshaar RC, Speckens AE. Cognitive-behavioural therapy for depression in people with a somatic disease: meta-analysis of randomised controlled trials. Br J Psychiatry. 2010;197:11–29.
135. Malins S, Kai J, Atha C, Avery A, Guo B, James M, Patel S, Sampson C, Stubley M, Morriss R. Cognitive behaviour therapy for long-term frequent attenders in primary care: a feasibility case series and treatment development study. Br J Gen Pract. 2016;66:e729–36.
136. Fallon BA. Pharmacotherapy of somatoform disorders. J Psychosom Res. 2004;56:455–60.
137. Fallon BA, Javitch JA, Hollander E, Liebowitz MR. Hypochondriasis and obsessive compulsive disorder: overlaps in diagnosis and treatment. J Clin Psychiatry. 1991;52:457–60.
138. Kellner R, Fava GA, Lisansky J, Perini GI, Zielezny M. Hypochondriacal fears and beliefs in DSM-III melancholia: changes with amitriptyline. J Affect Disord. 1986;10:21–6.
139. Noyes R, Reich J, Clancy J, O'Gorman TW. Reduction in hypochondriasis with treatment of panic disorder. Br J Psychiatry. 1986;149:631–5.
140. Kamlana SH, Gray P. Fear of AIDS (letter). Br J Psychiatry. 1988;15:1291.
141. Stone AB. Treatment of hypochondriasis with clomipramine (letter). J Clin Psychiatry. 1993;54:5.
142. Fallon BA, Schneier FR, Marshall R, Campeas R, Vermes D, Goetz D, Liebowitz MR. The pharmacotherapy of hypochondriasis. Psychopharmacol Bull. 1996;32:607–11.
143. Viswanathan R, Paradis C. Treatment of cancer phobia with fluoxetine. Am J Psychiatry. 1991;148:1090.
144. Fallon BA, Feinstein S. Hypochondriasis. In: Phillips KA, editor. Somatoform and fictius disorders. Washington, DC: America Psychiatric Publishing; 2001.

145. Fallon BA, Liebowitz MR, Salomon E, Schneier FR, Jusino C, Holander E, Klein DF. Fluoxetine for hypochondriacal patients without major depression. J Clin Psychopharmacol. 1993;13:438–41.
146. Fallon BA, Qureshi AI, Schneier FR, Sanchez-Lacay A, Vermes D, Feinstein R, Connelly J, Liebowitz MR. An open trial of fluvoxamine for hypochondriasis. Psychosomatics. 2003;44:298–303.
147. Oosterbaan DV, van Balkom AJ, van Boijen CA, De Metji TG, van Dick R. An open study of paroxetine in hypochondriasis. Prog Neuropsychopharmacol Biol Psychiatry. 2001;25:1023–33.
148. Kjernisted KD, Enns MW, Lander M. An open clinical trial of nefazodone in hypochondriasis. Psychosomatics. 2002;43:290–4.
149. Wesner RB, Noyes R. Imipramine: an effective treatment for illness phobia. J Affect Disord. 1992;22:43–8.
150. Fallon BA. Pharmacotherapy of hypochondriasis. Marburg: International Congress on Somatoform Disorders; 2002.
151. Escobar JR, Rubio-Stipec R, Canino G, Karno M. Somatic Symptom Index (SSI): a new and abridged somatization construct. J Nerv Ment Dis. 1989;177:140–6.
152. Noyes R, Happel R, Muller B, Holt CS, Kathol RG, Sieren L, Amos J. Fluvoxamine for somatoform disorders: an open trial. Gen Hosp Psychiatry. 1998;20:339–44.

The Reality Distortion and Thought Disorganisation Dimensions

4

Lorenzo Tarsitani and Annalisa Maraone

4.1 The Reality Distortion Dimension

4.1.1 Introduction

The Reality Distortion dimension is characterised by both the erroneous perception of reality and the inability to discriminate reality from fantasy. *Reality* is a concept that essentially refers to social consensus and to the relationship between an individual's inner and outer world. What is shared and accepted by collectivity can be considered as real; on the other hand, what differs from collective consensus is considered to be not real or incomprehensible.

The Reality Distortion dimension includes a variety of symptoms that share a failure to distinguish between reality and imagination and the attribution of unusual sense, meaning, or relevance to stimuli. The Reality Distortion dimension can include disorders of the content of thought, ranging from prevalent ideas to delusions, as well as disorders of perception, including illusions and hallucinations. A synthesis of this concept can be found in the item "Reality Distortion" of the SVARAD (or the English version, RADAS; see Chap. 2): *Difficulty distinguishing between reality and fantasy; tendency to attribute unusual and unshared meanings to events or experiences; presence of delusions or hallucinations.*

The word delusion (*Wahn* in German) derives from Latin (*delirium*) and it means *jumping out of the furrow* (*lira*). Kraepelin [1] described delusion as pathologically derived mistakes that are not responsive to correction despite logical evidence to the contrary. Jaspers [2] considered a delusional idea to be a distorted view of reality (an erroneous idea), incorrigibly held with absolute conviction. These qualities were crucial in discriminating delusion from other beliefs, from a categorical perspective

L. Tarsitani (✉) · A. Maraone
Department of Human Neurosciences, Sapienza University of Rome, Rome, Italy
e-mail: lorenzo.tarsitani@uniroma1.it; annalisa.maraone@uniroma1.it

M. Biondi et al. (eds.), *Dimensional Psychopathology*,
https://doi.org/10.1007/978-3-319-78202-7_4

[3]. A delusion could be a way to explain a hallucinatory experience [3]. The latter was defined by Esquirol [4] as "the intimate conviction of actually perceiving a sensation for which there is no external object" [5, 6].

Delusions and hallucinations are characterised by some sense of distorted reality and are generally defined as psychotic symptoms—or, specifically, "positive symptoms"—in the context of psychotic disorders.

At the beginning, the term psychosis was used by Canstatt [7] and, then, by von Feuchtersleben [8], meaning "psychic neurosis". Afterwards, "psychosis" was usually used as a synonym for mental disorder or illness, as well as for insanity [9]. Later, the term psychosis was used referring to mental disorders with established, or with supposed, organic basis [10]. Then, the difference between endogenous and exogenous was described. Endogenous psychoses included the spectrum of hysteria, melancholy, mania, and paranoia [11]. Exogenous mental diseases were instead caused by any extraneous influence, somatic or psychological in nature. Kraepelin [12] and Bleuler [13] divided major psychoses into manic-depressive and schizophrenic disorders, on the basis of the clinical course of the disease [14, 15].

Jaspers [16], in 1913, first described the dichotomy of psychosis and neurosis (or nonpsychosis). Psychoses were always considered as resulting from somatic disease, while neuroses had a psychological etiology. This was followed by the hypothesis that psychoses had an explanation of biological cause, while neuroses could be comprehended in terms of psychology (behaviour, cognition, and social factors) alone.

Moreover, psychosis was assumed to be a brain disease identifiable by psychological symptomatology. Consequently, psychopathology has always considered psychotic symptoms as clearly distinguished from normality. Delusions had to be clearly different from normal ideas and beliefs. This view of psychosis as a category defined by the presence of psychotic symptoms has influenced psychiatric nosography for almost a century.

Over the past few decades, the dichotomous definition of psychosis has been questioned with new proposed descriptions in favour of a continuum view [17–21]. In fact, psychotic experiences and symptoms of psychosis can be observed in patients with nonpsychotic mental disorders and in non-clinical samples [22].

A systematic review of studies on non-clinical samples suggested that symptoms typical of schizophrenia and related disorders, such as paranoid delusional thinking and auditory hallucinations, occur in 5–8% of individuals [23]. In particular, a study reported that auditory hallucinations have a prevalence of 10–15% in persons without mental disorders [24]. Similar figures were reported in surveys of college students [25, 26]. Those data presented a correlation with age and gender [24], as well as culture and ethnicity [27, 28], with specific correlation with ethnic minority groups [22, 29]. For example, in a large-scale community study of diverse populations in England and Wales, hallucinations were found to be more frequent in Caribbean, and less frequent in South Asian, as compared with White, British residents [30].

According to this new perspective, the Reality Distortion dimension extends into a continuum between normality and psychopathology (Fig. 4.1) in which quantitative rather than qualitative differences can be observed.

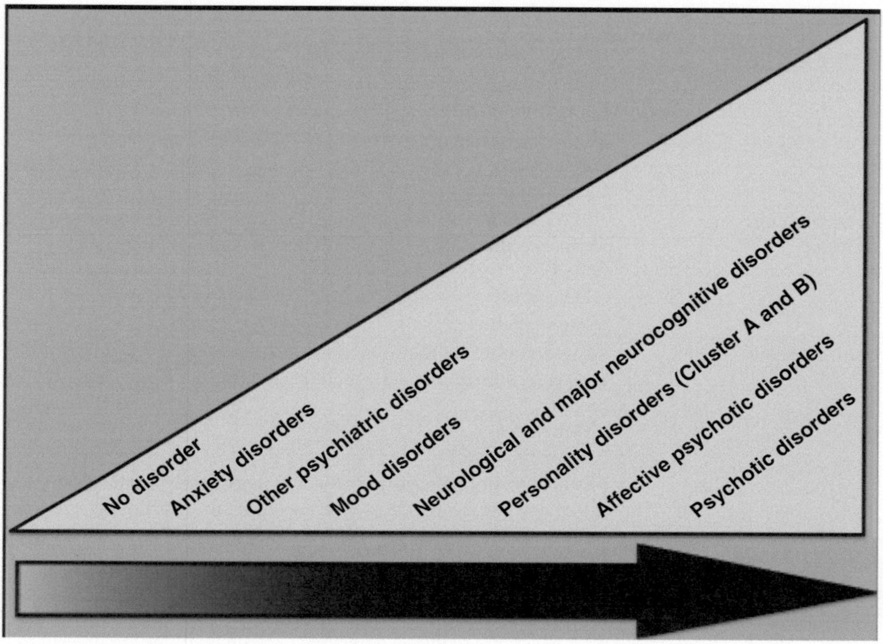

Fig. 4.1 The continuum of Reality Distortion across clinical conditions

The continuum of psychosis is defined as a distribution along a continuum on which nonpsychotic affective disorders and affective psychosis constitute an intermediate point that connect normal psychological experiences and psychotic disorders [31, 32].

However, although subclinical psychotic experiences in healthy individuals are usually transitory in about 80% of cases, some subjects may develop persistent psychotic experiences (20%) or a psychotic disorder (7%) [22, 33]. These figures suggested the hypothesis of an "extended psychosis phenotype" (including demographic, environmental, familial, and psychopathological features) that is both phenomenologically and temporally continuous with clinical psychotic disorder [34]. In other words, psychotic experiences are not exclusive to patients with a categorically defined psychotic disorder ("phenomenological continuity"), but these can persist in some individuals, resulting in a de facto psychotic disorder ("temporal continuity") [22]. It is possible to observe psychotic symptoms in a variety of nonpsychotic mental disorders such as mood or anxiety disorders [35, 36].

The Diagnostic and Statistical Manual of Mental Disorders 5 (DSM-5) [37] confirms the presence of reality distortion across mental disorders, including delusions and/or hallucinations outside the psychotic disorders, as summarised in Table 4.1.

Using the Community Assessment of Psychic Experiences (CAPE), a scale based on a 40-item self-report instrument with positive, negative, and depressive symptom dimensions, Hanssen and colleagues [35] observed that subjects with

Table 4.1 The Reality Distortion dimension in DSM-5 diagnostic criteria [37]

Mental disorder	Reality distortion-related diagnostic criteria
Delusional disorder	– The presence of one (or more) delusions with a duration of 1 month or longer
Brief psychotic disorder	– Presence of one (or more) of the following symptoms: 1. Delusions 2. Hallucinations
Schizophreniform disorder	– Two (or more) of the following, each present for a significant portion of time during a 1-month period (or less if successfully treated): 1. Delusions 2. Hallucinations
Schizophrenia	– Two (or more) of the following, each present for a significant portion of time during a 1-month period (or less if successfully treated): 1. Delusions 2. Hallucinations
Schizoaffective disorder	– Delusions or hallucinations for 2 or more weeks in the absence of a major mood episode (depressive or manic) during the lifetime duration of the illness
Substance/medication-induced psychotic disorder	– Presence of one or both of the following symptoms: 1. Delusions 2. Hallucinations
Bipolar I disorder	With psychotic features, delusions or hallucinations are present at any time in the episode: – With mood-congruent psychotic features – With mood-incongruent psychotic features
Major depressive disorder	With psychotic features, delusions and/or hallucinations are present: – With mood-congruent psychotic features – With mood-incongruent psychotic features
Anorexia nervosa	– Disturbance in the way in which one's body weight or shape is experienced, undue influence of body weight or shape on self-evaluation, or persistent lack of recognition of the seriousness of the current low body weight
Delirium	– An additional disturbance in cognition (e.g. disturbed visuospatial ability or perception)
Major or mild neurocognitive disorder with Lewy bodies	– Recurrent visual hallucinations that are well formed and detailed
Other hallucinogen intoxication	– Clinically significant changes (e.g. ideas of reference, paranoid ideation, impaired judgment) that developed during hallucinogen use – Perceptual changes (e.g. subjective intensification of perceptions, illusions, hallucinations, synesthesias) that developed during hallucinogen use
Hallucinogen persisting perception disorder	– The re-experiencing of perceptual symptoms that were experienced while intoxicated with a hallucinogen (e.g. geometric hallucinations, flashes of colour, intensified colours, trails of images of moving objects, macropsia, and micropsia)

anxiety and mood disorders show high scores on positive psychosis items. The authors suggest that common psychotic experiences may be found in nonpsychotic disorders and that these disorders represent intermediate points along the continuum of the psychosis phenotype. Wigman et al. [38] reported a higher prevalence of psychotic experiences in patients with depression or anxiety disorder as compared with subjects without these disorders.

Almost half of soldiers with post-traumatic stress disorder (PTSD) may present psychotic symptoms [39]. Psychotic features in PTSD are rarely due to a primary psychotic disorder but could occur in a distinct subtype of PTSD. These types of experiences are clinically different from typical psychotic symptoms and might be distinct in their etiology and pathogenesis, but they can be included in a Reality Distortion dimension. Similarly, some authors have suggested that patients with poor-insight obsessive-compulsive disorders, who firmly believe that compulsions serve to prevent bad thoughts from manifesting as reality and actually happening, might be considered delusional [40]. Although challenging from a phenomenological viewpoint, some obsessions might be included in the Reality Distortion dimension [41].

Usually, psychosis is common in the elderly, and multiple etiologies for late-life psychosis are recognisable, as any pathological cerebral process may present as a psychosis. Moreover, neuropsychiatric etiologies of psychosis comprise both chronic disorders such as dementias or other chronic neurological conditions and acute, sometimes reversible, conditions such as delirium, alcohol use or withdrawal, and some acute neurological diseases. Ropacki and Jeste [42] reviewed 55 studies on psychosis during Alzheimer's disease and observed that psychotic symptoms had a prevalence of 41%, delusions had a prevalence of 36%, and hallucinations had a prevalence of 18%. Furthermore, psychotic symptoms may complicate other non-Alzheimer's dementias like vascular dementia, dementia associated with Parkinson's disease, and Lewy-Body dementia. In fact, Ostling [43] found that delusions were more frequent in vascular dementia than in Alzheimer's disease. Patients with Parkinsonism and psychosis, including both dementia associated with Parkinson's disease and Lewy-Body dementia, may experience hallucinosis, which can be defined as benign or complex psychotic symptoms. In fact, subjects with Parkinson's disease without dementia can present with benign hallucinosis that occurs as mild visual perceptual disturbances for which patients usually preserve insight and which generally do not need any pharmacological treatment. On the other hand, a study of 239 community-based patients with Parkinson's disease (followed up for over 12 years) observed that 60% of these patients had psychotic symptoms [44].

4.1.2 Neurobiology of the Reality Distortion Dimension

Isolated positive psychotic symptoms are not easily accessible to neuroscience and are very difficult to model in animals because of their subjective (beliefs, perceptions) and dimensional nature. In fact, the neurobiology of Reality Distortion can mainly be inferred from studies in patients with schizophrenia.

Studies using structural brain imaging have described a subtle, nearly universal decrease in grey matter, enlargement of ventricles, and focal alteration of white matter tracts. However, despite the large body of data from studies and meta-analysis, the role of the abnormalities typical of schizophrenia in the pathophysiology of psychotic symptoms is not completely understood.

One of the most consistent neuroimaging findings in schizophrenia is lateral ventricular enlargement [45, 46]. Enlarged lateral ventricles in patients with schizophrenia were first observed in the 1970s by computed tomography (CT) [47], and this finding was later confirmed by magnetic resonance imaging (MRI) studies. In particular, recent meta-analytic reviews have indicated that, compared with healthy controls, patients with schizophrenia present reduced brain size [48, 49], enlarged lateral and third ventricles [49, 50], reduced frontal lobe volume [49], reduced volumes of temporo-limbic structures [48, 49, 51] and corpus callosum [52], and increased volume of the basal ganglia [49].

A meta-analysis of MRI studies conducted on brain morphology in first-episode schizophrenic patients showed that, while some brain abnormalities were already present during the first episode, other abnormalities found in chronic schizophrenic patients were absent from these first-episode patients. These data support the hypothesis that schizophrenia presents with diverse involvement of different cerebral areas over time [53].

Csernansky and Cronenwett [54] suggested that in schizophrenia, although brain changes may involve abnormalities in a network of grey and white matter regions, these changes have been more specifically mapped in grey matter than in white matter. In particular, grey matter reductions have been observed in limbic, paralimbic, and frontal cortical regions, as well as the thalamus [55–57]. In contrast, the distribution of white matter changes was less defined [58, 59]. Ellison-Wright and Bullmore [60] conducted a meta-analysis of studies using diffusion tensor imaging (DTI) to test the implications of white matter changes in schizophrenia. They identified two consistent locations of fractional anisotropy reduction: one in the left frontal lobe, with white matter tracts interconnecting the frontal lobe, thalamus, and cingulate gyrus, and the other in the temporal lobe, with tracts interconnecting the frontal lobe, insula, hippocampus-amygdala, temporal lobe, and occipital lobe. These results suggest that two networks of white matter tracts may be implicated in schizophrenia and that changes in these networks may lead to the potential for disconnection of the associated grey matter regions. In fact, it has been hypothesised that the pathogenesis of auditory hallucinations (AH) may be related to alterations in connectivity between frontal and parietotemporal speech-related areas. Moreover, it has also been proposed that schizophrenia involves altered frontoparietotemporal connectivity [61, 62]. Hubl et al. [63] hypothesised that altered neuronal activity during AH is related to changes in structural interconnections between the frontal and parietotemporal speech-related areas in subjects with frequent AH. Using diffusion tensor imaging (DTI), significant alteration of white matter fibres in patients who had AH was observed in an important area of connection between language-related frontal and temporal regions.

A recent DTI study revealed a positive correlation between the Scale for the Assessment of Positive Symptoms (SAPS) hallucination score [64] and fractional anisotropy in the left uncinate fasciculus and left corticospinal tract [65]. These results are in line with those from other recent studies showing significantly altered fractional anisotropy within the left corticospinal tract in patients with current hallucinations [66].

An interesting hypothesis on the development of specific psychotic symptoms is the failure of a mechanism called corollary discharge, in psychotic states and during the experience of psychotic symptoms [67]. Corollary discharge is a neurophysiological mechanism that allows the recognition of perceptions resulting from self-generated movements or speech. An impairment of corollary discharge throughout auditory or motor sensory systems could lead to external misattribution of individual movements or speech. Such a mechanism could underlie psychotic symptoms, such as delusions of alien control and auditory hallucinations. Although thus far only demonstrated in patients with schizophrenia [68], a failure of corollary discharge could also lead to attenuated psychotic symptoms related to external misattribution of self-generated actions in individuals without a diagnosis of psychotic disorder [67].

Imaging studies of schizophrenia show an increase in dopamine synthesis and dopamine release as related to acute symptoms, mainly positive psychotic symptoms. A few decades ago, it was proposed that hyperactivity of dopaminergic transmission is associated with schizophrenia [69]. In fact, it was observed that dopamine D2 receptor antagonists ameliorate symptoms of schizophrenia, mainly positive symptoms; on the other hand, dopamine agonists are able to induce psychotic states similar to some seen in schizophrenia [70]. This may suggest a dopamine system dysfunction underlying psychotic symptoms in schizophrenia. Since D2 receptors are mainly expressed in the striatum, some authors have proposed that psychosis may be associated with hyperactivity of dopaminergic systems in the limbic striatum [71, 72]. Moreover, the effect of D2 receptor antagonists in treating positive symptoms indicates a putative basis for the positive symptoms in the dopamine hypothesis of schizophrenia. The increased striatal dopamine transmission in both first-episode neuroleptic naive patients and previously treated chronically ill patients with acute psychosis was observed using single-photon emission computed tomography (SPECT) [73, 74]. Furthermore, postmortem studies found an increased dopamine D2 receptor density in the striatum in schizophrenia [75]. Nevertheless, this increased receptor density is usually attributed to a compensatory upregulation in response to chronic neuroleptic treatment.

Dopamine-enhancing drugs like amphetamine, methylphenidate, and L-dopa represent a cornerstone of the classical dopamine hypothesis of schizophrenia; in fact, high doses of psychostimulants in subjects without schizophrenia may gradually induce paranoid psychosis. In addition, low doses of psychostimulants, usually not psychotogenic in healthy subjects, may produce or even worsen psychotic symptoms in patients with schizophrenia. In the early 1970s, several authors studied the clinical pattern of amphetamine psychosis in non-schizophrenic amphetamine abusers [70, 76, 77] and formally recognised that sustained psychostimulant exposure can produce paranoid psychosis in non-schizophrenic individuals. Subsequently, a review by Lieberman et al. [78] provided evidence that in schizophrenia, there is an increased sensitivity to the psychotogenic effects of acute psychostimulant use. Laruelle [79] measured amphetamine-induced dopamine release in the striatum using SPECT and observed an association of amphetamine with the emergence or worsening of positive psychotic symptoms in patients with schizophrenia compared with healthy controls, suggesting an exaggerated stimulation of dopaminergic transmission. Those observations support an abnormal responsiveness of dopaminergic neurons in positive symptoms of schizophrenia.

4.1.3 From Neurochemical Alterations to Reality Distortion

An interesting model proposed by Kapur 15 years ago [80, 81] tried to explain the neurobiology of positive symptoms using the concept of motivational salience, which could link an aberrant dopamine mesolimbic release found in schizophrenia to delusions. Dopamine mediates rewards from pleasant stimuli and is physiologically aroused by rewarding experiences, like sex, food, and substances of abuse. According to this hypothesis, neurons in the mesolimbic dopamine system fire in response to novel rewards in the environment, and released dopamine leads to a switch in attention and behaviour towards a rewarding situation, thus driving the individual towards the stimulus. Abnormal functioning of the dopamine system might lead to the wrong assignment of motivational salience to external stimuli. Neutral stimuli appear salient and this situation leads to an emotional activation. The person then tries to interpret this new aberrant experience and develop an explanation to understand the environment. A cognitive scheme that attempts to explain these experiences leads to the development of a delusion. From this perspective, a delusion develops from an unfounded attribution of salience to normal events. This model is dimensional in nature because it can be applied to a variety of clinical conditions, beyond schizophrenia, in which aberrant dopamine firing suggests salience, and the patient develops a delusion. For example, delusions in cocaine-dependent patients can be transient during the intoxication, or persistent in chronic users [82]. Also, high prevalence rates of delusions in patients with Parkinson's disease on dopaminergic treatment confirm the role of dopamine, possibly mediated by aberrant salience, across categories [83].

Mania shares dopaminergic hyperactivity with psychotic disorders; positive psychotic symptoms occur frequently during manic episodes and might be related to aberrant dopamine firing [84] (Fig. 4.2).

Another interesting model to explain the development of delusional beliefs in a neurobiological framework was developed by Corlett and colleagues [85]. Delusion may result from an abnormal specification of hierarchical predictions by the brain circuits and from how they compute and respond to prediction errors. In particular, the study focused on a specific parameter—prediction error—"that involves a computational mechanism common to cortical hierarchies, frontostriatal circuits, and

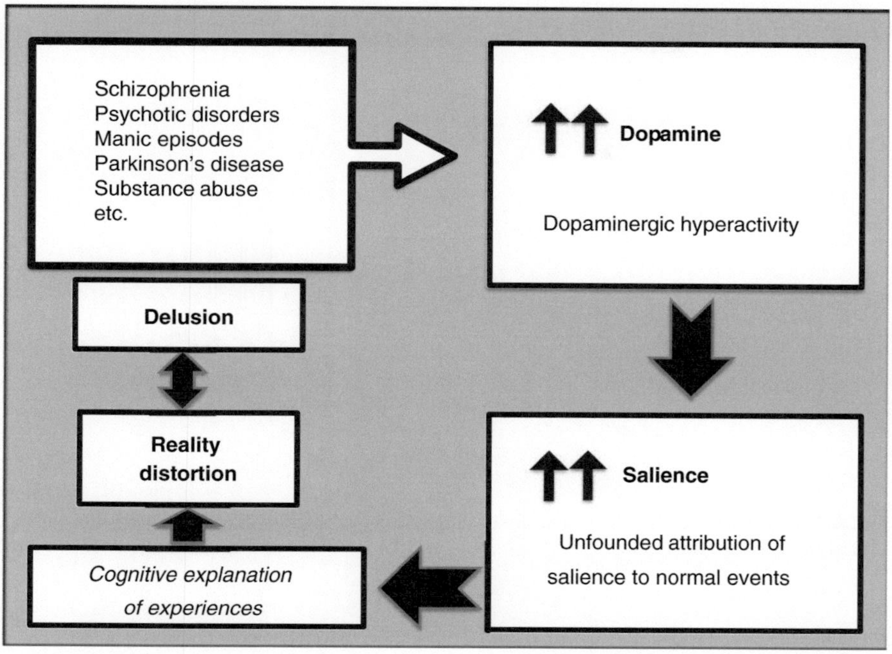

Fig. 4.2 Schematic relationship between dopaminergic hyperactivity, salience, and delusion in mental disorders with psychotic symptoms (see text)

the amygdala. Defects in these fundamental brain mechanisms can vitiate perception, memory, and bodily and social learning such that individuals with delusions experience an internal and external world that healthy individuals would find difficult to understand" [85].

4.1.4 Application of the SVARAD Score for Reality Distortion

SVARAD mean scores across categorical diagnoses in psychiatric outpatients confirm high levels of Reality Distortion in schizophrenia, delusional disorder, and psychotic disorder NOS. As expected, lower but significant levels were found also in bipolar disorder-manic episode. Scores for reality distortion are considerably lower, but above zero, in depressive disorders, obsessive-compulsive disorder, eating disorders, and borderline personality disorder (Fig. 4.3).

In the same study, the SVARAD detected a significant proportion of outpatients with obsessive-compulsive disorder (16%) with at least a mild score for reality distortion. A similar proportion was found in outpatients with borderline personality disorder (16%) (Figs. 4.4 and 4.5).

These figures are in line with studies showing a significant Reality Distortion dimension across nonpsychotic mental disorders.

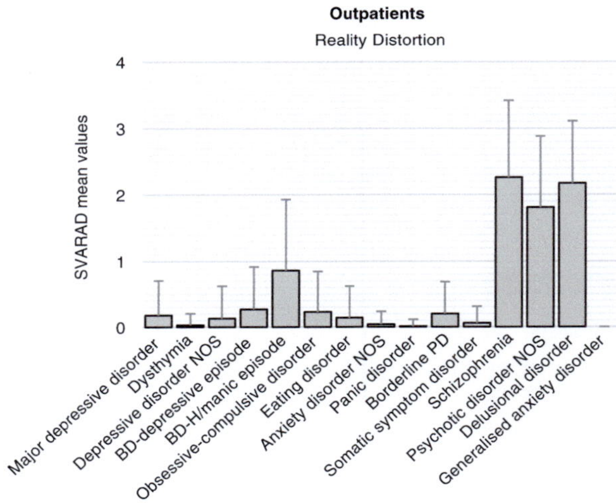

Fig. 4.3 SVARAD Reality Distortion dimension across outpatients' diagnostic categories: mean scores and standard deviations

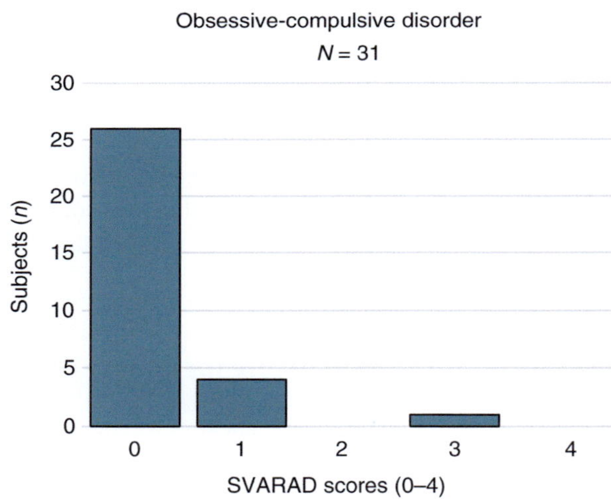

Fig. 4.4 Scores of SVARAD Reality Distortion dimension across outpatients with obsessive-compulsive disorder ($n = 31$)

In psychiatric inpatients, SVARAD Reality Distortion scores are higher than those in outpatients, as expected. Acute phase mental disorders, despite categorical diagnosis, appear to be characterised by significant levels of reality distortion. Data show that severe mood disorders and borderline personality disorder also have significant levels of reality distortion (Fig. 4.6).

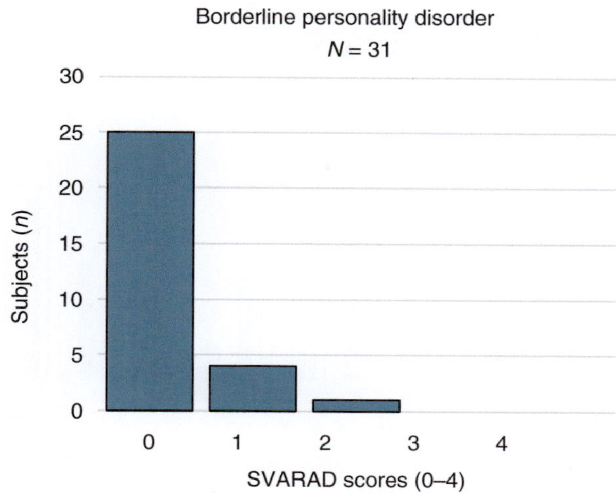

Fig. 4.5 Scores of SVARAD Reality Distortion dimension across outpatients with borderline personality disorder ($n = 31$)

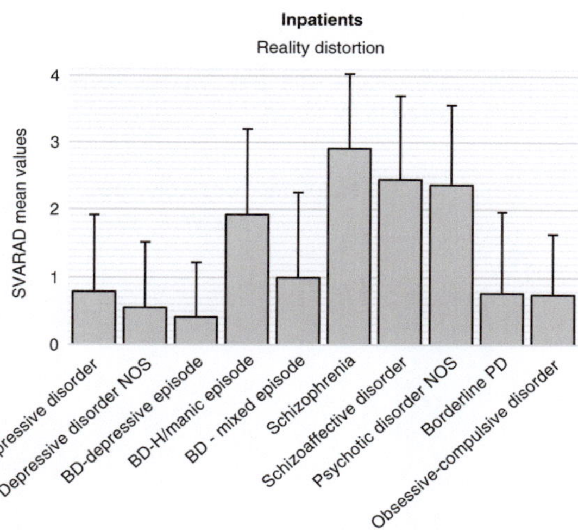

Fig. 4.6 SVARAD Reality Distortion dimension across inpatients' diagnostic categories: mean values and standard deviations

In inpatients with psychotic disorders, Reality Distortion is the dimension with the highest score. Schizophrenia shows the highest scores among the psychotic disorders (Fig. 4.7).

Forty-three percent of inpatients with major depression have at least a mild level of Reality Distortion. This percentage was found to be 31.7 in inpatients with depressive disorder NOS (Figs. 4.8 and 4.9). These data confirm clinically significant levels of Reality Distortion in depressive disorders that are severe enough to lead to hospital admission.

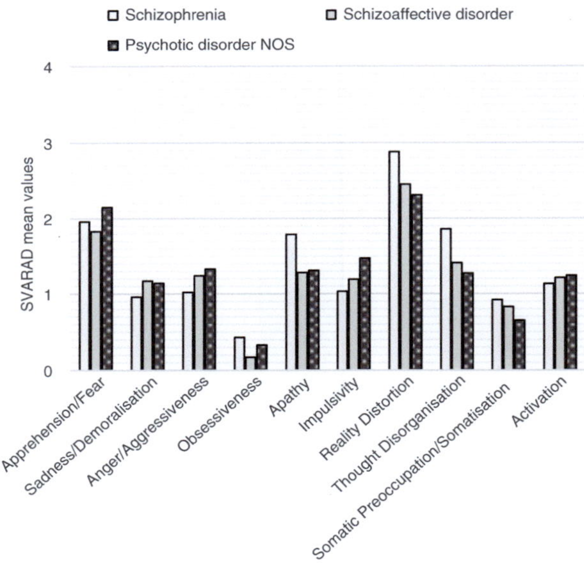

Fig. 4.7 SVARAD profile of inpatients with schizophrenia (N = 85), schizoaffective disorder (N = 53), and psychotic disorder NOS (N = 226)

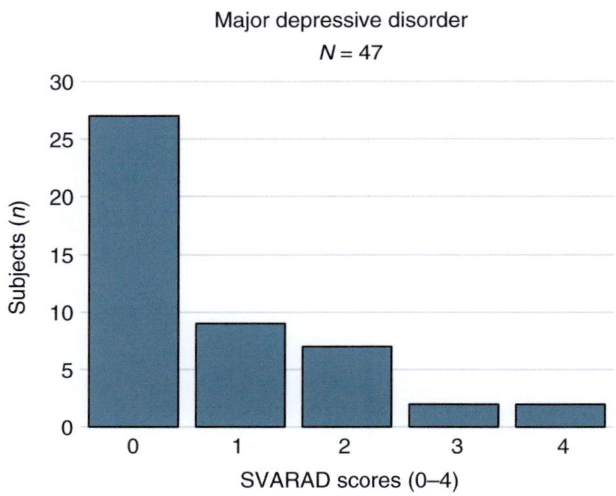

Fig. 4.8 Scores of SVARAD Reality Distortion dimension across inpatients with major depressive disorder (n = 47)

The majority of inpatients with hypomanic or manic episodes (85%) scored at least 1 (mild) for Reality Distortion on the SVARAD, confirming that reality distortion in severe mania is the rule rather than the exception. Among inpatients with borderline personality disorder, 36% showed significant levels of reality

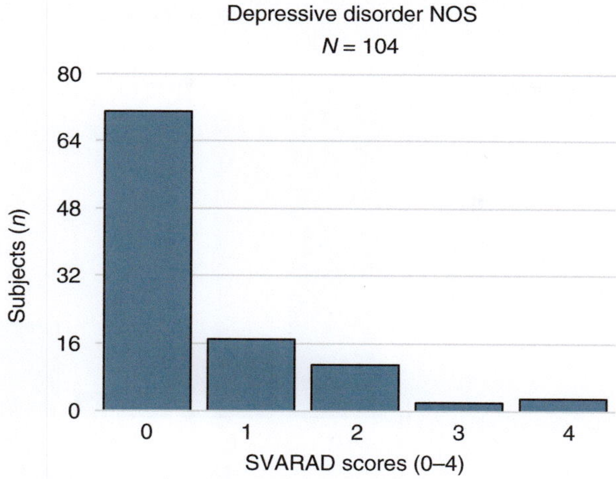

Fig. 4.9 Scores of SVARAD Reality Distortion dimension across inpatients with depressive disorder NOS ($n = 104$)

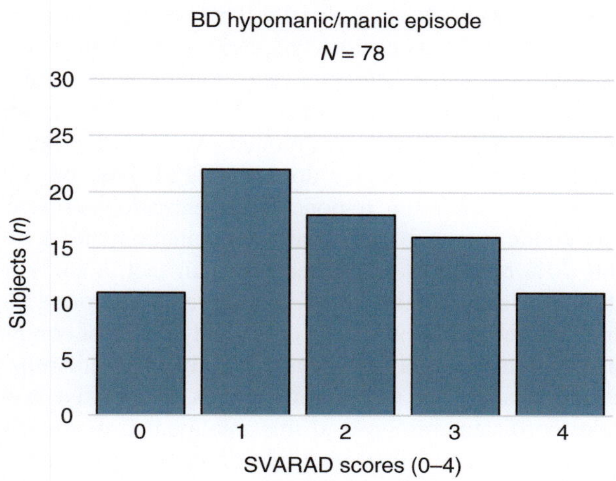

Fig. 4.10 Scores of SVARAD Reality Distortion dimension across inpatients with BD hypomanic/manic episode ($n = 78$)

distortion and 18% showed high levels. High levels of reality distortion might increase the likelihood of admission in a psychiatric intensive care unit (Figs. 4.10 and 4.11).

As described in Chap. 5, the Reality Distortion dimension represents one of the most important independent predictors of both voluntary and compulsory admission to a psychiatric intensive care unit.

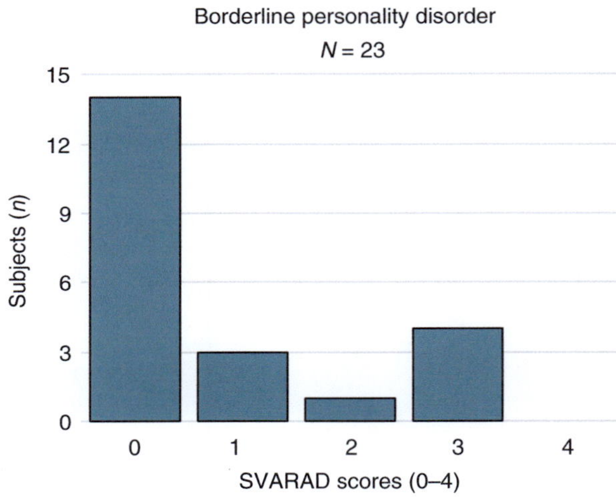

Fig. 4.11 Scores of SVARAD Reality Distortion dimension across inpatients with borderline personality disorder ($n = 23$)

4.1.5 Treatment of the Reality Distortion Dimension

The era of modern antipsychotics began in the 1950s with the introduction of chlorpromazine. Subsequently, fluphenazine, haloperidol, perphenazine, and thioridazine have been developed. In the following years, the role of dopamine in psychosis was confirmed using neuroimaging studies [86], and it became understood that these first-generation antipsychotics ameliorated psychosis by diminishing abnormal dopamine transmission by blocking dopamine D2/D3 receptors in the striatum and in frontal cortical and in limbic regions. First-generation antipsychotic medications improve psychotic symptoms, especially hallucinations and delusions, through inhibition of 60–70% of dopamine D2 receptors. In fact, a study led by the US National Institute of Mental Health (NIMH) observed that around 60% of patients treated with first-generation antipsychotic drugs had an almost complete resolution of acute positive symptoms within 6 weeks of treatment [87]. Antipsychotic treatment improves both positive and negative symptoms, but positive symptoms have a greater and more consistent response to this treatment than negative symptoms. Subsequently, it was described that after remission through antipsychotic drugs, long-term treatment is more effective in preventing new episodes [88]. Inhibition of dopamine D2 receptors by more than 80% can provoke extrapyramidal side effects and hyperprolactinemia, especially with high D2 affinity antipsychotics that do not block 5HT2 receptors. An exception to the role of D2 receptors in the treatment of reality distortion is clozapine, which has a low D2 affinity but is highly effective in treating positive psychotic symptoms. A possible explanation is the clozapine blockade of D1 receptors and other non-D2 receptors, which could be involved in reality distortion and antipsychotic action.

After clozapine, around ten new drugs (e.g. olanzapine, risperidone, quetiapine, and ziprasidone, among others) were approved and used as second-generation (or atypical) antipsychotics. All these new medications block D2 receptors with different affinity, and they have an antagonist action on other receptors, like 5-HT2A. A D2 partial agonist, aripiprazole, was also approved for the treatment of schizophrenia and bipolar disorder. As a group, all antipsychotics have been shown to be effective in reducing psychotic positive symptoms (versus placebo), with small differences in comparative efficacy [89]. Trials have indicated that antipsychotics differ substantially in side effects and that this is what should inform drug choice in clinical practice.

Studies on reality distortion in nonpsychotic mental disorders, as well as SVARAD scores for mood disorders, personality disorders, and others, suggest a role for antipsychotics (in association with other compounds) in the treatment of these conditions. The wide use of D2 blockers in clinical practice, in association with standard treatments for nonpsychotic disorders, confirms this concept.

As described, these pharmacotherapies can contribute greatly to reality distortion symptom relief, but mainly during the acute phase and during maintenance treatment, with a significant risk of relapse after discontinuation, especially in schizophrenia. Evidence-based therapy should be personalised, integrating both pharmacotherapies and psychosocial interventions [90]. In fact, medications are not able to preserve or restore premorbid levels of social and vocational functioning and do not lead to normal functioning. Approaches to individual therapy, including cognitive-behavioural therapy (CBT), aim to improve residual psychotic symptoms and to prevent relapse by modifying individual patterns of stress and response to the illness. A randomised study by Kuipers and colleagues [91], comparing CBT plus usual care with usual care alone in patients with schizophrenia, observed that CBT led to a significant reduction in overall symptom scores, delusional distress, and hallucinations. These effects remained significant for up to 9 months after treatment ended. Another study by Tarrier and colleagues [92] compared CBT with supportive counselling and routine care alone and found significant reductions in delusions and hallucinations in the CBT group. At 12-month follow-up [93], CBT was still superior to the other treatment conditions. In a meta-analysis of psychotherapy in psychotic disorders, cognitive-behavioural therapy was significantly more efficacious than other interventions (pooled) in reducing positive symptoms.

Several studies examined individual characteristics at baseline as predictors of response to CBT for psychosis. One of the first studies of CBT for treatment-resistant psychosis found that, among the patients with delusions, the acceptance of the "possibility of being mistaken" (in relation to their delusional belief) was associated with a good response to therapy [94]. CBT (based on a specific cognitive model) may change cognitive mechanisms and lead to good outcomes [95]. Morrison [96] proposed an integrative cognitive model of hallucinations and delusions, which focuses on the culturally unacceptable interpretations that patients with psychosis make for events, in addition to their responses to such events. This approach to the treatment of psychosis involves normalising the interpretations that patients make, helping them to create alternative explanations, helping them test out such appraisals using behavioural experiments, and helping them to identify and modify unhelpful cognitive and behavioural responses [95]. This procedure of the cognitive treatment of delusional beliefs is in line with the dimensional concept of reality distortion on a continuum. In fact, a categorical view of a delusion as an incorrigible belief would contraindicate any attempt towards normalisation. Therapy is aimed at reducing the levels of delusional reality distortion, towards a normal reality exam.

It is also important to emphasise that a study conducted by Kumari and colleagues [97], using functional magnetic resonance imaging, reported that patients who received CBT for psychosis, along with conventional treatment, showed significantly better clinical improvement compared with patients who received only conventional treatment. After treatment, the CBT with conventional treatment for psychosis group displayed decreased activation of frontal regions, insula, thalamus, putamen, and occipital areas in response to fearful and angry expressions. The angry expressions signal a direct and immediate threat, while fearful expressions indicate the presence of a significant, but uncertain, source of threat in the environment [98]; paranoia is fundamentally a threat response [99]. These neurophysiological results correlated directly with symptom improvement.

4.2 Thought Disorganisation Dimension

4.2.1 Introduction

Using the term "dissociation" as a core function in schizophrenia, Eugen Bleuler first described the disorganisation dimension in 1911 [13]. In fact, Bleuler recognised the concept of loosening of associations by the results seen on word association tests, considering some disorders of association as symptoms of dementia praecox. This concept of loosening of associations indicates a core, organically based psychological deficit that could be the basis for other symptoms of schizophrenia [100].

Besides the first categorical models proposed by Kraepelin [12] and Bleuler [13], due to the heterogeneity of schizophrenia, new classifications have been proposed within which the disorganisation dimension was described. Fifty years ago, in many studies on schizophrenia, conceptual disorganisation was combined with reality distortion symptoms, such as delusions and hallucinations, to form a positive symptom factor [101]. Eventually, other dimensional models, including the tridimensional positive-negative-disorganisation model, have been used [102, 103]. Factor analysis of studies conducted using the Scale for the Assessment of Negative Symptoms (SANS) [104] and the Scale for the Assessment of Positive Symptoms (SAPS) [64] demonstrated the existence of disorganisation as a separate dimension worthy of consideration. In 1996, Pancheri [105] and colleagues described three basic elements of disorganisation: "(a) disintegration of fundamental characteristics of communication, with related inability to decode a message; (b) loss of logical connecting associations between communication concepts; (c) disconnection between verbal and nonverbal, emotionally hued communication" [105, 106]. In a factor-analytic study on symptom structure of schizophrenia [107], a disorganisation factor of the Brief Psychiatric Rating Scale (BPRS) [108] was found to significantly discriminate among three heterogeneous groups of patients.

Key features of disorganisation symptoms include formal thought disorders (FTDs), which could lead to both disorganised speech and behaviour, and inappropriate affect. In classical psychopathology, this "dimensional" alteration is described with the terms "disorganisation of thought" and/or FTDs. The latter is considered a disorder of thought processes that is characterised by faulty organisation of thought into a definite logical sequence for a specific aim. In 1994, Thomas and Frazer [109] described FTD as a multidimensional impairment that included disorders in thought, language processing, and social cognition. FTDs encompass symptoms such as derailment, incoherence, loss of goal, illogicality, paucity of speech, perseveration, tangentiality, circumstantiality, thought blocking, and the novelty of neologisms [110]. Usually, thought disorders are split into two subtypes: negative, with inhibited speech (e.g. paucity of speech, strained speech, and/or language derailment), and positive, with distorted language (e.g. tangential association, illogical patterns, and/or word substitutions). In schizophrenia, derailment, loss of goal, poverty of content, and tangentiality were the most commonly observed [111].

Dwyer et al. [112] suggested that difficulty in determining context for language, as well as difficulty in differentiating between emotionally negative and neutral sentences by context, is observed in subjects who experience thought disorders. Afterwards, they indicated that the failure of aspects of higher-order semantic processing could produce word substitutions, as well as incorrect association of context, to contextless sentences [112]. This may not be related to the dysfunction of separate language processes but may instead be the result of higher cognitive dysfunction. In fact, thought disorders have been associated with semantic and executive dysfunction [113, 114] and, to a minor extent, with working memory and attentional impairments as well [115].

Another aspect of disorganised behaviour, inappropriate or bizarre comportments, including gestures, may be either related to environmental stimuli or completely incongruous to the situation. In particular, disorganised behaviour refers to difficulty with any type of purposeful behaviour, which includes personal self-care, bizarre or inappropriate dressing, sexual self-stimulation in public, or agitated shouting or cursing. Inappropriate or incongruous affect includes exhibiting incorrect emotional responses for a given context.

Over the last few years, in patients affected by Axis I mental disorders, factor analyses conducted on psychopathological symptoms and signs have identified factors as disorganisation, disorganisation syndrome, or FTD. Moreover, those showed symptom covariance with poor attention, stereotyped thinking, thought disorder, disorientation, and other symptoms [116, 117]. The disorganisation dimension can be detected in a variety of mental disorders. Disorders for which the disorganisation dimension is included in diagnostic criteria of the DSM-5 [37] are summarised in Table 4.2.

Table 4.2 The Thought Disorganisation dimension in DSM-5 [37] diagnostic criteria

Mental disorder	Disorganisation-related diagnostic criteria
Attention-deficit/ hyperactivity disorder (ADHD)	– Often has difficulty organising tasks and activities – Often runs about or climbs in situations where it is inappropriate
Brief psychotic disorder	– Disorganised speech (e.g. frequent derailment or incoherence) – Grossly disorganised or catatonic behaviour
Schizophreniform disorder	– Disorganised speech (e.g. frequent derailment or incoherence) – Grossly disorganised or catatonic behaviour
Schizophrenia	– Disorganised speech (e.g. frequent derailment or incoherence) – Grossly disorganised or catatonic behaviour
Schizoaffective disorder	– Disorganised speech (e.g. frequent derailment or incoherence) – Grossly disorganised or catatonic behaviour
Delirium	– A disturbance in attention (i.e. reduced ability to direct, focus, sustain, and shift attention) and awareness (reduced orientation to the environment) – An additional disturbance in cognition (e.g. memory deficit, disorientation, language, visuospatial ability, or perception)
Major neurocognitive disorder	– Evidence of significant cognitive decline from a previous level of performance in one or more cognitive domains (e.g. complex attention, executive function, learning and memory, language, perceptual-motor, or social cognition)
Stimulant intoxication	Clinically significant changes (e.g. stereotyped behaviours) that developed during use of a stimulant
Other (or unknown) substance intoxication	Clinically significant changes that are attributable to the effect of the substance on the central nervous system (e.g. cognitive impairment) and develop during use of the substance

Despite the diagnostic criteria of DSM-5, other disorders, such as bipolar I disorder and major depressive disorder, often include disorganisation dimension symptoms in the clinical picture. Such features have also been described also in the paragraph Diagnostic Features of Bipolar I Disorder of the DSM-5 [37]:

> Often the individual's thoughts race at a rate faster than they can be expressed through speech. Frequently there is flight of ideas evidenced by a nearly continuous flow of accelerated speech, with abrupt shifts from one topic to another. When flight of ideas is severe, speech may become disorganized, incoherent, and particularly distressful to the individual. Sometimes thoughts are experienced as so crowded that it is very difficult to speak.

Similarly, two factor-analytic studies performed with the BPRS in patients with bipolar disorder found a relevant disorganisation factor [118, 119], but two similar studies did not find a factor with disorganisation items [120, 121]. It is likely that the BPRS-E [122] simply does not contain sufficient items to adequately define disorganisation as a consistent independent factor. This hypothesis seems to be supported by the findings of factor-analytic studies on the Positive and Negative Syndrome Scale (PANSS) [123], in which disorganisation is consistently found as a factor including most of the following items: disorientation, conceptual disorganisation, and mannerisms and posturing. Dazzi and colleagues [124] suggested that BPRS-E would probably benefit from an increase in the number of items used to specifically describe the disorganisation dimension, such as inappropriate affect, impoverished thinking, and poor attention, as well as from two refined versions of the existing BPRS-E items (conceptual disorganisation and bizarre behaviour), to create a version of the BPRS with 26 items (BPRS-26)—which should capture the existing four major dimensions, as well as the fifth disorganisation dimension of schizophrenia.

In the past several years, in addition to the items on conceptual disorganisation in the Positive And Negative Syndrome Scale (PANSS) [123] and Brief Psychiatric Rating Scale (BPRS) [108], different psychometric scales have been proposed to assess the disorganisation dimension, such as the Scale for the Measurement of Disorganisation (SCADIS) [105] and 3-THREE, a brief scale for the assessment of psychosis [105, 125].

The Thought Disorganisation item of the SVARAD is *Disruption of connection between ideas and principles governing the organisation of thought, which thus turns out to be altered in its logical organisation and impaired in its communicative functions.*

4.2.2 Neurobiology of the Thought Disorganisation Dimension

The neurobiology underlying the Thought Disorganisation dimension (mainly studied in schizophrenia) is not totally understood. Symptoms of disorganisation could be the results of failure in the working memory functions of the prefrontal cortex [126], as well as other cerebral structures, such as the hippocampus [127]. A recent review of 97 studies on structural neuroimaging in patients with thought disorders found a major role for the left superior temporal gyrus. Also, associations between thought disorders and structural changes within the orbitofrontal cortex, cerebellum, nucleus accumbens, and amygdala-hippocampal region were described [128].

Symptoms of disorganisation seemed to be strictly correlated with cognitive impairment [129], which is believed to be the central deficit in the disorganisation syndrome of schizophrenia [130], and with an involvement of several functions of working memory [131].

In schizophrenia, the disorganisation syndrome has long been considered to be a disorder of impaired cognitive association [132], corresponding to impaired phase synchronisation between cortical areas. This deficiency of anatomical connectivity between cortical areas has been considered to be a cause of disorganisation in schizophrenia [133]. Data obtained from both neuroimaging studies [134] and visual psychophysics studies [135, 136] suggest a reduced functional connectivity that stems from anatomical disconnection of cortical areas [137].

Cognitive problems—such as deficits in areas of executive functioning, including working memory and inhibition, as well as memory and attention—are a core feature of schizophrenia and of disorganisation [138–140].

A recent study showed a negative correlation between disorganisation scores on the Scale for the Assessment of Positive Symptoms (SAPS) [64] and fractional anisotropy, using diffusion tensor imaging (DTI), along the right cingulum bundle within the dorsal cingulate gyrus. The cingulum bundle connects all parts of the limbic system [141], and it is involved in executive control. This study showed significant white matter alterations in a region involved in cognitive control and executive function related to the thought disorganisation dimension [65].

To better understand the molecular mechanisms underlying cognitive dysfunction in schizophrenia, some authors have focused their attention on potential genetic pathways. The glutamate metabolism pathway has been described as a potential molecular mechanism influencing cognition in schizophrenia, known as the "glutamate hypothesis" [142]. Glutamate is a primary excitatory neurotransmitter, controlled by *N*-methyl-D-aspartate (NMDA) receptors. NMDA receptors control synaptic plasticity and memory function, and their antagonists can simulate cognitive impairment and negative symptoms of schizophrenia [143, 144]. The "glutamate hypothesis" has been proposed by observing both positive and negative symptoms of schizophrenia in response to administration of the NMDA receptor antagonists phencyclidine (PCP) and ketamine in healthy subjects [145, 146]. Both PCP and ketamine are able to increase glutamate release in the cortex [147, 148], suggesting that inhibition of NMDA receptors, leading to abnormal glutamate transmission, may be correlated with schizophrenia.

Moreover, Phillips and Silverstein stated that schizophrenia is associated with impaired cognitive coordination caused by reduced ion flow through NMDA channels [130]. Later, studies using magnetic resonance spectroscopy (MRS) confirmed increased glutamatergic metabolites in cortical and subcortical areas of the brain, indicating that excess glutamatergic neurotransmission is associated with schizophrenia [149].

Moreover, several investigations have studied the hypothesis of association between glutamate and cognitive deficits in schizophrenia. Recently, a systematic review [150] conducted on pharmacological, candidate gene, and neuroimaging studies concluded that the glutamatergic pathway is likely to be involved in different domains of cognition, mostly memory and working memory. In fact, in 2005 a meta-analysis of 124 studies concluded that there is a significant deficit of working memory in individuals with schizophrenia [151].

4.2.3 Application of the Thought Disorganisation Dimension SVARAD Score

SVARAD mean scores across categorical diagnoses in psychiatric outpatients confirm high levels of disorganisation in schizophrenia, delusional disorder, and psychotic disorder NOS. Very low levels were also found in bipolar disorder (Fig. 4.12)

Accordingly, one out of seven outpatients with hypomanic or manic episodes (14%) showed mild levels of disorganisation (Fig. 4.13). In outpatients, the SVARAD did not detect significant disorganisation in nonpsychotic mental disorders.

SVARAD disorganisation scores in psychiatric inpatients are higher than those in outpatients. The severity of the mental disorder might explain the significant levels of disorganisation in mood disorders and borderline personality disorders in addition to psychotic disorders (Fig. 4.14).

Twenty-three percent of inpatients with major depressive disorder had at least a mild level of disorganisation, as did 18.3% of inpatients with depressive disorder NOS (Figs. 4.15 and 4.16). These data confirm clinically significant levels of disorganisation in a subgroup of patients admitted to the hospital for a depressive disorder.

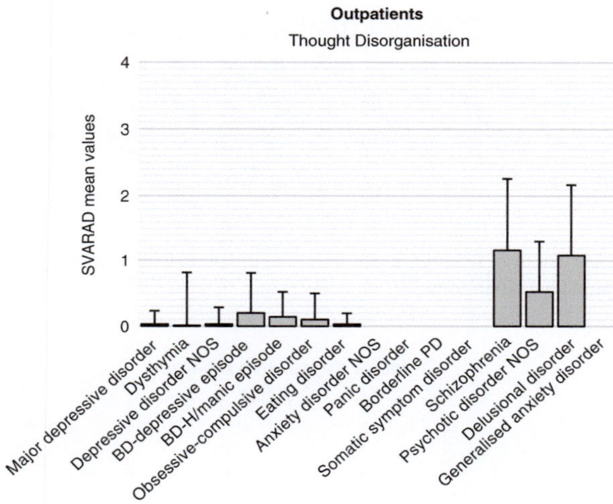

Fig. 4.12 SVARAD Thought Disorganisation dimension across outpatients' diagnostic categories: mean scores and standard deviations

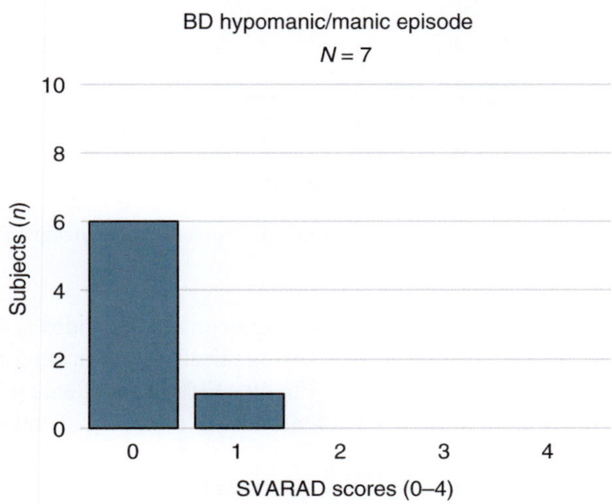

Fig. 4.13 Scores of SVARAD Thought Disorganisation dimension across outpatients with BD hypomanic/manic episode (n = 7)

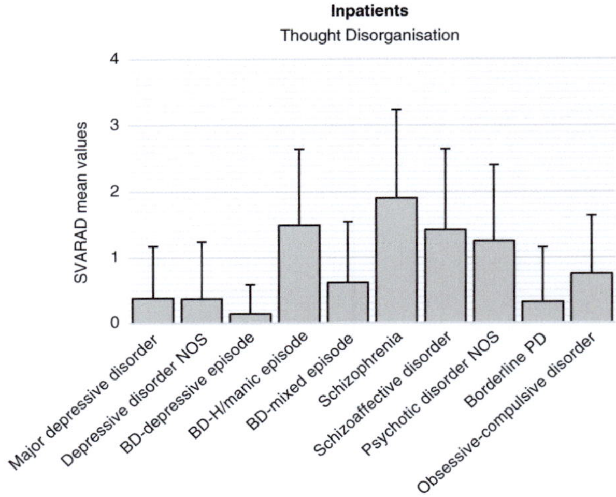

Fig. 4.14 SVARAD Thought Disorganisation dimension across inpatients' diagnostic categories: mean values and standard deviations

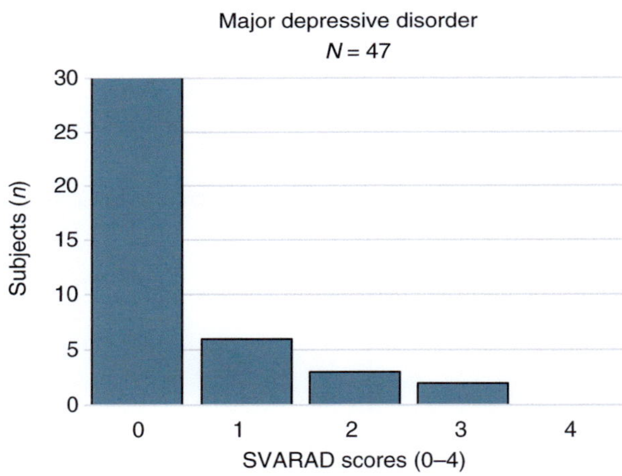

Fig. 4.15 Scores of SVARAD Thought Disorganisation dimension in inpatients with major depressive disorder ($n = 47$)

The majority of inpatients with hypomanic or manic episodes (74%) scored at least mild on the SVARAD for Thought Disorganisation, confirming that disorganisation in severe mania is very frequent. Finally, 13% of inpatients with borderline personality disorder also showed significant levels of disorganisation (Figs. 4.17 and 4.18).

Subgroups of patients with clinically significant disorganisation, as detected by the SVARAD, were seen in patients admitted to a psychiatric intensive care unit for

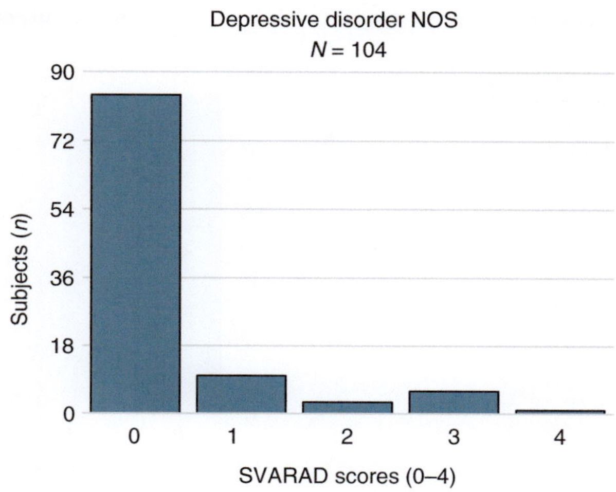

Fig. 4.16 Scores of SVARAD Thought Disorganisation dimension in inpatients with depressive disorder NOS ($n = 104$)

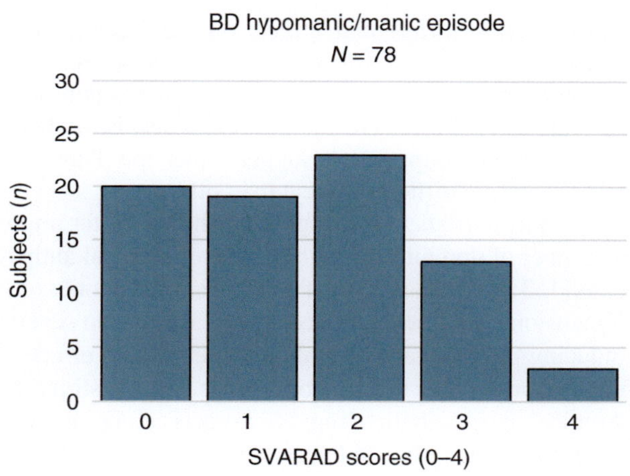

Fig. 4.17 Scores of SVARAD Thought Disorganisation dimension in inpatients with BD hypomanic/manic episode ($n = 78$)

different categorical diagnoses, but not in outpatients. Disorganisation is a severe psychopathological sign, and it might be an important reason for hospital admission.

4.2.4 Treatment for the Thought Disorganisation Dimension

Treatment for the Thought Disorganisation dimension presents some controversies. Studies of antipsychotics deal with mechanisms of action and the efficacy on positive and negative symptoms. However, pro-cognitive effects have not been fully explained. Antipsychotics are effective in treating all psychotic symptoms in patients with schizophrenia, with small differences in efficacy [89], but trials do not

Fig. 4.18 Scores of SVARAD Thought Disorganisation dimension in inpatients with borderline personality disorder (n = 23)

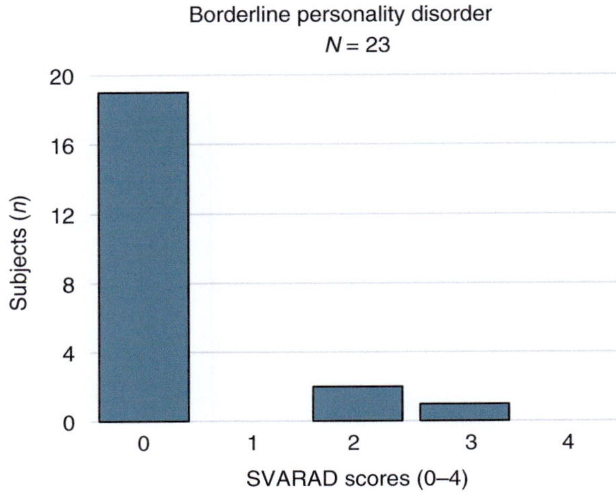

show specific differential effects on the Thought Disorganisation dimension. A reanalysis of a large study on antipsychotic treatment found a lower symptom reduction in patients with disorganised schizophrenia as compared with the paranoid group [152]. This result was found despite higher baseline BPRS scores in the disorganised group, and it suggests that the Thought Disorganisation dimension might be more difficult to treat than others.

Data from trials in schizophrenia show a better improvement in cognition across a number of domains through the use of atypical antipsychotics, rather than typical ones [153, 154]. The action of atypical antipsychotics on the Thought Disorganisation dimension could be explained by the activity on several receptors, in particular by inducing the blockade of glutamate neurotoxic effects through modulation of gene expression in specific cerebral areas, or by stimulation of neurogenesis [155]. Atypical antipsychotics improve aspects of cognition, probably due to their ability to increase dopamine and acetylcholine in the prefrontal cortex. In this area, dopamine activity is critical for cognitive functioning.

Patients with disorganised thought and speech may be considered not suitable for psychotherapies, because of their conversational problems. In fact, no data are available on the effectiveness of psychological interventions for disorganisation. Conversely, a study showed that thought and language disorders interfered significantly with therapeutic alliance in outpatients with schizophrenia and schizoaffective disorder [156].

Recently, Hamm and Firmin [157], in a case report, suggested an emergent integrative psychotherapy (called Metacognitive Reflection and Insight Therapy) could be useful in patients with disorganisation.

Symptoms of disorganisation and cognitive problems are associated with poorer quality of life and poor social functioning [158]. Impaired social functioning is closely correlated with the severity of the psychopathology and particularly with disorganised thought [106]. Although studies have shown limited effectiveness in improving quality of life, social, or cognitive functioning, interventions such as community treatment, skills training, cognitive remediation, and supported employment are potentially very helpful for individuals with schizophrenia, but do not specifically address disorganisation [159].

4.3 Reality Distortion and Thought Disorganisation

A study involving 1114 outpatients was conducted to test the correlation between the SVARAD dimensions of Reality Distortion and Thought Disorganisation and found a Spearman correlation coefficient of 0.59 ($p = 0.0001$) between the two dimensions. A similar correlation was found among inpatients (0.55; $p = 0.0001$, $N = 846$). This correlation between reality distortion and thought disorganisation does not contradict several studies suggesting that these two dimensions are independent [160]. Clinical symptoms and signs are indeed distinct, without significant overlaps. Although reality distortion and disorganisation are treated with the same types of medications, they show distinct neurobiological correlates, as described in this chapter. However, the correlation might reflect the frequent co-occurrence of both dimensions in psychotic disorders and in manic episodes.

In conclusion, Reality Distortion and Thought Disorganisation are clinically important dimensions, because they are related to the severity of mental disorders, and also, even when mild, they lead to significant subjective distress and functional impairment. The data described above suggest that routine assessment and detection of these dimensions, such as by administering the SVARAD, should be mandatory in all patients, regardless of categorical diagnosis. In a subgroup of patients with nonpsychotic mental disorders, disorganisation and reality distortion might be present, and they need to be addressed in clinical decision-making, as they should represent an important target for treatment.

References

1. Kraepelin E. Psychiatrie. Nine editions. Leipzig: Abel; 1883.
2. Jaspers K. Allgemeine psychopathologie (7th edn). Berlin: Springer; 1959.
3. Kiran C, Chaudhury S. Understanding delusions. Ind Psychiatry J. 2009;18(1):3–18.
4. Esquirol E. Mental maladies; a treatise on Insanity. Philadelphia, PA: Lea and Blanchard; 1845.
5. Berrios GE. The history of mental symptoms: descriptive psychopathology since the nineteenth century. Cambridge: Cambridge University Press; 1996.
6. Waters F, Fernyhough C. Hallucinations: a systematic review of points of similarity and difference across diagnostic classes. Schizophr Bull. 2017;43(1):32–43.
7. Canstatt C. Handbuch der Medizinischen klinik. Stuttgart: Enke; 1841.
8. Feuchtersleben E. v. Lehrbuch der Ärztlichen Seelenkunde. Wien: Gerold; 1845.
9. Bürgy M. The Concept of psychosis: historical and phenomenological aspects. Schizophr Bull. 2008;34(6):1200–10.
10. Flemming CF. Pathologie und Therapie der Psychosen. Berlin: Hirschwald; 1859.
11. Möbius PJ. Über die Einteilung der Krankheiten. Neurologische Betrachtungen Centralbibliothek Nervenheilkd Psychiatr. 1892;15:289–301.

12. Kraepelin E. Psychiatrie: Ein Lehrbuch für Studirende und Aerzte. 5th ed. Leipzig: JA Barth; 1896.
13. Bleuler E. Dementia praecox oder Gruppe der Schizophrenien. Leipzig: Deuticke; 1911.
14. Hoenig J. The concept of Schizophrenia. Kraepelin-Bleuler-Schneider. Br J Psychiatry. 1983;142:547–56.
15. Hoff P. Emil Kraepelin und die Psychiatrie als Klinische Wissenschaft. Ein Beitrag zum Selbstverständnis Psychiatrischer Forschung. Berlin: Springer; 1994.
16. Jaspers K. Allgemeine psychopathologie. Berlin: Springer; 1913.
17. Strauss JS. Hallucinations and delusions as points on continua function. Arch Gen Psychiatry. 1969;21:581–6.
18. Claridge G. Final remarks and future directions. In: Claridge G, editor. Schizotypy. Implications for Illness and Health. Oxford: Oxford University Press; 1997. p. 301–17.
19. Peters ER, Joseph SA, Garety PA. Measurement of delusional ideation in the normal population: introducing the PDI (Peters et al. Delusions Inventory). Schizophr Bull. 1999;25(3):553–76.
20. Stefanis NC, Hanssen M, Smirnis NK, Avramopoulos DA, Evdokimidis IK, Stefanis CN, et al. Evidence that three dimensions of psychosis have a distribution in the general population. Psychol Med. 2002;32(2):347–58.
21. Verdoux H, van Os J. Psychotic symptoms in non-clinical populations and the continuum of psychosis. Schizophr Res. 2002;54(1-2):59–65.
22. Linscott RJ, van Os J. An updated and conservative systematic review and meta-analysis of epidemiological evidence on psychotic experiences in children and adults: on the pathway from proneness to persistence to dimensional expression across mental disorders. Psychol Med. 2013;43:1133–49.
23. Van Os J, Schizophrenia KS. Lancet. 2009;374:635–45.
24. Tien AY. Distributions of hallucinations in the population. Soc Psychiatry Psychiatr Epidemiol. 1991;26:287–92.
25. Posey TB, Losch ME. Auditory Hallucinations of Hearing Voices in 375 Normal Subjects. Imag Cogn Personal. 1983;3(2):99–113.
26. Barrett TR, Etheridge JB. Verbal hallucinations in normals. I: People who hear voices. Appl Cogn Psychol. 1992;6:379–87.
27. Al-Issa I. Social and cultural aspects of hallucinations. Psychol Bull. 1977;84:570–87.
28. Schwab MEA. study of reported hallucinations in a Southeastern county. Ment Health Soc. 1977;4:344–54.
29. Morgan C, Fisher H, Hutchinson G, Frissa S, study team SELCH, Hotopf M, et al. Ethnicity, social disadvantage and psychotic-like experiences in a healthy population based sample. Acta Psychiatr Scand. 2009;119:226–35.
30. Johns LC, Nazroo JY, Bebbington P, Kuipers E. Occurrence of hallucinatory experiences in a community sample and ethnic variations. Br J Psychiatry. 2002;180:174–8.
31. van Os J, Hanssen M, Bijl R-V, Ravelli A. Straus (1969) revisited: a psychosis continuum in the general population? Schizophr Res. 2000;45:11–20.
32. van Os J, Hanssen M, Bijl RV, Vollebergh W. Prevalence of psychotic disorder and community level of psychotic symptoms: an urban–rural comparison. Arch Gen Psychiatry. 2001;58:663–8.
33. Zammit S, Kounali D, Cannon M, et al. Psychotic experiences and psychotic disorders at age 18 in relation to psychotic experiences at age 12 in a longitudinal population-based cohort study. Am J Psychiatry. 2013;170:742–50.
34. van Os J, Linscott RJ. Introduction: the extended psychosis phenotype – relationship with schizophrenia and with ultrahigh risk status for psychosis. Schizophr Bull. 2012;38:227–30.
35. Hanssen M, Peeters F, Krabbendam L, Radstake S, Verdoux H, van Os J. How psychotic are individuals with non-psychotic disorders? Soc Psychiatry Psychiatr Epidemiol. 2003;38:149–54.
36. Honings S, Drukker M, Groen R, et al. Psychotic experiences and risk of self-injurious behaviour in the general population: a systematic review and meta-analysis. Psychol Med. 2015;30:1–15.

37. American Psychiatric Association. Diagnostic and statistical manual of mental disorders. 5th ed. Arlington, VA: American Psychiatric Publishing; 2013.
38. Wigman JT, van Nierop M, Vollebergh WA, Lieb R, Beesdo-Baum K, Wittchen HU, et al. Evidence that psychotic symptoms are prevalent in disorders of anxiety and depression, impacting on illness onset, risk, and severity implications for diagnosis and ultra-high risk research. Schizophr Bull. 2012;38:247–57.
39. Hamner MB, Frueh BC, Ulmer HG, Arana GW. Psychotic features and illness severity in combat veterans with chronic posttraumatic stress disorder. Biol Psychiatry. 1999;45(7):846–52.
40. Foa EB, Kozak MJ, Goodman WK, Hollander E, Jenike MA, Rasmussen SA. DSM-IV field trial: obsessive-compulsive disorder. Am J Psychiatry. 1995;152:90–6.
41. Raveendranathan D, Narayanaswamy JC, Reddi SV. Response rate of catatonia to electroconvulsive therapy and its clinical correlates. Eur Arch Psychiatry Clin Neurosci. 2012;262(5):425–30.
42. Ropacki SA, Jeste DV. Epidemiology of and risk factors for psychosis of Alzheimer's disease: a review of 55 studies published from 1990 to 2003. Am J Psychiatry. 2005;162(11):2022–30.
43. Ostling S, Gustafson D, Blennow K, Börjesson-Hanson A, Waern M. Psychotic symptoms in a population-based sample of 85-year-old individuals with dementia. J Geriatr Psychiatry Neurol. 2011;24(1):3–8.
44. Forsaa EB, Larsen JP, Wentzel-Larsen T, Goetz CG, Stebbins GT, Aarsland D, et al. A 12-year population-based study of psychosis in Parkinson disease. Arch Neurol. 2010;67(8):996–1001.
45. DeLisi LE, Sakuma M, Tew W, Kushner M, Hoff AL, Grimson R. Schizophrenia as a chronic active brain process: a study of progressive brain structural change subsequent to the onset of schizophrenia. Psychiatry Res. 1997;74:129–40.
46. Lieberman J, Chakos M, Wu H, Hoffman E, Robinson D, Bilder R. Longitudinal study of brain morphology in first episode schizophrenia. Biol Psychiatry. 2001;49:487–99.
47. Johnstone EC, Crow TJ, Frith CD, Husband J, Kreel L. Cerebral ventricular size and cognitive impairment in chronic schizophrenia. Lancet. 1976;2:924–6.
48. Ward KE, Friedman L, Wise A, Schulz SC. Metaanalysis of brain and cranial size in schizophrenia. Schizophr Res. 1996;22:197–213.
49. Wringht IC, Rabe-Hesketh S, Woodruff PW, David AS, Murray RM, Bullmore ET. Meta-analysis of regional brain volumes in schizophrenia. Am J Psychiaty. 2000;157(1):16–25.
50. Lawrie SM, Abukmeil SS. Brain abnormalities in schizophrenia: a systematic and quantitative review of volumetric magnetic resonance imaging studies. Br J Psychiatry. 1998;172:110–20.
51. Nelson MD, Saykin AJ, Flashman LA, Riordan HJ. Hippocampal volume reduction in schizophrenia as assessed by magnetic resonance imaging: a meta-analytic study. Arch Gen Psychiatry. 1998;55:433–40.
52. Woodruff PW, McManus IC, David AS. Meta-analysis of corpus callosum size in schizophrenia. J Neurol Neurosurg Psychiatry. 1995;58:457–61.
53. Vita A, De Peri L, Silenzi C, Dieci M. Brain morphology in first-episode schizophrenia: a meta-analysis of quantitative magnetic resonance imaging studies. Schizophr Res. 2006;82:75–88.
54. Csernansky JG, Cronenwett WJ. Neural networks in schizophrenia. Am J Psychiatry. 2008;165:937–9.
55. Ellison-Wright I, Glahn DC, Laird AR, Thelen SM, Bullmore E. The Anatomy of first-episode and chronic schizophrenia: an anatomical likelihood estimation meta-analysis. Am J Psychiatry. 2008;165:1015–23.
56. Glahn DC, Laird AR, Ellison-Wright I, Thelen SM, Robinson JL, Lancaster JL, et al. Meta-analysis of gray matter anomalies in schizophrenia: application of anatomic likelihood estimation and network analysis. Biol Psychiatry. 2008;64:774–81.
57. Shenton ME, Dickey CC, Frumin M, McCarley RWA. review of MRI findings in schizophrenia. Schizophr Res. 2001;49:1–52.
58. Kubicki M, McCarley R, Westin CF, Park HJ, Maier S, Kikinis R, et al. A review of diffusion tensor imaging studies in schizophrenia. J Psychiatr Res. 2007;41:15–30.

59. Kanaan RA, Kim JS, Kaufmann WE, Pearlson GD, Barker GJ, McGuire PK. Diffusion tensor imaging in schizophrenia. Biol Psychiatry. 2005;58:921–9.
60. Ellison-Wright I, Bullmore E. Meta-analysis of diffusion tensor imaging studies in schizophrenia. Schizophr Res. 2009;108:3–10.
61. Schizophrenia ANC. the fundamental questions. Brain Res Brain Res Rev. 2000;31:106–12.
62. Liddle PF. Functional imaging—schizophrenia. Br Med Bull. 1996;52:486–94.
63. Hubl D, Koenig T, Strik W, Federspiel A, Kreis R, Boesch C, et al. Pathways that make voices: white matter changes in auditory hallucinations. Arch Gen Psychiatry. 2004;61:658–68.
64. Andreasen N. The scale for the assessment of positive symptoms (SAPS). Iowa City, IA: University of Iowa; 1984.
65. Bopp MHA, Zöllner R, Jansen A, Dietsche B, Krug A, Kircher TTJ. White matter integrity and symptom dimensions of schizophrenia: a diffusion tensor imaging study. Schizophr Res. 2017;184:59–68.
66. Curcic-Blake B, Nanetti L, van der Meer L, Cerliani L, Renken R, Pijnenborg GH, et al. Not on speaking terms: hallucinations and structural network disconnectivity in schizophrenia. Brain Struct Funct. 2015;220(1):407–18.
67. Poletti M, Gebhardt E, Raballo A. Corollary discharge, self-agency, and the neurodevelopment of the psychotic mind. JAMA Psychiat. 2017;74:1169.
68. Ford JM, Mathalon DH, Heinks T, Kalba S, Faustman WO, Roth WT. Neurophysiological evidence of corollary discharge dysfunction in schizophrenia. Am J Psychiatry. 2001;158(12):2069–71.
69. Rossum V. Arch Int Pharmacodyn Ther. 1966;160:492–4.
70. Angrist BM, Gershon S. The phenomenology of experimentally induced amphetamine psychosis—preliminary observation. Biol Psychiatry. 1970;2:95–107.
71. Matthysse S. Dopamine and the pharmacology of schizophrenia: the state of the evidence. J Psychiatr Res. 1974;11:107–13.
72. Stevens J. An anatomy of schizophrenia? Arch Gen Psychiatry. 1973;29:177–89.
73. Abi-Dargham A, Gil R, Krystal J, Baldwin RM, Seibyl JP, Bowers M, et al. Increased striatal dopamine transmission in schizophrenia: confirmation in a second cohort. Am J Psychiatry. 1998;155(6):761–7.
74. Abi-Dargham A, Rodenhiser J, Printz D, Zea-Ponce Y, Gil R, Kegeles LS, et al. Increased baseline occupancy of D2 receptors by dopamine in schizophrenia. Proc Natl Acad Sci U S A. 2000;97(14):8104–9.
75. Zakzanis KK, Hansen KT. Dopamine D2 densities and the schizophrenic brain. Schizophr Res. 1998;32(3):201–6.
76. Bell DS. The experimental reproduction of amphetamine psychosis. Arch Gen Psychiatry. 1973;29:35–40.
77. Griffith JJ, Oates J, Cavanaugh J. Paranoid episodes induced by drugs. JAMA. 1968;205:39.
78. Lieberman JA, Kane JM, Sarantakos S, et al. Prediction of relapse in schizophrenia. Arch Gen Psychiatry. 1987;44:597–603.
79. Laruelle M, Abi-Dargham A, van Dyck CH, Gil R, D'Souza CD, Erdos J, et al. Single photon emission computerized tomography imaging of amphetamine-induced dopamine release in drug-free schizophrenic subjects. Proc Natl Acad Sci U S A. 1996;93(17):9235–40.
80. Kapur S. Psychosis as a state of aberrant salience: a framework linking biology, phenomenology, and pharmacology in schizophrenia. Am J Psychiatry. 2003;160(1):13–23.
81. Kapur S, Mizrahi R, Li M. From dopamine to salience to psychosis--linking biology, pharmacology and phenomenology of psychosis. Schizophr Res. 2005;79(1):59–68.
82. Roncero C, Comín M, Daigre C, Grau-López L, Martínez-Luna N, Eiroa-Orosa FJ, et al. Clinical differences between cocaine-induced psychotic disorder and psychotic symptoms in cocaine-dependent patients. Psychiatry Res. 2014;216(3):398–403.
83. Fénelon G, Alves G. Epidemiology of psychosis in Parkinson's disease. J Neurol Sci. 2010;289(1-2):12–7.
84. Park SY, Kang UG. Hypothetical dopamine dynamics in mania and psychosis its pharmacokinetic implications Original Research Article. Prog Neuro-Psychopharmacol Biol Psychiatry. 2013;43:89–95.

85. Corlett PR, Taylor JR, Wang XJ, Fletcher PC, Krystal JH. Toward a neurobiology of delusions. Prog Neurobiol. 2010;92(3):345–69.
86. Healy D. The creation of psychopharmacology. Cambridge, MA: Harvard University Press; 2002.
87. Guttmacher MS. Phenothiazine treatment in acute schizophrenia; effectiveness: the National Institute of Mental Health Psychopharmacology Service Center Collaborative Study Group. Arch Gen Psychiatry. 1964;10:246–61.
88. Hogarty GE, Ulrich RF, Mussare F, Aristigueta N. Drug discontinuation among long term, successfully maintained schizophrenic outpatients. Dis Nerv Syst. 1976;37:494–500.
89. Leucht S, Cipriani A, Spineli L, Mavridis D, Orey D, Richter F, et al. Comparative efficacy and tolerability of 15 antipsychotic drugs in schizophrenia: a multiple-treatments meta-analysis. Lancet. 2013;382(9896):951–62.
90. Lehman AF, Kreynbuhl J, Buchanan RW, Dickerson FB, Dixon LB, Goldberg R, et al. The Schizophrenia Patient Outcomes Research Team (PORT): updated treatment recommendations 2003. Schizophr Bull. 2004;30:193–217.
91. Kuipers E, Fowler D, Garety P, Chisholm D, Freeman D, Dunn G, et al. London-East Anglia randomized controlled trial of cognitive-behavioural therapy for psychosis, III: follow-up and economic evaluation at 18 months. Br J Psychiatry. 1998;173:61–8.
92. Tarrier N, Yusupoff L, Kinney C, McCarthy E, Gledhill A, Haddock G, et al. Randomised controlled trial of intensive cognitive behaviour therapy for patients with chronic schizophrenia. BMJ. 1998;317:303–7.
93. Tarrier N, Wittkowski A, Kinney C, McCarthy E, Morris J, Humphreys L. Durability of the effects of cognitive-behavioural therapy in the treatment of chronic schizophrenia: 12-month follow-up. Br J Psychiatry. 1999;174:500–4.
94. Garety P, Fowler D, Kuipers E, Freeman D, Dunn G, Bebbington P, et al. London-East Anglia randomised controlled trial of cognitive-behavioural therapy for psychosis. II: predictors of outcome. Br J Psychiatry. 1997;171:420–6.
95. Morrison AP, Turkington D, Wardle M, Spencer H, Barratt S, Dudley R, et al. A preliminary exploration of predictors of outcome and cognitive mechanisms of change in cognitive behaviour therapy for psychosis in people not taking antipsychotic medication. Behav Res Ther. 2012;50(2):163–7.
96. Morrison AP. The interpretation of intrusions in psychosis: an integrative cognitive approach to hallucinations and delusions. Behav Cogn Psychother. 2001;29:257–76.
97. Kumari V, Fannon D, Peters ER, ffytche DH, Sumich AL, Premkumar P, et al. Neural changes following cognitive behaviour therapy for psychosis: a longitudinal study. Brain. 2011;134(8):2396–407.
98. Fridlund AJ. Human facial expression: an evolutionary view. San Diego, CA: Academic Press; 1994.
99. Freeman D, Garety PA, Kuipers E, Fowler D, Bebbington PEA. cognitive model of persecutory delusions. Br J Clin Psychol. 2002;41(Pt 4):331–47.
100. Hardy-Baylé MC, Sarfati Y, Passerieux C. The cognitive basis of disorganization symptomatology in schizophrenia and its clinical correlates: toward a pathogenetic approach to disorganization. Schizophr Bull. 2003;29(3):459–71.
101. Guy W. Assessment manual for psychopharmacology ECDEU. Rockville, MD: U.S. Department of Health and Human Services; 1976.
102. Crow TJ. Molecular pathology of schizophrenia: more than one disease process? Br Med J. 1980;280:66–8.
103. Liddle PF. Schizophrenic syndromes, cognitive performance and neurological dysfunction. Psychol Med. 1987;17:49–57.
104. Andreasen N. The scale for the assessment of negative symptoms (SANS). Iowa City, IA:University of Iowa; 1983.
105. Pancheri P, Romiti R, Maselli P, Marconi PL. SCADIS: una nuova scala per la valutazione della disorganizzazione. 2. Omogeneità, struttura fattoriale, validità concorrente, attendibilità. Giorn Ital Psicopat. 1996;3:232–46.
106. Brugnoli R, Pacitti F, Tarsitani L, Troisi A, Rossi A, Pancheri P. Dimensione disorganizzazione in corso di schizofrenia e funzionamento sociale. Giorn Ital Psicopat. 2006;12:208–16.

107. Picardi A, Viroli C, Tarsitani L, Miglio R, de Girolamo G, Dell'Acqua G, et al. Heterogeneity and symptom structure of schizophrenia. Psychiatry Res. 2012;198:386–94.
108. Overall JE, Gorham DR. The brief psychiatric rating scale. Psychol Rep. 1962;10:799–812.
109. Thomas P, Fraser W. Linguistics, human communication, and psychiatry. Br J Psychiatry. 1994;165:585–92.
110. Yudofsky SC, Hales RE. The American Psychiatric publishing textbook of clinical psychiatry. Washington, DC: American Psychiatric Association; 2002.
111. Covington MA, He C, Brown C, Naçi L, McClain JT, Fjordbak BS, et al. Schizophrenia and the structure of language: the linguist's view. Schizophr Res. 2005;77:85–98.
112. Dwyer K, Peters E, McKenna P, David A, McCarthy R. Verbatim recall in formal thought disorder in schizophrenia: a study of contextual influences. Cogn Neuropsychiatry. 2014;19(4):337–58.
113. Stirling J, Hellewell J, Blakey A, Deakin W. Thought disorder in schizophrenia is associated with both executive dysfunction and circumscribed impairments in semantic function. Psychol Med. 2006;36:475–84.
114. Tan EJ, Rossell SL. Building a neurocognitive profile of thought disorder in schizophrenia using a standardized test battery. Schizophr Res. 2014;152:242–5.
115. Docherty NM, Hawkins KA, Hoffman RE, Quinlan DM, Rakfeldt J, Sledge WH. Working memory, attention and communication disturbances in schizophrenia. J Abnorm Psychol. 1996;105:212–9.
116. Higashima M, Urata K, Kawasaki Y, Maeda Y, Sakai N, Mizukoshi C, et al. P300 and the thought disorder factor extracted by factor-analytic procedures in schizophrenia. Biol Psychiatry. 1998;44:115–20.
117. Drake RJ, Dunn G, Tarrier N, Haddock G, Haley C, Lewis S. The evolution of symptoms in the early course of non-affective psychosis. Schizophr Res. 2003;63:171–9.
118. Picardi A, Battisti F, de Girolamo G, Morosini P, Norcio B, Bracco R, et al. Symptom structure of acute mania: a factor study of the 24-item Brief Psychiatric Rating Scale in a national sample of patients hospitalized for a manic episode. J Affect Disord. 2008;108:183–9.
119. Perugi G, Medda P, Swann AC, Reis J, Rizzato S, Mauri M. Phenomenological subtypes of severe bipolar mixed states: a factor analytic study. Compr Psychiatry. 2014;55:799–806.
120. Pacchiarotti I, Nivoli AM, Mazzarini L, Kotzalidis GD, Sani G, Koukopoulos A, et al. The symptom structure of bipolar acute episodes: in search for the mixing link. J Affect Disord. 2013;149:56–66.
121. Perlick DA, Rosenheck RA, Clarkin JF, Sirey J, Raue P. Symptoms predicting inpatient service use among patients with bipolar affective disorder. Psychiatr Serv. 1999;50:806–12.
122. Ventura J, Lukoff D, Nuechterlein KH, Liberman RP, Green MF, Shaner A. Brief Psychiatric Rating Scale (BPRS) expanded version: scales, anchor points, and administration manual. Int J Methods Psychiatr Res. 1993;3:227–43.
123. Kay SR, Fiszbein A, Opler LA. The Positive And Negative Syndrome Scale (PANSS) for schizophrenia. Schizophr Bull. 1987;13:261–76.
124. Dazzi F, Shafer A, Lauriola M. Meta-analysis of the Brief Psychiatric Rating Scale e Expanded (BPRSE) structure and arguments for a new version. J Psychiatr Res. 2016;81:140–51.
125. Pancheri P. La 3-TRE una scala per la valutazione rapida della sintomatologia schizofrenica. I: basi teoriche, descrizione, campi di applicazione. Riv Psichiatr. 1995;30:205–23.
126. Goldman-Rakic PS, Selemon LD. Functional and anatomical aspects of prefrontal pathology in schizophrenia. Schizophr Bull. 1997;23:437–58.
127. Olypher AV, Klement D, Fenton AA. Cognitive disorganization in hippocampus: a physiological model of the disorganization in psychosis. J Neurosci. 2006;26(1):158–68.
128. Sumner PJ, Bell IH, Rossell SL. A systematic review of the structural neuroimaging correlates of thought disorder. Neurosci Biobehav Rev. 2018;84:299.
129. Castner SA, Vosler PS, Goldman-Rakic PS. Amphetamine sensitization impairs cognition and reduces dopamine turnover in primate prefrontal cortex. Biol Psychiatry. 2005;57(7):743–51.
130. Phillips WA, Silverstein SM. Convergence of biological and psychological perspectives on cognitive coordination in schizophrenia. Behav Brain Sci. 2003;26(1):65–82. discussion 82–137.

131. Bertolino A, Sciota D, Brudaglio F, Altamura M, Blasi G, Bellomo A, et al. Working memory deficits and levels of N-acetylaspartate in patients with schizophreniform disorder. Am J Psychiatry. 2003;160(3):483–9.
132. Haig AR, Gordon E, De Pascalis V, Meares RA, Bahramali H, Harris A. Gamma activity in schizophrenia: evidence of impaired network binding. Clin Neurophysiol. 2000;111:1461–8.
133. Hoffman RE, McGlashan TH. Reduced corticocortical connectivity can induce speech perception pathology and hallucinated 'voices'. Schizophr Res. 1998;30:137–41.
134. Josin GM, Liddle PF. Neural network analysis of the pattern of functional connectivity between cerebral areas in schizophrenia. Biol Cybern. 2001;84:117–22.
135. Silverstein SM, Kovacs I, Corry R, Valone C. Perceptual organization, the disorganization syndrome, and context processing in chronic schizophrenia. Schizophr Res. 2000;43:11–20.
136. Izawa R, Yamamoto S. Spatio-temporal disintegration of visual perception in schizophrenia as revealed by a novel cognitive task, the Searchlight Test. Schizophr Res. 2002;53(1-2):67–74.
137. Bressler SL. Cortical coordination dynamics and the disorganization syndrome in schizophrenia. Neuropsychopharmacology. 2003;28(Suppl 1):S35–9.
138. Kiehl KA, Smith AM, Hare RD, Liddle PF. An event-related potential investigation of response inhibition in schizophrenia and psychopathy. Biol Psychiatry. 2000;48(3):210–21.
139. Minzenberg MJ, Laird AR, Thelen S, Carter CS, Glahn DC. Meta-analysis of 41 functional neuroimaging studies of executive function in schizophrenia. Arch Gen Psychiatry. 2009;66(8):811–22.
140. Weickert TW, Goldberg TE, Gold JM, Bigelow LB, Egan MF, Weinberger DR. Cognitive impairments in patients with schizophrenia displaying preserved and compromised intellect. Arch Gen Psychiatry. 2000;57(9):907–13.
141. Schmahmann J, Pandya D. Fiber Pathways of the Brain. Oxford: Oxford University Press; 2006.
142. Jia P, Wang L, Meltzer HY, Zhao Z. Common variants conferring risk of schizophrenia: a pathway analysis of GWAS data. Schizophr Res. 2010;122(1):38–42.
143. Coyle JT, Tsai G. The NMDA receptor glycine modulatory site: a therapeutic target for improving cognition and reducing negative symptoms in schizophrenia. Psychopharmacology (Berl). 2004;174(1):32–8.
144. Neill JC, Barnes S, Cook S, Grayson B, Idris NF, McLean SL, Harte MK. Animal models of cognitive dysfunction and negative symptoms of schizophrenia: focus on NMDA receptor antagonism. Pharmacol Ther. 2010;128(3):419–32.
145. Adler CM, Malhotra AK, Elman I, Goldberg T, Egan M, Pickar D, et al. Comparison of ketamine-induced thought disorder in healthy volunteers and thought disorder in schizophrenia. Am J Psychiatry. 1999;156(10):1646–9.
146. Allen RM, Young SJ. Phencyclidine-induced psychosis. Am J Psychiatry. 1978;135(9):1081–4.
147. Stone JM, Dietrich C, Edden R, Mehta MA, De Simoni S, Reed LJ, et al. Ketamine effects on brain GABA and glutamate levels with 1H-MRS: relationship to ketamine-induced psychopathology. Mol Psychiatry. 2012;17(7):664.
148. Stone JM, Pepper F, Fam J, Furby H, Hughes E, Morgan C, et al. Glutamate, N-acetyl aspartate and psychotic symptoms in chronic ketamine users. Psychopharmacology (Berl). 2014;231(10):2107–16.
149. Merritt K, Egerton A, Kempton MJ, Taylor MJ, McGuire PK. Nature of glutamate alterations in schizophrenia: a meta-analysis of proton magnetic resonance spectroscopy studies. JAMA Psychiat. 2016;73(7):665–74.
150. Thomas EHX, Bozaoglu K, Rossell SL, Gurvich C. The influence of the glutamatergic system on cognition in schizophrenia: a systematic review. Neurosci Biobehav Rev. 2017;77:369–87.
151. Lee J, Park S. Working memory impairments in schizophrenia: a meta-analysis. J Abnorm Psychol. 2005;114(4):599–611.
152. Corves C, Engel RR, Davis J, Leucht S. Do patients with paranoid and disorganized schizophrenia respond differently to antipsychotic drugs? Acta Psychiatr Scand. 2014;130(1):40–5.
153. Harvey PD, Keefe RSE. Studies of cognitive change in patients with schizophrenia following novel antipsychotic treatment. Am J Psychiatry. 2001;158:176–84.

154. Weiss EM, Bilder RM, Fleischhacker WW. The effects of second-generation antipsychotics on cognitive functioning and psychosocial outcome in schizophrenia. Psychopharmacology (Berl). 2002;162:11–7.
155. Stip E, Chouinard S, Boulay LJ. On the trail of a cognitive enhancer for the treatment of schizophrenia. Prog Neuropsychopharmacol Biol Psychiatry. 2005;29(2):219–32.
156. Cavelti M, Homan P, Vauth R. The impact of thought disorder on therapeutic alliance and personal recovery in schizophrenia and schizoaffective disorder: an exploratory study. Psychiatry Res. 2016;239:92–8.
157. Hamm J, Firmin RL. Disorganization and individual psychotherapy for schizophrenia: a case report of metacognitive reflection and insight therapy. J Contemp Psychother. 2017;46:227.
158. Norman RM, Malla AK, Cortese L, Cheng S, Diaz K, McIntosh E, et al. Symptoms and cognition as predictors of community functioning: a prospective analysis. Am J Psychiatry. 1999;156:400–5.
159. Pilling S, Bebbington P, Kuipers E, Garety P, Geddes J, Orbach G, et al. Psychological treatments in schizophrenia: I. Meta-analysis of family intervention and cognitive behaviour therapy. Psychol Med. 2002;32(5):763–82.
160. Ventura J, Thames AD, Wood RC, Guzik LH, Hellemann GS. Disorganization and reality distortion in schizophrenia: a meta- analysis of the relationship between positive symptoms and neurocognitive deficits. Schizophr Res. 2010;121(1-3):1–14.

Psychopathological Dimensions in Emergency Psychiatry: Determinants of Admission, Compulsory Treatment, and Therapeutic Intervention

5

Federico Dazzi, Luigi Orso, Angelo Picardi, and Massimo Biondi

5.1 Definition

A psychiatric emergency is a complex condition involving individual (e.g. biochemical and psychopathological processes), interpersonal, and environmental dynamic factors [1]. A number of definitions have been proposed over time, emphasising either the urgency or the context, i.e. a mismatch of needs and resources. Providing an exhaustive definition is difficult since the emergency is often self-determined by the patient, the community, or law enforcement. In our experience, severity of psychopathology, urgency, individual or environmental (e.g. family) degree of distress, and abnormal behaviours can all be part of a psychiatric emergency and also act as potential triggers for presentation in an emergency department.

The American Psychiatric Association (APA) Task Force on Psychiatric Emergency Services [2] comprehensively defines a psychiatric emergency as "an acute disturbance of thought, mood, behavior, or social relationship that requires an immediate intervention as defined by the patient, family, or community" and as "a set of circumstances in which (a) the behavior or condition of an individual is

F. Dazzi (✉)
Marconi University, Rome, Italy
e-mail: federicodazzi@hotmail.com

L. Orso
Department of Human Neurosciences, Sapienza University of Rome, Rome, Italy
e-mail: l.orso@hotmail.it

A. Picardi
Centre for Behavioural Sciences and Mental Health, Italian National Institute of Health, Rome, Italy
e-mail: angelo.picardi@iss.it

M. Biondi
Department of Human Neurosciences, Policlinico Umberto I Hospital, Sapienza University of Rome, Rome, Italy
e-mail: massimo.biondi@uniroma1.it

M. Biondi et al. (eds.), *Dimensional Psychopathology*,
https://doi.org/10.1007/978-3-319-78202-7_5

perceived by someone, often not the identified individual, as having the potential to rapidly eventuate in a catastrophic outcome, and (b) the resources available to understand and deal with the situation are not available at the time and place of the occurrence".

It is worth underscoring that, from a clinical perspective, a psychiatric emergency represents a crucial condition where a clinical intervention may drastically modify the trajectory of the clinical picture.

5.2 Features, Settings, Aims, and Criticalities

Emergency psychiatry is a challenging field of application, involving specific features, aims, settings, and criticalities. There are a variety of models for psychiatric emergency services (PES), largely varying in organisation and features from facility to facility but also sharing some common core characteristics. PES are 24-h operating services that aim to provide a prompt assessment of clinical conditions, to evaluate and prevent potentially life-threatening situations, to stabilise the current emergent clinical conditions, and to plan proper therapeutic interventions. Though PES can be community based, usually, they are hospital-based services, delivered in the form of a consultation with an emergency physician in the emergency department (ED).

Patients can reach the service either voluntarily or compulsorily, with the involvement of ambulance or police. Though there are a variety of different models, Italian hospital-based services are often organised as follows: all the patients reaching the ED share a common pathway of assessment, regardless of the main complaint or reason for the visit. First they undergo a triage evaluation, during which they are assigned a code, ranging from white to red, indicating the degree of emergency, and then they are assessed by an emergency physician and undergo blood and instrumental exams if needed. Upon request of the emergency physician, patients are assessed by a consultant psychiatrist from the psychiatric intensive care unit (PICU). After assessing mental conditions, the psychiatrist can require further exams, provide a medication, monitor the clinical conditions and re-evaluate, and decide whether to admit the patient to the PICU or to discharge and refer him or her to a community service (Fig. 5.1).

Emergency settings also share a number of potential criticalities. Compared with outpatient services, time is limited and spaces might sometimes be chaotic, especially in EDs serving crowded metropolitan areas, where admissions are very frequent. This makes it difficult to gather exhaustive information, to conduct a careful assessment of the clinical conditions, and to evaluate the potential risks.

Patients may often be frightened, agitated, uncooperative, or even hostile and thus unable to provide clinical information. Other sources of information, such as relatives and friends, may be unavailable, especially when the patient reaches the ED with the involvement of ambulance or police.

Medications are limited as well, with only a few drugs, usually antipsychotics and benzodiazepines, available to stabilise the clinical condition. Finally, and importantly, clinicians are essentially limited in planning therapeutic intervention to just two relevant choices: a decision to hospitalise or a referral of the patient to an outpatient service.

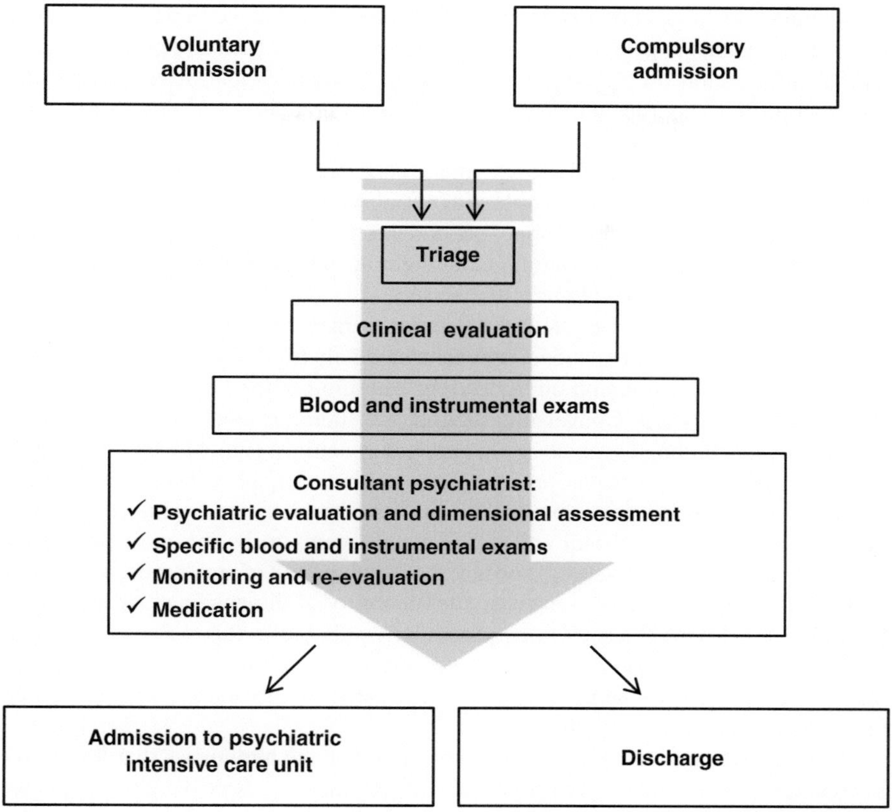

Fig. 5.1 Common pathway of assessment and therapy in the emergency department for patients with psychiatric complaints

Such complexity is challenging not only for clinicians but also for researchers and administrators. The conditions and criticalities aforementioned often prevent the comprehensive analysis of context that is needed to develop common guidelines, therapeutic strategies, and assessment instruments tailored to the needs of PES. This, in turn, gets reflected back as a problematic lack of standards [2].

Most of the available tools are indeed borrowed from general psychiatry and show limited usefulness in emergency settings.

5.3 The Diagnostic Approach in Emergency Psychiatry

A relevant concern is raised by the diagnostic model in emergency psychiatry. The general aims of a diagnostic assessment are to provide synthetic information on the patient's current and past condition, to predict a prognostic trajectory, and, above all, to guide the therapeutic intervention. The categorical approach matches most of these requirements in general psychiatric settings, as it synthetically provides longitudinal information on the patient's history and prognosis and drives therapeutic intervention. This approach, though, may be less accurate in describing current psychopathological conditions and predicting an immediate or short-term risk of life-threatening behaviours, which are among the core matters to focus on in emergency psychiatry. For example, two people diagnosed with schizophrenia might present very different, sometimes even opposite, clinical pictures, with negative symptoms or psychomotor agitation prevailing. One might be at high risk of self-harm behaviour; the other might be catatonic.

From this perspective, a categorical approach may show limited usefulness in emergency settings. A dimensional approach may help fill this gap, being more suitable to providing information on the current condition of the patient and guiding a short-term intervention. Conversely, though, the literature underscores that inter-rater agreement among psychiatrists in emergency settings is low for psychopathology, impulse control problems, and danger to self but adequate for some categorical diagnoses [3–5]. This brings up two relevant considerations: first, a mixed diagnostic approach, both categorical and dimensional, seems to be the most promising model for the emergency setting; second, the development of standardised diagnostic tools is needed in order to increase the reliability of dimensional assessment in the ED. The SVARAD seems to be particularly suitable to this purpose, as it is a quick, reliable, and easy-to-use instrument that allows clinicians to easily collect information on most of the psychopathological dimensions routinely evaluated during a mental state examination.

Recently, we conducted a clinical study [6] to test the hypothesis that SVARAD assessment can be predictive of psychiatric hospitalisation. Specifically, we sought to evaluate whether and to what extent a standardised dimensional assessment, as assessed with the SVARAD, can predict the need for hospitalisation in acute psychiatric patients presenting to the ED.

5.4 The Role of Dimensional Assessment in the ED: A Clinical Study

According to the considerations mentioned above, we recently conducted a clinical study [6] to assess the role of the dimensional assessment in the ED. Our main goal (a) was to evaluate whether and to what extent a standardised dimensional evaluation, as assessed with the SVARAD (both alone and in a mixed diagnostic model), could predict the need for hospitalisation in acute psychiatric patients presenting to the ED. The other major goal of this study (b) was to identify which, if any, psychopathological dimensions can independently predict the need for hospitalisation. Secondary objectives included (c) identifying the psychopathological dimensions differentiating between voluntarily and compulsorily admitted patients and (d) analysing the relationship between psychopathological dimensions and medications.

The results will be reported and discussed later in the chapter, but, for better comprehension of the results in the context of the study, we will first describe the main characteristics and methods of the study (for details see Dazzi et al. [6]).

5.4.1 Setting: The Policlinico Umberto I

The study was conducted at the Policlinico Umberto I in Rome in 2008 over a 6-month period.

The Policlinico Umberto I is the biggest university hospital in Italy. Located in downtown Rome, it serves a large and crowded catchment area with over 600,000 inhabitants. The ED admits about 140,000 patients each year, including over 900 emergency psychiatric consultations and almost 400 admissions to the PICU. The population presenting to the ED is particularly varied, as only one third of the patients come from the downtown area, one third come from the rest of the city or from the regional area, and one third from other Italian regions or from other countries.

Patients presenting with a psychiatric complaint follow specific pathways that were developed by the psychiatric department and the ED together. Specifically, three pathways are defined according to the presenting condition:

Group A. Patients presenting with a psychiatric condition due to a disorder already diagnosed, usually in treatment in an outpatient service, and referred for voluntary admission. These patients move along a fast track: psychiatric consultation is immediately requested after triage, and they can be admitted to the PICU after routine medical assessment.

Group B. Patients presenting with a condition to be diagnosed, potentially due to either a psychiatric disorder or a medical condition. A careful evaluation of the medical and psychopathological condition, including specific blood and instrumental exams, is performed first by the emergency physician and then by the emergency psychiatrist in order to make a correct diagnosis. If the condition is primarily due to a psychiatric disorder, the emergency psychiatrist can eventually admit the patient to the PICU or refer him or her to an outpatient service.

Group C. Patients presenting for self-harm behaviour. Such patients require a multi-professional approach, with different figures involved, including an emergency psychiatrist, according to the type and severity of self-harm behaviour. The primary aim is the support and stabilisation of the vital functions and damage control; later, once stabilisation has been achieved, the patient can be admitted to the intensive care unit, the PICU, or another ward, according to the highest priority treatment required.

When recommended, the patient is admitted to the PICU of Policlinico Umberto I or, if a bed is not available, transferred to another PICU, within 12 h. Patients requiring compulsory admission are immediately admitted to the PICU regardless of bed availability (Fig. 5.2).

Conversely, when admission is not recommended, the patient is usually referred to a community-based outpatient service or even to the psychiatric services of the Policlinico Umberto I, which include a day hospital and an outpatient service delivering both pharmacological and psychotherapeutic treatment.

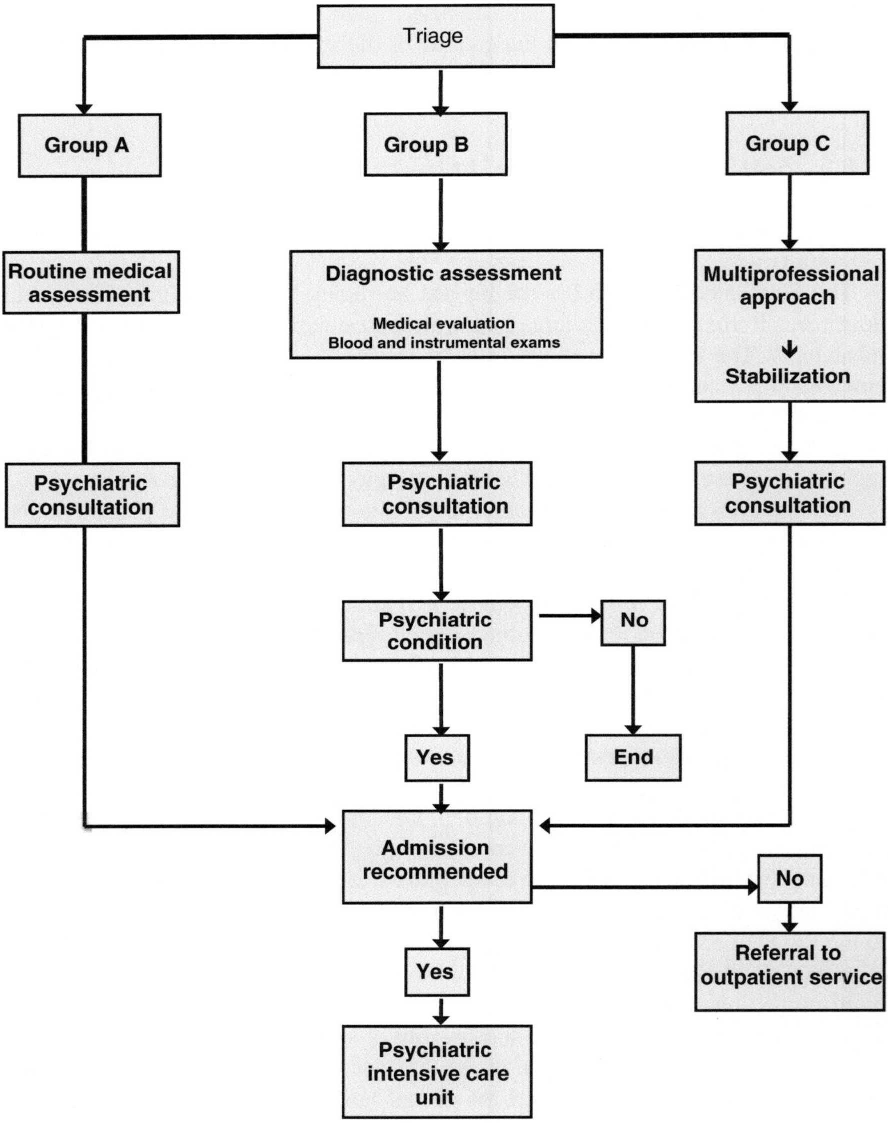

Fig. 5.2 Pathways for evaluation and admission in the emergency department for acute psychiatric patients

5.4.2 Procedure and Sample

The subjects of the study were composed of all the patients presenting to the ED for whom an acute psychiatric evaluation was required by an ED physician. For each patient, we collected socio-demographic and clinical data, including a dimensional evaluation as assessed with the SVARAD by a senior psychiatrist. For our main goal (A), all the variables were tested as potential predictors of the need for hospitalisation, which was chosen as the outcome of interest, rather than the actual rate of hospitalisation, as this could be affected by many factors, such as bed availability, presence of other clinical priorities, or patient's refusal.

We recruited a total of 312 patients, mainly Italian, with a balanced gender distribution, a mean age of 40 years (SD = 14.3), and a history of psychiatric admission in half of the cases. The most common diagnoses were psychotic, depressive, or bipolar disorder. Most of the patients reached the ED voluntarily, whereas 12.6% of them had a proposal for compulsory admission (see below for a description of the procedure for compulsory admission in Italy). Almost 40% of the sample was recommended for hospitalisation in the acute psychiatric ward, and half of the patients received one or more medications, typically a benzodiazepine or an antipsychotic drug.

Later, a replication study was conducted with the same methods in a new setting (San Filippo Neri Hospital, Rome, Italy), in order to test the predictors of recommended admission that were identified in the original study on a new sample.

5.5 Psychopathological Dimensions and Need for Hospitalisation

As we mentioned above, a crucial issue for emergency psychiatrists assessing an acute psychopathological picture is whether or not to hospitalise the patient, in order to prevent short-term risk and to plan a therapeutic intervention. From this perspective, emergency clinicians should carefully evaluate all the conditions that commonly may lead to hospitalisation. A large variety of socio-demographic, logistic, and clinical predictors for admission were identified. Age, gender, family and social support, marital status, employment status, homelessness, and ethnicity were all suggested to be potential socio-demographic risk factors. In addition, logistic and contextual factors were proposed, such as bed availability or day of the week. For these non-clinical factors, the role played by most of them is controversial. Stronger evidence was observed for clinical predictors, such as severity of illness, psychotic or bipolar disorder, and previous hospitalisation. Also, a number of symptoms and conditions were found to predict hospitalisation, including (1) psychotic symptoms such as hallucination, delusion, lack of insight, and odd behaviour; (2) suicidality and danger to self or others; (3) agitation, destructive behaviour, or psychomotor inhibition; and (4) confusion and abnormal consciousness (for a detailed overview, please see Dazzi et al. [6]).

Though a variety of psychopathological alterations were described, we found a lack of studies adopting a systematic dimensional approach. In our study, as expected, we observed that patients who were recommended for admission scored significantly higher on most of the SVARAD dimensions, such as Anger/Aggressiveness, Apathy, Impulsivity, Reality Distortion, Thought Disorganisation, and Activation, whereas patients who were recommended for admission scored lower for Somatic Preoccupation/Somatisation (Fig. 5.3).

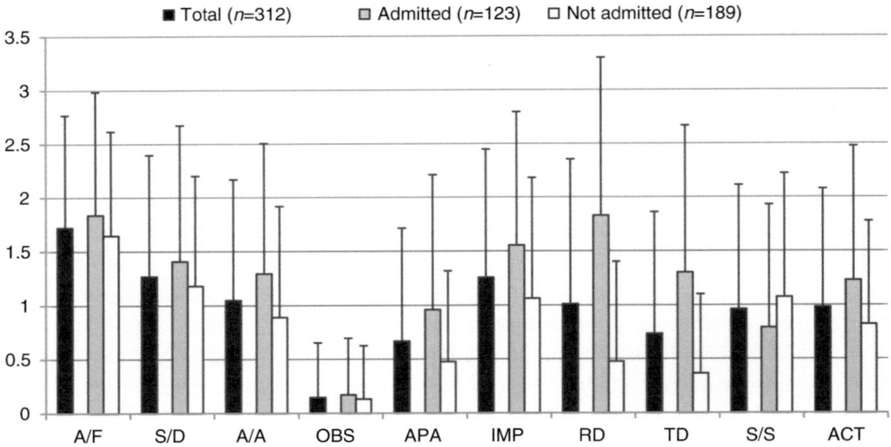

Fig. 5.3 SVARAD dimensions in the total sample and comparison between patients recommended for admission and patients discharged. Data are displayed as mean ± standard deviation. A/F, Apprehension/Fear; S/D, Sadness/Demoralisation; A/A, Anger/Aggressiveness; OBS, Obsessiveness; APA, Apathy; IMP, Impulsivity; RD, Reality Distortion; TD, Thought Disorganisation; S/S, Somatic preoccupation/Somatisation; ACT, Activation

More specifically, though, we identified only three dimensions that independently predicted the need for hospitalisation: Reality Distortion, Impulsivity, and Apathy. Such alterations can be commonly observed in a wide range of clinical pictures, commonly falling within the psychotic area, as well as affective disorders, including both manic and depressive episode, and personality disorders (e.g. borderline personality disorder). Our findings therefore suggest that it is useful for emergency clinicians to focus on these trans-diagnostic dimensions, independently from the categorical disorder, in order to evaluate whether to recommend admission or to discharge the patient in an emergency setting.

It is also interesting to compare the dimensional profiles by diagnostic group (Fig. 5.4). In the schizophrenic group, the patients who were recommended for admission showed significantly higher degrees of Reality Distortion, Thought Disorganisation, and Impulsivity; similarly, in the bipolar subgroup, they showed significantly higher degrees of Reality Distortion. The psychopathological differences between the patients recommended for admission and those discharged within the same disorder suggest that the need for hospitalisation is mainly determined by the current psychopathological alterations, rather than the categorical diagnosis.

This raises particular interest, both from theoretical and clinical perspectives, as to which diagnostic model is more suitable for the emergency setting. To this aim, which was the main goal of our study, we compared the categorical and the dimensional approaches as predictive models for the need for hospitalisation, using a hierarchical regression analysis. Our results suggested that 43% of recommended admission can be correctly predicted according to a dimensional evaluation alone, as assessed with the SVARAD, whereas the categorical model predicted only 29% of the recommended admissions. Also, when both socio-demographic (age, sex,

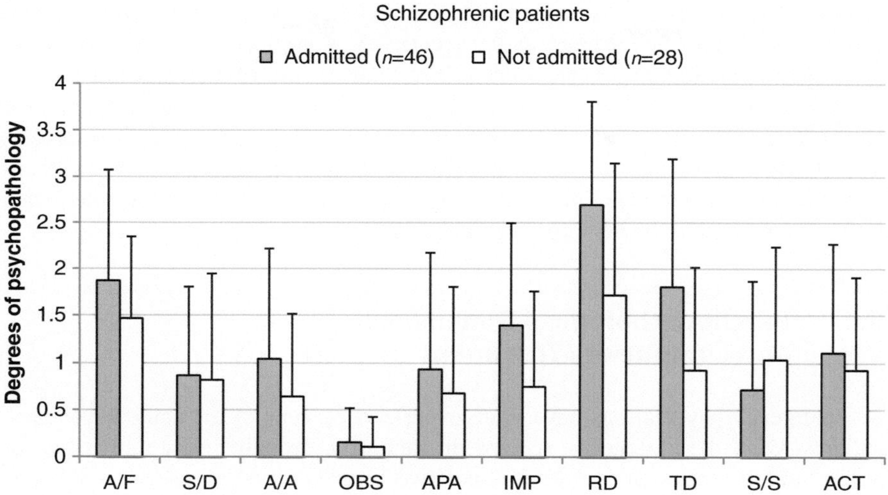

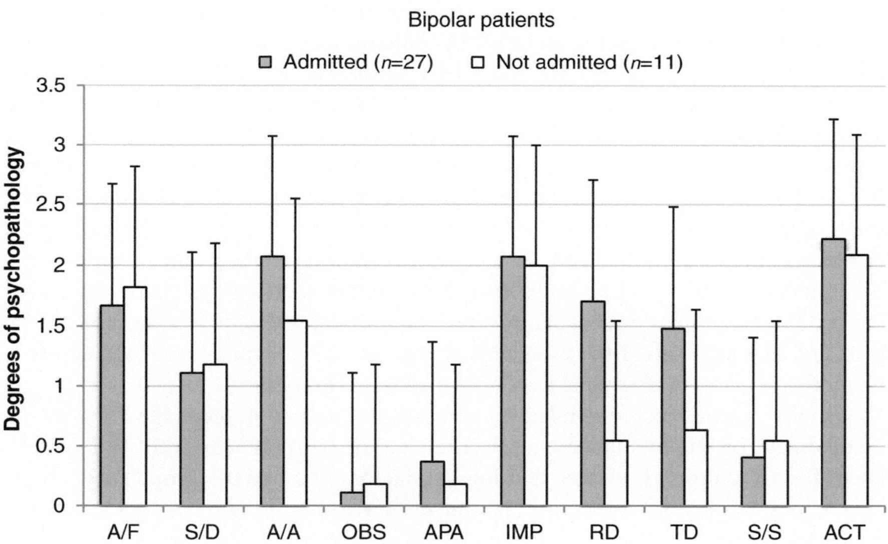

Fig. 5.4 Comparison of the mean SVARAD dimensions between patients recommended for admission and patients discharged in the schizophrenic ($n = 74$) and bipolar ($n = 38$) samples. Data are displayed as mean ± standard deviation. A/F, Apprehension/Fear; S/D, Sadness/Demoralisation; A/A, Anger/Aggressiveness; OBS, Obsessiveness; APA, Apathy; IMP, Impulsivity; RD, Reality Distortion; TD, Thought Disorganisation; S/S, Somatic preoccupation/Somatisation; ACT, Activation

marital status) and general clinical (previous admission, proposal for compulsory admission) predictors of admission were included, the two models predicted, respectively, 54% and 41% of the total rate of recommended admissions. Our findings suggest thus that dimensional assessment, globally, is the most important determinant of psychiatric admission. On the other hand, however, we observed that adopting both the dimensional and the categorical approach, in addition to the socio-demographic and general clinical information, further increased the predictive power of the model, which then correctly predicted 58% of total recommended admissions. Overall, such findings, which were globally confirmed in our replication study, suggest that a hybrid model, including both approaches plus socio-demographic and general clinical information, is the best solution in the emergency setting to predicting the need for hospitalisation.

5.6 Psychopathological Dimensions and Compulsory Treatment

As emergency psychiatrists have often experienced, some of the acute psychiatric patients reach the emergency service involuntarily, brought in by ambulance or police, and they are uncooperative, oppositive, or even hostile. Lack of compliance with treatment is a critical issue in psychiatry, in terms of both ethical and clinical implications: emergency psychiatrists are commonly required to decide whether or not to compulsorily admit a patient who is not compliant with treatment. The regulations for civil commitment, though, differ across countries. In Italy, unlike other countries, the patient's dangerousness to self or others is not included as a criterion for compulsory admission; instead, three criteria need to be met concurrently: (1) the patient shows psychic alterations requiring urgent treatment; (2) the patient refuses the required treatment; and (3) the patient cannot be adequately treated by nonhospital-based services. Also, the Italian law stipulates a specific procedure for ordering compulsory admission: first, compulsory admission must be recommended by a physician (proposal of compulsory admission); then a physician working for the National Health Service, usually a psychiatrist, must decide whether to confirm it or not after evaluating the case; finally, the city mayor must order the commitment. Compulsory admissions may only take place in a PICU.

Usually patients recommended for compulsory admission reach the ED with the involvement of an ambulance and, in a few cases, the police as well. With differences in law, culture, tradition, and logistic factors, the rate of compulsory admission varies greatly across countries and even regions, ranging across European countries between 3.2% in Portugal and 30% in Sweden [7]. In Italy, it is 9% [8].

Though such differences are cause for caution when interpreting the findings from different countries, it is presumable that a variety of clinical conditions play a similar role in determining the need for compulsory admission across different countries. The literature reports many socio-demographic and clinical risk factors, such as male gender, immigrant status, low social support, previous compulsory admission, premature termination of treatment, psychotic and bipolar disorder, severity and duration of illness, lack of insight, positive symptoms, and excitement (as measured with the Positive and Negative Syndrome Scale Excited Component) [7, 9–13]. It is clear that most of these studies describe socio-demographic and previous clinical risk factors, without focusing on the current psychopathological

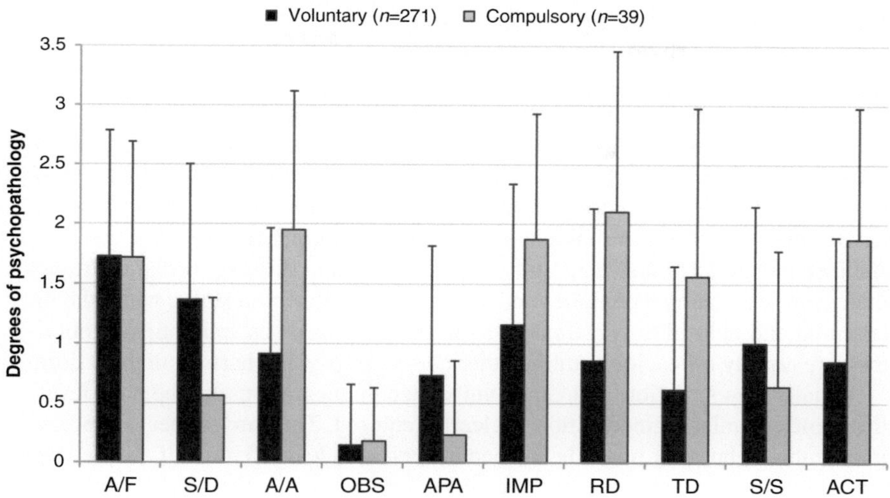

Fig. 5.5 Comparison of the mean SVARAD dimensions between patients admitted voluntarily and compulsorily. Data are displayed as mean ± standard deviation. A/F, Apprehension/Fear; S/D, Sadness/Demoralisation; A/A, Anger/Aggressiveness; OBS, Obsessiveness; APA, Apathy; IMP, Impulsivity; RD, Reality Distortion; TD, Thought Disorganisation; S/S, Somatic preoccupation/Somatisation; ACT, Activation. Two patients were not included because of missing data

dimensions. In our study, we observed that patients presenting to the ER with a proposal for compulsory admission were more likely to show higher degrees of Anger/Aggressiveness, Reality Distortion, Activation, Thought Disorganisation, and Impulsivity, and lower degrees of Sadness/Demoralisation, Apathy, and Somatic Preoccupation/Somatisation (Fig. 5.5). Though the difference in approach prevents us from comparing such findings with previous studies, some of our results seem to be consistent with the literature in key aspects. Indeed, Reality Distortion may be framed in the broader positive symptom dimension, whereas Impulsivity, Activation, and Anger/Aggressiveness can reflect, to a certain degree, the PANSS Excited Component. Both positive symptoms and excitement have been described as risk factors for compulsory admission [10]. Conversely, to our knowledge, the role played by other dimensions has not been clearly assessed in previous studies and will need to be investigated through future research. Interestingly, though, we observed in a recent study [14] that the Brief Psychiatric Rating Scale (BPRS) dimensions of Disorganisation and Activation, as well as Resistance (as measured with the expanded version of the BPRS), were independent risk factors for the use of restraint in acute psychiatric patients, whereas Negative Affect, which represents a broad dimension including depression, guilt, anxiety, and suicidality, and Negative Symptoms, which partially resemble the SVARAD Apathy dimension, were associated with a lower risk of restraint.

In conclusion, our findings suggest that a systematic evaluation of psychopathological dimensions may be usefully adopted in emergency settings to discriminate between psychopathological profiles of compulsorily vs. voluntarily admitted patients.

5.7 Psychopathological Dimensions and Therapeutic Intervention

The psychiatric examination in the ED often represents a crucial step for many patients, as it can drastically change the trajectory of their clinical history. Although it shows several relevant clinical implications, as we discussed above, first and foremost, it is a therapeutic intervention per se. To this purpose, emergency clinicians have a few instruments available, both verbal and pharmacological, that can help to stabilise the clinical conditions and mitigate the mental distress. Verbal instruments include de-escalation techniques and specific communication skills, while the most common and helpful medications used are benzodiazepines and antipsychotics, as they are rapidly efficacious, unlike other classes of psychopharmacological drugs.

Usually de-escalation and other verbal techniques represent the first level of intervention, unless a medication is clearly required. They are proper and practical interventions, but they might be unspecific, and thus less effective, if adopted without taking into consideration the prevailing psychopathological dimensions of the patient. Rather, in our experience, the evaluation of the dimensional profile is fundamental as it helps clinicians to tailor their approach to the patient, which promotes patient compliance and, more importantly, increases the therapeutic efficacy [15]. For example, when the patient shows high degrees of Activation, the psychiatrist should avoid dictating or competitive/dominant behaviours; rather, he should favour a calm, firm, and helpful style of communication. When managing a patient with Reality Distortion, behaviours best avoided by the therapist include minimising the importance of the current condition, giving false reassurance in order to relieve the patient, and questioning the patient's delusion. Instead, it is best for the therapist to empathically listen to such a patient.

In a number of cases, acute psychiatric patients also require a pharmacological intervention. Unlike in general psychiatry, where medications are recommended according to a categorical diagnosis, in emergency psychiatry, therapeutic guidelines are developed according to a number of specific conditions, e.g. psychomotor agitation, that may be symptoms of various medical and psychiatric disorders. In order to stabilise the psychopathological and behavioural condition and to choose the best medication, clinicians are thus required to focus on the current picture and the underlying causes, rather than carefully discriminating the primary psychiatric disorder, e.g. between schizophrenia and mania when managing an agitated patient. To this purpose, a dimensional approach seems particularly suitable and useful, especially when all the main psychopathological dimensions are systematically assessed, helping clinicians to easily identify a specific condition and plan the proper treatment.

In our experience, the choice of medication in emergency settings is strongly driven by the dimensional assessment. To test this hypothesis, we analysed the relationship between SVARAD dimensions and the medications most frequently used in the ED, such as benzodiazepines alone (BDZ), antipsychotics alone (AP), or a combination of benzodiazepines and antipsychotics (BDZ/AP). In our study, half of the patients did not receive a medication in the ED, 27.9% received BDZ, 6%

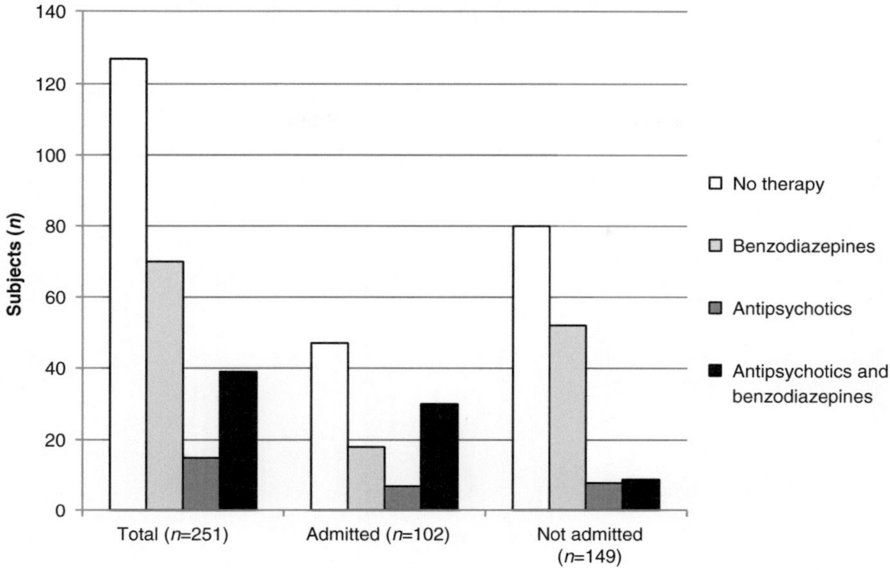

Fig. 5.6 Medication administered in the emergency department in the total sample and in subsamples of patients who were admitted and patients who were discharged. BDZ, benzodiazepines; AP, antipsychotics. Sixty-one patients were not included because of missing data

received AP, and 15% received BDZ/AP (Fig. 5.6). The logistic regression analysis showed that 42.4% of the medications were chosen according to the SVARAD assessment alone. This finding supports our initial hypothesis that emergency psychiatrists would highly value the dimensional profile when selecting medications in the ED. It is important, however, to emphasise that, since in more than half of the cases, the medication was not chosen according to the SVARAD dimensions. Emergency psychiatrists also clearly utilise a variety of other factors that affect the choice of proper medication.

The most relevant dimensions involved in choice of medication were Apprehension/Fear and Reality Distortion. The former independently predicted the use of BDZ and AP, whereas the latter independently predicted the use of AP and AP/BDZ. Conversely, other dimensions were less likely to be treated in the ED with one or more classes of medication, e.g. patients with Sadness/Demoralisation were less likely to receive BDZ or AP.

The mean levels of Activation, Impulsivity, Anger/Aggressiveness, and Thought Disorganisation were significantly higher in the patients who were treated with AP/BDZ, compared with those who did not receive any medication or BDZ only. However, such dimensions did not independently predict the medication. While the relationship between a high level of Apprehension/Fear and the use of BDZ is obvious and well known, AP might also be used when the patient is experiencing extreme apprehension or anguish, mainly under psychotic conditions. The administration of AP, either alone or in association with BDZ, to patients with psychotic conditions,

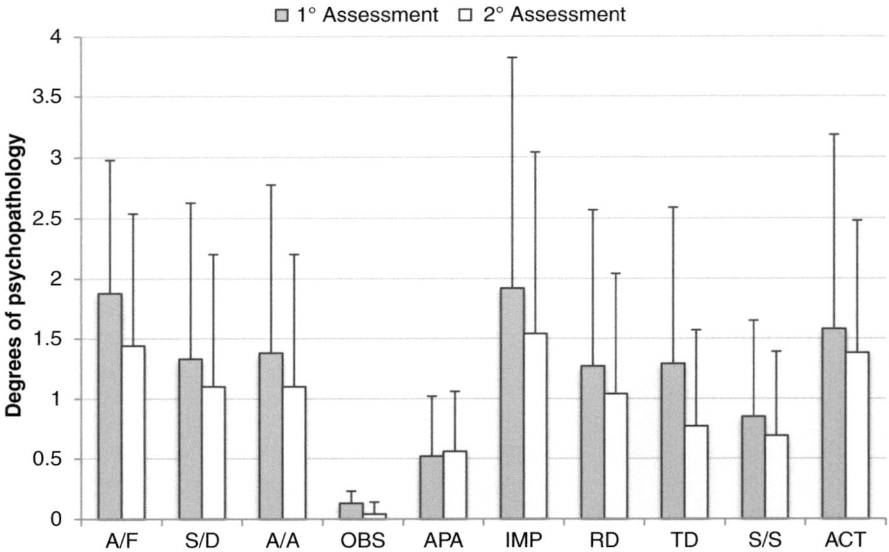

Fig. 5.7 Comparison of SVARAD mean profiles between the first and second psychiatric assessment in a sample of 48 patients who were monitored in the emergency department. . Data are displayed as mean ± standard deviation. A/F, Apprehension/Fear; S/D, Sadness/Demoralisation; A/A, Anger/Aggressiveness; OBS, Obsessiveness; APA, Apathy; IMP, Impulsivity; RD, Reality Distortion; TD, Thought Disorganisation; S/S, Somatic preoccupation/Somatisation; ACT, Activation

where Reality Distortion is typically observed, is not only aimed at achieving rapid sedation but also seems to generate early therapeutic effects, improving the psychotic condition even within the first 24 h of treatment [16].

To evaluate the therapeutic efficacy of psychiatric interventions in the ED, we compared the SVARAD dimensions over time in a group of patients who were monitored and re-evaluated in the ED. The sample included 48 patients who were monitored in the ED after their initial psychiatric assessment and evaluated again within 12 h. These patients had been monitored and re-evaluated for different reasons, e.g. because they required a further investigation to determine the diagnosis, such as a neurological evaluation or a toxicology screen. In these patients, we observed a significant reduction of Apprehension/Fear, Thought Disorganisation, and Activation (Fig. 5.7). In addition, the SVARAD total score decreased significantly. This suggests that a psychiatric intervention in the ED based on dimensional assessment is indeed therapeutic and effective even in the short term, decreasing the global degree of distress and mitigating some specific dimensions.

5.8 Conclusions, Limitations, and Future Perspective

Amid a growing concern for the current psychiatric diagnostic system, the dimensional approach has been gaining importance in the last 20 years as a potential complement to the categorical system. According to our findings, a systematic evaluation of psychopathological dimensions is a valid approach to emergency psychiatry: focusing on the current psychopathological condition, it allows clinicians to collect a large amount of valuable information with relevant implication for clinical practice. The dimensional approach to psychopathology serves to guide emergency psychiatrists through crucial but delicate decisions regarding hospitalisation, medication, and compulsory treatment. Indeed, our findings suggest that dimensional assessment represents the main determinant of admission. It also discriminates between patients requiring compulsory vs. voluntary treatment and guides the therapeutic intervention.

The dimensional assessment thus represents a useful approach in emergency psychiatry, but it obviously needs to be integrated into a larger paradigm. Indeed, a variety of other conditions, e.g. the risk of self-harm behaviours or the presence or absence of family and social support, need to be considered, as well as the underlying causes of the psychiatric condition. From this perspective, even in emergency psychiatry, the dimensional model should not represent an alternative to the current diagnostic system but rather a valuable complementary approach.

Considering the recent development of the dimensional approach, further investigation is needed to determine its validity and potential implications. As our results suggest, emergency psychiatry seems to be a promising field of interest, but further research is needed to corroborate these findings over time, as well as to extend the potential application of the dimensional approach and to tailor it to the emergency setting.

With no rating tools specifically developed to systematically assess the psychopathological dimensions in the emergency setting, the SVARAD represents, in our experience, a useful instrument guiding emergency psychiatrists through a systematic and standardised dimensional evaluation of acute psychopathological conditions. Indeed, though the SVARAD still needs to be extensively tested in emergency psychiatry, it seems particularly suitable to the emergency setting, as it is brief and easy to use and evaluates the dimensions that are routinely assessed during the mental state examination.

Acknowledgements We are grateful to Drs. Pier Luigi Andreassi, Pier Luca Bandinelli, Maria Caredda, Piero Cipriano, Roberto Delle Chiaie, Massimo Di Genio, Michele Di Nunzio, Giuseppe Ducci, Fabio Di Fabio, Nicoletta Giacchetti, Alessandro Grispini, Michele Majorana, Martino Mistretta, Donato Paciello, Mauro Pallagrosi, Massimo Pasquini, Daria Piacentino, Michele Risso, Massimo Salviati, Lorenzo Tarsitani, and Valeria Trincia for their help with data collection.

References

1. Bassuk EL, Birk AW, editors. Emergency psychiatry: concepts, methods, and practices. New York, NY: Plenum Press; 1984.
2. Allen MH, Forster P, Zealberg J, Currier G. Report and recommendations regarding psychiatric emergency and crisis services. A review and model program descriptions. APA Task Force on Psychiatric Emergency Services. Washington, DC: APA; 2002.
3. Warner MD, Peabody CA. Reliability of diagnoses made by psychiatric residents in a general emergency department. Psychiatr Serv. 1995;46:1284–6.

4. Way BB, Allen MH, Mumpower JL, Stewart TR, Banks SM. Interrater agreement among psychiatrist in psychiatric emergency assessment. Am J Psychiatry. 1998;155:1423–8.
5. Brown JF. Psychiatric emergency services: a review of the literature and a proposed research agenda. Psychiatry Q. 2005;76:139–65.
6. Dazzi F, Picardi A, Orso L, Biondi M. Predictors of inpatient psychiatric admission in patients presenting to the emergency department: the role of dimensional assessment. Gen Hosp Psychiatry. 2015;27(6):587–94.
7. Salize HJ, Dressing H. Epidemiology of involuntary placement of mentally ill people across the European Union. Br J Psychiatry. 2004;184(2):163–8.
8. De Girolamo G, Rucci P, Gaddini A, Picardi A, Santone G. Compulsory admission in Italy: results of a national survey. Int J Mental Health. 2008;37(4):48–61.
9. Webber M, Huxley P. Social exclusion and risk of emergency compulsory admission. A case-control study. Soc Psychiatry Psychiatr Epidemiol. 2004;39(12):1000–9.
10. Montemagni C, Badà A, Castagna F, Frieri T, Rocca G, Scalese M, Villari V, Rocca P. Predictors of compulsory admission in schizophrenia-spectrum patients: excitement, insight, emotion perception. Prog Neuro-Psychopharmacol Biol Psychiatry. 2011;35(1):137–45.
11. Myklebust LH, Sørgaard K, Røtvold K, Wynn R. Factors of importance to involuntary admission. Nord J Psychiatry. 2012;66(3):178–82.
12. Van der Post L, Visch I, Mulder C, Schoevers R, Dekker J, Beekman A. Factors associated with higher risks of emergency compulsory admission for immigrants: a report from the ASAP study. Int J Soc Psychiatry. 2012;58(4):374–80.
13. De Wit MA, Tuinebreijer WC, van Brussel GH, Selten JP. Ethnic differences in risk of acute compulsory admission in Amsterdam, 1996-2005. Soc Psychiatry Psychiatr Epidemiol. 2012; 47(1):111–8.
14. Dazzi F, Tarsitani L, Di Nunzio M, Trincia V, Scifoni G, Ducci G. Psychopathological Assessment of Risk of Restraint in Acute Psychiatric Patients. J Nerv Ment Dis. 2017;205(6):458–65.
15. Biondi M. Comunicazione con il paziente. Il metodo END. Empatia, normalizzazione, descalation in crisi ed emergenza. Roma: Alpes Italia; 2014.
16. Kapur S, Arenovich T, Agid O, Zipursky R, Lindborg S, Jones B. Evidence for onset of antipsychotic effects within the first 24 hours of treatment. Am J Psychiatry. 2005;162(5):939–46.

6 Sadness/Demoralisation and Apathy

Angelo Picardi, Paola Gaetano, and Elisa Fabi

6.1 Sadness/Demoralisation

6.1.1 Psychopathology and Assessment

The Sadness/Demoralisation dimension concerns the core symptoms of low mood and depression and is described in the English version of the SVARAD (RADAS) as follows: "distrust in oneself and one's own abilities; decreased creativity and energy; pessimism; decreased interests and pleasure". The symptoms of depression are varied and multifaceted, to the point that in his famous *The Anatomy of Melancholy*, published in the seventeenth century, Robert Burton considered the chaos created by the sheer variety of symptoms of melancholia to be greater than the confusion of tongues yielded by the tower of Babel [1]. However, it should be recognised that deep sadness and its variants (e.g., hopelessness, sorrow, dejection, despondency, emptiness, despair, discouragement) have been mentioned as core features of depression from the earliest medical texts in ancient Greece to the modern DSM and ICD classification systems, along with other symptoms alluded to in the SVARAD dimension, such as fatigue, repetitive focus on a few negative ideas, and lack of pleasure or interest in usual activities [2].

The key importance in the depressive syndrome of the symptoms covered by the SVARAD "Sadness/Demoralisation" dimension is corroborated by the fact that depressed mood and anhedonia are considered to be essential requirements for the

A. Picardi (✉)
Centre for Behavioural Sciences and Mental Health, Italian National Institute of Health, Rome, Italy
e-mail: angelo.picardi@iss.it

P. Gaetano
Italian Society of Cognitive and Behavioural Therapy, Rome, Italy

E. Fabi
Department of Human Neurosciences, Sapienza University of Rome, Rome, Italy

M. Biondi et al. (eds.), *Dimensional Psychopathology*,
https://doi.org/10.1007/978-3-319-78202-7_6

diagnosis of a major depressive episode in both the DSM-5 and ICD-10. Furthermore, in the ICD-10, decreased energy or increased fatigability can also be considered as a "core feature" in the definition of a depressive episode, together with depressed mood and anhedonia. The relevance of this dimension is also supported by the examination of the post-Kraepelin Western psychiatric tradition as described in textbooks. In fact, all authors of the 19 major textbooks of psychiatry or psychological medicine published from 1900 to 1960 mention mood disturbances, described by terms such as sadness, hopelessness, emptiness, gloominess, mental pain, and misery, as a key manifestation of depression. Also, all of these authors describe specific cognitive content, such as hopelessness, guilt, worthlessness, pessimism, self-accusation, self-derogation, and feelings of inadequacy and of being a failure, as characteristic of depressed patients. Twelve authors commented on loss of interest, while anhedonia was described by seven authors, and fatigue and exhaustion by eight authors [3].

Various aspects of anxiety were noted by most authors of classical textbooks as characteristic of depressed patients, and depression and anxiety show considerable overlap at every level that has been studied. However, a number of psychometric studies lend support to the hypothesis of a core depressive dimension, distinct from anxiety and from general symptoms of somatic and emotional distress. For instance, many previous studies, including one from our group [4], have identified a distinct depressive factor in the Hamilton Depression Rating Scale (HDRS), which is defined by items covering core symptoms of depression, such as depressed mood, psychomotor retardation, loss of interest in work and other activities, suicidal ideation, and guilt, although there are some differences among studies regarding the specific variables included in this factor and their relative importance as estimated by the size of factor loadings. Indeed, four separate meta-analyses of factor analytic studies of the Beck Depression Inventory (BDI), the Center for Epidemiological Studies Depression Scale (CES-D), the HDRS, and the Zung Self-Rating Depression Scale (SDS) identified relatively robust and well-established specific depression symptom factors within each scale [5]. A factor analytic study of the SVARAD itself identified a core depression factor loaded by the Sadness/Demoralisation item and the Apathy item, which covers the other dimension examined in this chapter [6]. A pattern that is seen with some frequency in symptom factor analyses is a tripartite division of the items relevant to depression and anxiety: a general "neurotic" factor and two smaller factors specific for depression and anxiety. Items that consistently load on the depression factor include depressed mood, loss of interest and pleasure, crying easily, hopelessness, loneliness, and suicidal ideation. This led to the proposal of a "tripartite model" of symptoms of depressive and anxiety disorders, which posits three symptom groups: nonspecific symptoms of general distress, interpreted as high negative affect; symptoms of anhedonia and low positive affect that are relatively unique to depression; and manifestations of somatic tension and arousal that are relatively specific to anxiety [7]. While the model might need further refinement and specifications [8], it has received some empirical support, with the depression factor showing no overlap with any of the anxiety symptoms [9].

Clinimetric studies also corroborate the hypothesis of a core depressive dimension, similar to that described in the SVARAD. When the prevalence and sensitivity to change of the symptoms included in the HDRS were evaluated using item analysis or factor analysis, a few core symptoms were found to have greater sensitivity to change and less distortion by treatment-emergent side effects. This led to the derivation from the HDRS of smaller scales measuring core depression, such as the six-item HAM-D6, which features symptoms such as mood, guilt, anhedonia, loss of energy, and psychomotor retardation, along with psychic anxiety that, as described above, is regarded by many influential authors as an important clinical feature of depressive states but can be psychometrically separated from a specific depressive dimension. The six items included in the HAM-D6 were found to be the only HDRS items that validly reflected a global clinical assessment of depression severity made by experienced psychiatrists [10]. In dose-response trials, the HAM-D6 was found to be much more sensitive in discriminating between antidepressants and placebo than the full version of the scale [11, 12], which lends further empirical support to this core depressive dimension and suggests it might also have biological validity.

The SVARAD Sadness/Demoralisation dimension is also deeply rooted in the phenomenological tradition, as it grasps the most characteristic features of the experience of being depressed. Classical psychopathologists have underscored how the vital sadness that permeates the life of the depressed patient is so deep and prolonged that it borders stagnation of being and hampers their basic vital movement. The perception of time is altered in such a way that the patient's attention is directed to the past, while the present is experienced as empty, and the future is perceived as blocked. This creates a painful experience of stagnation of time lived, characterised by a heavy past, an empty present, and a lost future, with poor or no prospects [13, 14]. Action seems difficult, impossible, or futile, because there is no sense of any possibility for positive change. The body is perceived as heavy and inert, as it is no longer drawn in by situations, solicited to act. It feels stuck where it is, incapable of projection and purposeful action. Space is experienced as empty, dull, flat, and closed, with no possibilities [15]. The patient might be unable to find anything practically significant and feels as if something is missing, painfully lacking. Nothing appears quite as it should, and nothing offers the potential for positive change. In the more severe cases, the patient does not just take all human life to be without value but cannot even contemplate the possibility of its being otherwise, as the experience has a feeling of irrevocable certainty to it [16]. It is in the midst of this painful experience of vital sadness, standstill of lived time, draining of energy, and loss of significance in life that pessimistic and self-devaluing thoughts breed in the patient's mind. From an intersubjective perspective, the clinician clearly perceives the patient's painful condition, and increased severity of this dimension was found to be correlated with higher clinician's feelings of impotence. However, severity of affective symptoms was also found to be correlated with a greater positive involvement with the patient, as if his or her vulnerability and help-seeking attitude were reassuring for the clinician, thereby promoting the creation of a classical doctor-patient relationship [17].

Although the SVARAD "Sadness/Demoralisation" dimension includes the term "demoralisation" in its name, it mainly refers to the spectrum of depressive phenomena, rather than to the clinical concept of demoralisation first elaborated by Jerome Frank. However, particularly at the lower end of severity, the item may capture the phenomenon of demoralisation, so it may be useful to provide a brief overview of this construct. According to Frank, demoralisation results from persistent failure to cope with a stressful event or situation that the person and their significant others expect them to handle. It is characterised by damaged self-esteem and by feelings of impotence, isolation, and despair. Demoralised persons are conscious of having failed to meet their own expectations or those of others, or of being unable to cope with some pressing problem. They feel powerless to change the situation and cannot extricate themselves from trouble [18]. At the core of demoralisation lies a feeling of being trapped—not knowing what to do, becoming helpless. There is also a breakdown in core assumptions and beliefs about the world, with resulting loss of meaning. Therefore, demoralisation is experienced as a persistent inability to cope, together with associated feelings of helplessness, hopelessness, meaninglessness, subjective incompetence, and reduced self-esteem [19]. Several studies have suggested that demoralisation can be distinguished from passing or transient distress, nonspecific distress, subthreshold depression or anxiety, and major depression [20, 21]. From a psychopathological perspective, it has been emphasised that, differently from depression, demoralisation is characterised not by a reduction in hedonic capacity and motivation but rather by feelings of subjective incompetence and helplessness [22].

6.1.2 Neurobiology

The neurobiology of this psychopathological dimension is inextricably linked to the neurobiology of depression. Persistent low mood and anhedonia are both believed to result from functional abnormalities in the neural circuitries underlying implicit emotion regulation and reward processing. Persistent low mood mainly involves the amygdala and the prefrontal cortex, as well as the hippocampus, while anhedonia centres on the striatum and the prefrontal cortex. A substantial body of evidence suggests that serotonin modulates activity in the implicit emotion regulation neural circuitry, and a smaller number of studies indicate that other neurotransmitters, such as norepinephrine, may also modulate functioning in this circuitry. A large body of research highlights the modulating role of dopamine in the neural circuitry underlying reward processing. The symptoms of depression are hypothesised to reflect changes in the activity of frontal brain areas, related to decreased activity in the hippocampus. These changes represent either a heightening (the ventral system) or an inhibition (the dorsal system) of the usual role of these areas in the processing of, and coping with, affective information. Specific disruptions in these networks can be related to specific symptoms of depression, for instance, rumination and negative self-referential attributions would reflect hyperactivity in the amygdala and ventral prefrontal regions, while anhedonia would reflect hypoactivity in the nucleus accumbens and dorsal prefrontal cortex [23].

Indeed, both structural and functional brain imaging findings consistently indicate alterations in cortical and subcortical regions involved in emotion regulation and reward processing. Among the findings of altered brain structure in depression, the most consistent finding is the volume loss of the hippocampus, orbital, and ventral prefrontal cortex [24]. Also, functional abnormalities in both the implicit emotion regulation and reward circuitries have been observed in patients with depression. A number of studies have provided evidence of reduced functional connectivity between the amygdala and the medial prefrontal cortex and associated anterior cingulate cortex regions, while other studies have suggested increased amygdala response to emotional stimuli, especially those with negative valence [25]. In depressed patients with anhedonia, abnormal activity levels in the ventral and dorsal striatum, as well as in the orbital prefrontal cortex, have been observed [26].

An influential hypothesis, first put forward more than 50 years ago, posits that the underlying biological basis for depression is a deficiency of central noradrenergic and serotonergic systems [27]. While in its original form this hypothesis turned out to be clearly inadequate, the monoamine hypothesis has been of great importance in understanding depression and has evolved over the years, so that depressive pathophysiology is currently recognised as involving multiple levels. It is now thought that acute increases in the amount of synaptic monoamines induced by antidepressants produce secondary neuroplastic changes that are on a longer timescale and involve transcriptional and translational changes that mediate molecular and cellular plasticity [28].

Besides deficiencies in noradrenergic and serotonergic function, other potentially contributing neurobiological factors underlying depression have been proposed, such as deficits in other neurotransmitters and in neurotrophic factors, changes in hippocampal neurogenesis, and HPA dysregulation [29]. Recently developed frameworks for understanding the neurobiology of depression emphasise the role of chronic stress [23, 30]. Specifically, the role of stress in depression is postulated to be mediated by activation of the hypothalamic–pituitary–adrenal (HPA) axis, which leads to functional and, later, morphological damage to the hippocampus, which in turn causes widespread disruption of information processing in the forebrain. In rodents, chronic exposure to adrenal glucocorticoids decreases synaptic number and function and causes atrophy of neurons in the hippocampus and prefrontal cortex. The importance in depression of the hippocampal formation stems from its connections with other brain regions that are more directly involved in the psychological (amygdala, prefrontal cortex and other areas of association cortex, and nucleus accumbens) and physiological (hypothalamus) phenomena that characterise depression.

Stress-induced activation of the inflammatory response, particularly inflammatory cytokines, may also play a role, since inflammatory cytokines may influence all the pathways believed to be involved in the pathophysiology of depression, including dysfunction of monoaminergic systems and the HPA axis, alterations in glutamate neurotransmission, changes in growth factors, and decreased neurogenesis.

In fact, cytokines and inflammatory signalling pathways may affect dopaminergic neurotransmission and the synthesis of dopamine. Also, inflammatory cytokines have been shown to increase serotonin transporter expression and function. Moreover, they can activate indoleamine 2,3-dioxygenase (IDO), which, in turn, breaks down tryptophan (the primary amino acid precursor of serotonin) into kynurenine. On the one hand, this contributes to reduced serotonin availability. On the other hand, the activation of IDO and kynurenine pathways in the brain affects glutamate neurotransmission and may contribute to glutamate dysfunction. Furthermore, inflammatory cytokines may influence HPA axis function through effects on negative feedback regulation. They may also adversely influence neurogenesis and neuroplasticity, as they have been shown to decrease systemic brain-derived neurotrophic factor (BDNF) levels and to influence BDNF receptor phosphorylation, thereby further interfering with BDNF signalling [31, 32].

Therefore, there are several signalling pathways that influence synapse formation and stability and that could contribute to loss of synapses in depression. These pathways include neurotransmitters (such as glutamate), growth factors, and neurotrophic factors (such as BDNF), inflammatory cytokines, and the HPA axis. They affect multiple intracellular signalling cascades that regulate all aspects of neuronal function, including the protection and survival of neurons and the induction of synaptic plasticity. A key downstream convergence pathway for activity-dependent synaptic plasticity, and translation of synaptic proteins, is the mechanistic target of rapamycin complex 1 (mTORC1), which serves as a neuronal sensor of activity-dependent demand for new protein synthesis and synaptogenesis [33]. Figure 6.1 presents an overview of the mechanisms involved in the neurobiology of depressed mood.

Serotonergic and noradrenergic antidepressants act via different mechanisms, involving serotonergic and noradrenergic synapses, respectively. However, the various cortical or subcortical routes to antidepressant action are hypothesised to converge at a point downstream from the primary actions at serotonergic and noradrenergic synapses. For example, after chronic treatment, antidepressants of all classes increase the responsiveness of D2 dopaminergic receptors in the nucleus accumbens, the terminal integrative area of the mesolimbic dopaminergic system. Also, following chronic administration, antidepressants of all classes have been found to increase the expression of the nuclear transcription factor cAMP response element-binding protein (CREB) in the hippocampus. CREB regulates the activity of several genes, with resultant changes in the production of various proteins, among which are BDNF and vascular endothelial growth factor (VEGF). Chronic antidepressant treatment, not only with common monoaminergic antidepressants but also with putative antidepressants from very different pharmacological classes and with non-pharmacological treatments such as electroconvulsive therapy or vagus nerve stimulation, increases hippocampal neurogenesis, an effect that requires stimulation of monoamine transmission and is mediated by BDNF and, likely, also by VEGF. Therefore, the effects of currently used monoaminergic antidepressants are hypothesised to be mediated principally by stimulation of serotonergic and noradrenergic synapses in the hippocampus, which would be expected to lead to increased

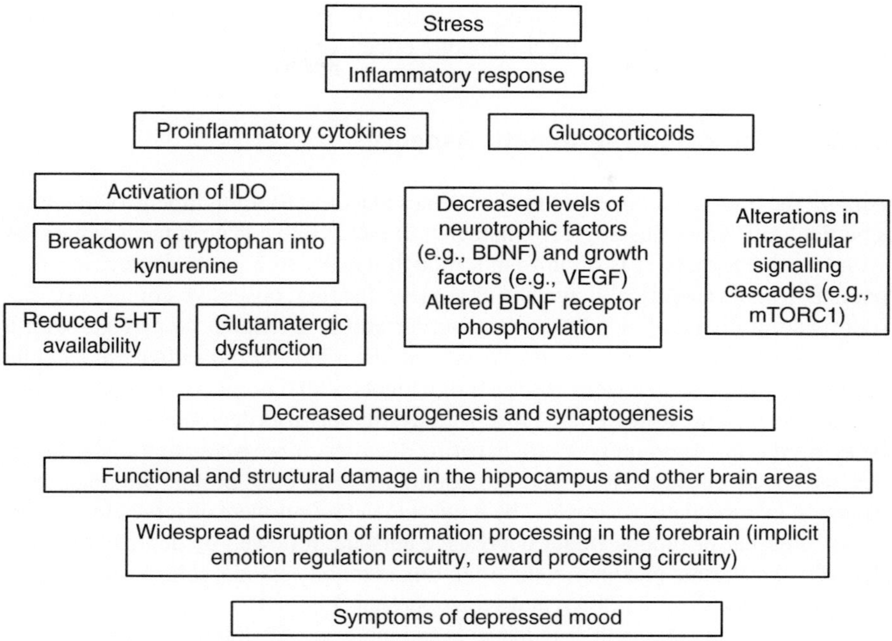

Fig. 6.1 Mechanisms involved in the neurobiology of depressed mood

production of neurotrophins and neurogenesis, and, consequently, to reorganisation and repair of the stress-induced morphological damage. This, in turn, would rebalance activity in forebrain circuits, with a normalising decrease in the impact of aversive events and an increase in the impact of rewards [23].

The changes in synaptic plasticity brought about by antidepressants through increased neurotrophic factor expression require some weeks, and this is hypothesised to account for the time lag in their clinical effectiveness. The rapid antidepressant action displayed by agents such as intravenous ketamine and scopolamine in some small-scale randomised clinical trials suggests a mechanism that results in fast changes in synaptic function and plasticity. The available data indicate that these agents stimulate mTORC1 signalling in the prefrontal cortex, which in turn leads to the increased synthesis of proteins that are required for synapse maturation and formation. Both ketamine and scopolamine cause a burst of glutamate, the former via blockade of N-methyl-D-aspartate (NMDA) receptors on GABAergic interneurons, the latter via blockade of acetylcholine muscarinic M1 receptors on GABAergic interneurons. The glutamate burst activates alpha-amino-3-hydroxy-5--methylisoxazole-4-propionic acid (AMPA) receptors, and this activation stimulates mTORC1 signalling. Interestingly, neither acute nor chronic administration of common antidepressants increases mTORC1 signalling, which corroborates the notion that its activation is involved in rapid antidepressant action. It may also be noted that the rapid antidepressant effect displayed by scopolamine in preliminary trials raises the possibility that the anticholinergic actions of tricyclic antidepressants might contribute to some degree to the therapeutic effects of these agents [33].

6.1.3 Clinical and Therapeutic Aspects

Among the SVARAD dimensions, the Sadness/Demoralisation dimension displays a particularly "cross-nosological" character. In fact, while it is prominent in patients with mood disorders, it is often found, though usually to a lesser degree, in many patients with a categorical diagnosis of anxiety disorder, obsessive-compulsive disorder, eating disorder, or adjustment disorder and in a number of patients with a diagnosis of schizophrenia or other psychotic disorders. In our outpatient sample, the highest mean scores were observed in patients with a unipolar or bipolar major depressive episode, followed by those with dysthymic disorder, depressive disorder NOS, borderline personality disorder, eating disorders, obsessive-compulsive disorder, psychotic disorder NOS, and schizophrenia (Fig. 6.2). Similar findings were observed in our inpatient sample. The highest mean scores were observed in patients with a unipolar or bipolar major depressive episode, followed by depressive disorder NOS, borderline personality disorder, mixed mood episode, schizoaffective disorder, psychotic disorder NOS, and schizophrenia (Fig. 6.3).

It is noteworthy that both in outpatients and inpatients the findings observed in the total group of patients with a given primary diagnosis were comparable to those observed in the subgroup of patients who did not receive any other Axis I diagnosis. This suggests that the prominence of the Sadness/Demoralisation dimension in diagnostic groups other than depressive disorders does not result from depressive comorbidity.

In patients with mood disorders, the symptoms making up the Sadness/ Demoralisation dimension are usually moderate to severe, and this dimension

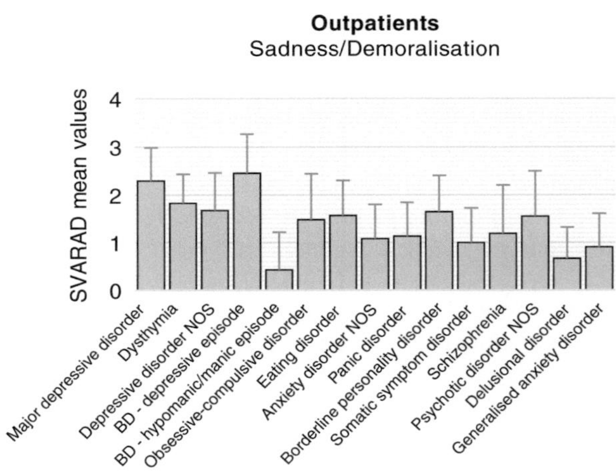

Fig. 6.2 SVARAD Sadness/Demoralisation dimension across outpatients' diagnostic categories: mean scores and standard deviations

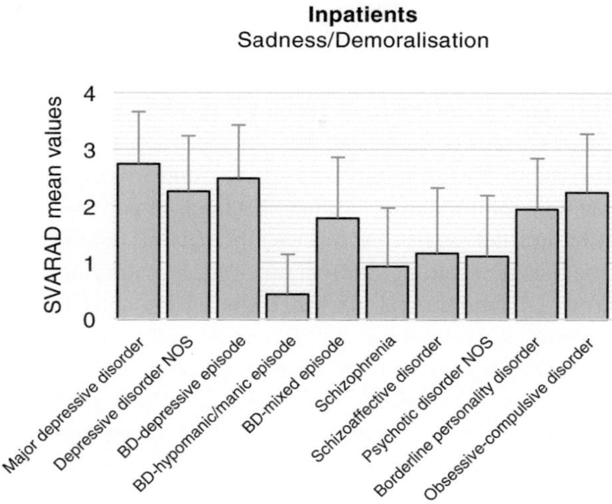

Fig. 6.3 SVARAD Sadness/Demoralisation dimension across inpatients' diagnostic categories: mean values and standard deviations

typically shows the highest mean score among the SVARAD dimensions. In patients with other mental disorders, these symptoms often take a milder form that does not justify a formal diagnosis of depressive disorder but might nevertheless be clinically important. In fact, the examination of mean scores across diagnoses shows that this dimension often ranks second or third among all SVARAD dimensions. Also, the proportion of outpatients without psychiatric comorbidity who were rated 2 ("moderate: poorly modifiable, extended to almost all areas of experience, mild reduction of the patient's social or occupational functioning") or more on this item was 59, 53, 42, 41, and 38% among patients with borderline personality disorder, psychotic disorder NOS, schizophrenia, obsessive-compulsive disorder, and eating disorders, respectively. Among inpatients with no psychiatric comorbidity, the proportion was 77%, 34%, and 27% for patients with borderline personality disorder, psychotic disorder NOS, and schizophrenia, respectively.

In outpatients, the Sadness/Demoralisation dimension displayed several significant ($p < 0.001$) correlations, such as a moderate to strong correlation with Apathy (rho = 0.48), a moderate correlation with Apprehension/Fear (rho = 0.28), a small correlation with Anger/Aggressiveness (rho = 0.11), and a small negative correlation with Activation (rho = −0.15). In inpatients, it similarly showed a moderate to strong correlation with Apathy (rho = 0.39) and a moderate correlation with Apprehension/Fear (rho = 0.32), and it showed several sizable negative correlations, such as with Activation (rho = −0.32), Reality Distortion (rho = −0.39), and Thought Disorganisation (rho = −0.37) (all $p < 0.001$).

Usually, specific treatment is warranted for patients with clinically significant levels of this dimension, independent of categorical diagnosis. A decrease in mood that is severe enough to cause substantial suffering or impairment in psychosocial functioning is usually indicated by a score of 2 or higher on the corresponding SVARAD item. Treatment can take the form of medication, psychotherapy, or a combination of both. From a pharmacological perspective, the underlying neurobiology suggests the use of agents enhancing noradrenergic or serotonergic transmission in the central nervous system, which is in fact an effect that is shared by most drugs classified as "antidepressants". Neurobiological considerations also underpin other kinds of pharmacological or neurophysiological interventions, which are mainly used as second-line treatments in patients with treatment-resistant depression. For instance, the link between depression and inflammation suggests the use of anti-inflammatory agents, and indeed cytokine inhibitors and nonsteroidal anti-inflammatory drugs (NSAIDs), in particular the selective cyclooxygenase 2 inhibitor celecoxib, have shown promising results in clinical trials. A recent meta-analysis, combining data on anti-inflammatory add-on treatment and monotherapy from 14 randomised clinical trials with a total of 6262 patients, found that anti-inflammatory treatment significantly reduced depressive symptoms as compared with placebo, without increased risks of adverse effects [34]. The anti-inflammatory actions of physical exercise, too, might be associated with its beneficial effect on depression, with recent data providing support for the involvement of IL-6 in the antidepressant effect of exercise [35]. Attenuation of cortisol release [36] has also been observed in depressed patients treated with bright light therapy, an intervention with preliminary evidence of effectiveness in seasonal as well as non-seasonal depression [37]. A further example is provided by repetitive transcranial magnetic stimulation over the left dorsolateral prefrontal cortex, another empirically supported treatment [38]. Although its effects are complex and only partially understood, they seem to involve normalisation of the HPA axis function and also activation of neuroprotective mechanisms in the brain [39, 40]. As far as electroconvulsive therapy is concerned, although there is evidence that it is a potent antidepressant treatment [41], its broad spectrum of effects ranging from depression to mania [42], and possibly catatonia [43], coupled with the great uncertainty surrounding its mechanism of action [44–47], makes it difficult for us to provide indications for its use in the framework of our theoretical perspective, which is based on psychopathological dimensions and their underlying neurobiology.

Biological interventions are not the only effective treatment options available. Empirically supported treatments for depression include cognitive behavioural therapy (CBT), interpersonal psychotherapy, behavioural therapy, and, to a lesser extent, short-term psychodynamic psychotherapy [48]. It should be underscored that psychotherapeutic interventions can not only profoundly influence patients' beliefs, ways of thinking, emotional states, and behavioural patterns but can also affect brain function.

Since the inception of psychotherapy, there has been interest in the brain as a mediator of its effects; as early as the end of the nineteenth century, Freud had attempted to translate psychotherapeutic concepts into the language of biological sciences [49]. However, until recently, the putative biological mechanisms of psychotherapy and the resulting, underlying changes in the brain have remained elusive. In the late 1990s, the significant advances in neuroscience that had occurred in the second half of the twentieth century enabled Kandel to suggest that the effectiveness of psychotherapy and its ability to produce long-term changes in behaviour are likely due to learning processes and related changes in gene expression that alter the strength of synaptic connections, as well as structural changes that alter the anatomical pattern of interconnections between neurons [50]. Then, in the last two decades, thanks to increases in the resolution of brain imaging techniques, the putative mechanisms of action of psychotherapy and the underlying changes in the brain have finally attracted the research attention they deserve [51]. Although the biological study of psychotherapy has barely begun, neuroimaging studies have clearly shown that psychotherapy leads to observable changes in the brain.

Neuroimaging studies of psychotherapy for major depression have reported changes in many brain regions, including the cingulate gyrus, medial prefrontal, orbitofrontal, dorsolateral, and dorsomedial prefrontal cortices, hippocampus, amygdala, and basal ganglia. The findings of these studies are complex and, at times, inconsistent. This is hardly surprising, given substantial differences between studies in neuroimaging technique (PET, SPECT, fMRI), experimental paradigm (resting state, symptom provocation task), therapeutic approach (cognitive behavioural therapy, behavioural therapy, interpersonal psychotherapy, psychodynamic therapy), duration of treatment (and the related time interval between pre- and post-treatment scan), inclusion of a control condition, and type of control subjects (healthy, waiting list). However, despite the methodological heterogeneity of the studies, the majority suggest that psychotherapy results in a normalisation of neural function in brain regions that showed abnormalities before treatment [52–54].

Across different therapeutic approaches, reductions in activity or metabolism have been observed in several prefrontal cortical regions. Although the precise regions differ to some extent across studies and approaches, reductions in one or more of these regions have been observed following treatments as diverse as interpersonal therapy, cognitive behavioural therapy, behavioural activation, and psychodynamic psychotherapy. Regarding changes in other brain regions, the results are more preliminary and mixed. Of particular interest is preliminary evidence from a single study that behavioural activation can affect the reward circuitry in the brain, as it was reported to normalise functioning in the dorsal striatum during reward anticipation, and the orbital prefrontal cortex during reward feedback [55]. On the one hand, the findings of these studies are broadly consistent with predictions regarding the neural substrates of psychotherapy. On the other hand, the nature of the effects in prefrontal cortical regions, with a decrease in activity following psychotherapy, may be surprising, as it appears to contradict the hypothesis that psychotherapy should strengthen the ability of higher cortical regions to regulate processing in lower brain areas. However, it should be underscored that, beyond understanding which regions are involved, the state of the field has not yet evolved sufficiently to draw specific conclusions regarding the direction of effects. For instance, conflicting directional observations may be due to differences in experimental protocols and tasks (resting state, symptom provocation, application of a specific therapy skill). Furthermore, regional increases and decreases observed in neuroimaging are currently subject to multiple interpretations. Increased activation in a given region might be interpreted as reflecting an improvement in the strength of the region's function. However, it could also be interpreted as the opposite: an impairment in the region's efficiency, reflecting a need for greater activity in order to accomplish the same effect [56].

Interestingly, antidepressants and psychotherapy may have distinct neural effects, as the literature suggests that their mechanisms are divergent and that the two types of treatment normalise brain function in different ways [53, 57]. Whereas psychotherapy appears to lead to changes in the activation patterns in cortical areas, antidepressants tend to be associated with changes in brain activity in the limbic system, other subcortical structures, and the insula [58]. This pattern of findings is consistent with the hypothesis that psychotherapy may exert its effects "top-down", targeting mainly frontal cortical regions and strengthening the cortical emotion regulatory processes, whereas medication may produce "bottom-up" changes by disengaging mainly subcortical regions that mediate attention to personally relevant emotional and environmental stimuli or are involved in representing our internal bodily states [53, 58]. Although the findings are still too preliminary to draw firm conclusions, the possibility that antidepressant medication and psychotherapy have different neural effects is intriguing, as it suggests that their effects might be, at least partially, complementary.

The possibly complementary effects of psychotherapy and medication on brain function suggest that the two kinds of treatment may act synergistically. Indeed, there has been increased interest recently in their combination for the treatment of depression, which may aid in adherence to both treatments. Medication may allow for a more effective use of psychotherapy by providing initial relief from depressive symptoms and increasing concentration and motivation. Psychotherapy may aid in adherence to pharmacological treatment, as it may allow for reduced doses and thus a lower side effect burden of medication [59]. Indeed, a number of meta-analyses suggest that combined treatment has small but significant advantages over each treatment modality alone [60–62] and may have a protective effect against depression relapse or recurrence [63]. Importantly, care should be applied not to merely add psychotherapy to antidepressant medication but to introduce it to the patient as a treatment that would work synergistically with pharmacotherapy. A psychobiological model overcoming the traditional brain-mind dichotomy by positing that both treatments affect mental phenomena and influence brain function may help the clinician give the patient a clear rationale for the combination of psychological and pharmacological treatment [64, 65].

In a dimensional perspective, it is important to underscore that in order to optimise treatment, the clinician should take into account the whole dimensional picture. Independent of categorical diagnosis, if dimensions other than Sadness/Demoralisation are present to a substantial degree, they, too, should be taken into due account. The most relevant dimensions are usually Apathy, the "anxious" dimensions (Apprehension/Fear and Somatic Preoccupation/Somatisation), or the "activation" dimensions (Anger/Aggressiveness, Activation, and Impulsivity). Indeed, in some previous studies using the SVARAD [6, 66], we found that patients with unipolar depression are typically characterised by a variable combination of three clusters of dimensions: "core depressive dimensions" (Sadness/Demoralisation, Apathy), "anxiety dimensions" (Apprehension/Fear, Somatic Preoccupation/Somatisation), and "activation dimensions" (Anger/Aggressiveness, Impulsivity, Activation).

The approach to patients with prominent apathy will be described in detail in the following section, which is specifically devoted to this dimension. Basically, the presence of substantial levels of both Sadness/Demoralisation and Apathy, with a SVARAD profile such as that depicted in Fig. 6.4, indicates the need to target not only the implicit emotion regulation circuitry but also the reward processing circuitry. Possible means to achieve this aim with pharmacological and psychological treatment would be the use of an antidepressant with noradrenergic properties and the inclusion of behavioural activation techniques, respectively.

When treating a patient with pronounced anxiety symptoms in addition to high levels of Sadness/Demoralisation, who has a SVARAD profile such as that illustrated in Fig. 6.5, the clinician should give such symptoms the attention that they deserve. There is, in fact, evidence that such patients are at greater risk for side

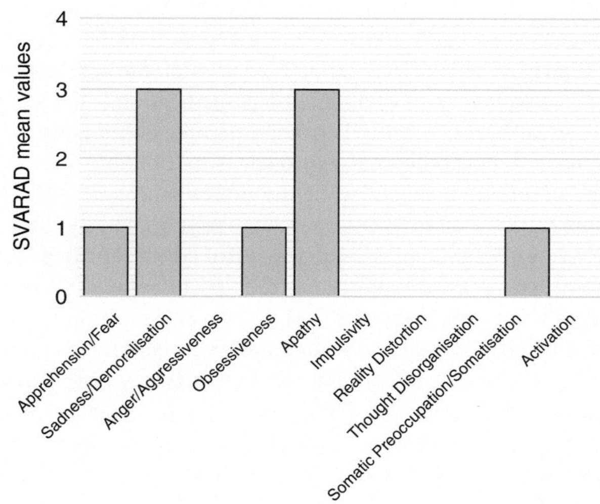

Fig. 6.4 A patient with prominent Sadness/Demoralization and Apathy

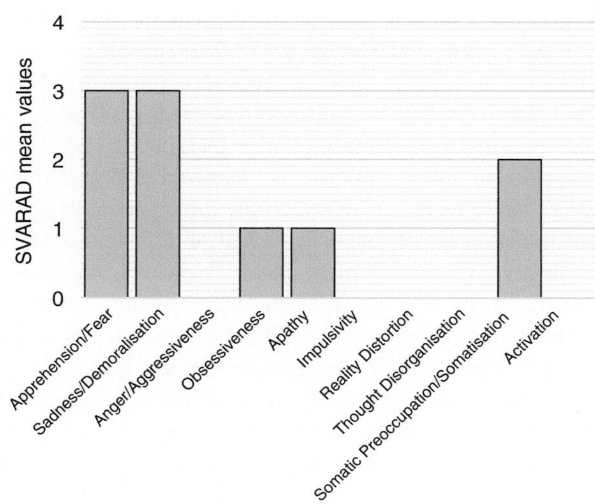

Fig. 6.5 A patient with prominent Sadness/Demoralization and Apprehension/Fear

effects and have poorer treatment outcomes than depressed patients without anxiety [67, 68]. If pharmacological therapy is chosen, neither the neurobiology of anxiety [69] nor the findings from clinical trials involving patients with anxious depression [70] reveal sufficient evidence to suggest choosing one class of antidepressant medication over another, providing that the serotonergic system is targeted, given its role in the regulation of anxiety-related behaviour and traits.

While some patients with this dimensional profile can be successfully treated with antidepressants alone, this clinical population is quite difficult to treat and may require additional measures. One possibility is to reduce dopaminergic transmission with low doses of D2 receptor blockers. However, this might be a double-edged sword, as the role of dopamine in anxiety states is complex. Dopaminergic pathways may affect anxiety states in several ways [71], either exacerbating anxiety or reducing it through increased feelings of self-efficacy and self-confidence. Another possibility is to aim for an increase in GABAergic transmission through a short-term course of benzodiazepines or by means of antiepileptic medication, such as gabapentin.

A further possibility is psychotherapy, either alone or in combination with medication. A psychological approach is particularly indicated for such patients, given the compelling evidence of the effectiveness of CBT in anxiety disorders. Other approaches, such as interpersonal psychotherapy, also show some promise [72]. Even psychotherapies used as control conditions in clinical trials of CBT in anxiety disorders, such as supportive therapy or relaxation protocols, are associated with significant improvements [73]. However, the empirical support is strongest for CBT, which has dominated research and treatment in this diagnostic area and even tends to have a greater effect than pharmacotherapy [74]. Therefore, even therapists using approaches other than CBT may wish to consider incorporating some of its elements, such as exposure, which can be encouraged without taking a very directive stance, and is considered by some scholars as a transdiagnostic component of successful psychotherapies [75].

When confronted with a patient who displays elevated levels of Anger/Aggressiveness, Impulsivity, or Activation, in addition to remarkable Sadness/Demoralisation, and who has a SVARAD profile such as that depicted in Fig. 6.6, care should be taken to address the "activation dimensions" in addition to depressed mood. This kind of profile is similar to that seen in some patients with bipolar depression, and indeed these patients require therapeutic measures similar to those indicated below, even if the severity of the "activation dimensions" is not pronounced. This is one of those instances in which the dimensional evaluation should take a back seat, and the categorical diagnosis based on a longitudinal perspective should carry the greatest weight in treatment choice. However, such a profile can be seen in patients with a number of diagnoses other than bipolar disorder, such as cluster B personality disorder, unipolar depression [76], and obsessive-compulsive disorder [77].

From a pharmacological perspective, for quite some time it was believed that the use of an antidepressant with serotonergic effects would be a simple option enabling the clinician to "kill two birds with one stone". However, in the past decade, the longstanding dogma that aggression and brain serotonergic activity are inversely related has been challenged on several levels [78]. Currently, it is recognised that

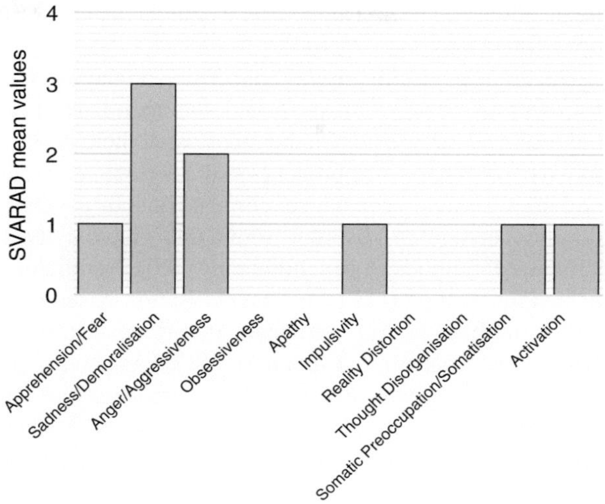

Fig. 6.6 A patient with prominent Sadness/Demoralization and Anger/Aggressiveness

serotonin is not the only relevant neurotransmitter underlying aggressive behaviour [79]; that there is only a small inverse correlation between central serotonergic function and aggression, anger, and hostility [80]; and that the anti-aggressive effect of serotonergic agents seems to be directly related to the intactness of serotonin synaptic function [81]. Therefore, although the clinician can try serotonergic monotherapy, other options should be considered. The first, which actually is still based on a single drug, would be to use one of the few tricyclic antidepressants that not only increase brain serotonergic function but also have considerable sedating properties, thanks to a marked action on histamine H1 receptors. The second option is the addition of another drug with the aim of decreasing dopaminergic activity or of shifting the balance of amino acid neurotransmission from excitatory (glutamatergic) towards inhibitory (GABAergic) transmission. The first aim can be achieved with antipsychotic medication and the second with several antiepileptic drugs, while lithium can achieve both aims as it reduces dopaminergic and glutamatergic neurotransmission while increasing GABAergic neurotransmission [82]. If there is a high risk of self-directed aggression due to the presence of very elevated levels of Sadness/Demoralisation, Impulsivity, and Anger/Aggressiveness, lithium deserves serious consideration due to the compelling evidence of its unique suicide prevention effect in patients with both unipolar and bipolar mood disorders, which sets it apart from other agents [83]. In patients without a pronounced risk of self-harm, or with uncertain compliance or contraindications to lithium, the use of an antiepileptic drug is a good alternative. In our experience, the combination of a serotonergic antidepressant with antiepileptic medication is quite effective in patients with high levels of Sadness/Demoralisation and one or more "activation dimensions". We have used a single-group, open-label design to test the effectiveness of this combination in patients with unipolar depressive disorder characterised by the presence of substantial levels of Anger/Aggressiveness as measured by the SVARAD. The findings, though preliminary, were promising, with an average percentage of improvement in the core depression items of the Hamilton Depression Rating Scale of 56% and in the SVARAD "activation dimensions" of 69% [84].

As far as psychotherapy is concerned, on the one hand, there is not much ground for recommending exclusive psychotherapeutic treatment for patients with substantial levels of Activation. On the other hand, there is plenty of evidence that a number of psychotherapeutic approaches can be effective in reducing anger, aggressiveness, and impulsivity. For instance, clinical trials of psychotherapeutic treatment of borderline personality disorder show that impulsivity can be reduced with well-structured methods of psychotherapy [85]. Prominent examples include dialectical behaviour therapy (DBT) and mentalisation-based therapy [86]. Also, several cognitive and behavioural treatments are effective in reducing impulsivity in children with various categorical diagnoses [87]. As far as anger and aggression are concerned, there is extensive literature supporting the effectiveness of psychological interventions, mainly cognitive and behavioural in nature, in reducing anger and, to a lesser degree, aggression, in both nonclinical and psychiatric populations [88]. Mindfulness meditation training, which can be incorporated into standard psychotherapeutic approaches [89], may also be effective in reducing anger [90]. Moreover, it has beneficial structural and functional effects on the brain and may reduce cortisol levels [91].

If psychotherapy is used as an alternative to, or in combination with, medication, the therapist should be careful not to neglect the patient's anger and difficulties in impulse control, because focusing exclusively on depressed mood would be a recipe for disaster. Some approaches, such as DBT, have an inherent modularity [92] and can thus be more easily adapted to the specific needs of each patient. However, therapists using less structured psychological interventions can also modify their approach in order to address the patient's feelings of anger, aggressive behaviour, and poor impulse control. In working with these patients, it is important to explore their experience and to guide them to recognise and make sense of their feelings of anger, in order to validate the experience and improve emotion regulation [93, 94]. Promoting forgiveness can also play a significant role in resolving the anger associated with depressive disorders [95].

6.2 Apathy

6.2.1 Psychopathology and Assessment

Apathy is a condition that presents itself in a wide variety of mental disorders and also in a number of neurological diseases. As far as mental disorders are concerned, apathy is more frequently associated with mood disorders and psychotic conditions. Among neurodegenerative diseases, apathy can be observed in Alzheimer's disease, extrapyramidal system diseases such as Parkinson's disease (PD), corticobasal ganglionic degeneration, Huntington's disease, stroke, brain traumas, some brain tumours (such as those involving the frontal lobe), and a number of infectious diseases, such as progressive multifocal leukoencephalopathy and human immunodeficiency virus and herpes simplex virus infections.

In the SVARAD, Apathy is defined as "indifference, detachment, affective flattening and blunting; decreased planning and initiative". Given its relevance in a wide variety of health conditions, this dimension is conceptualised and described in many other assessment instruments, some of which focus specifically on it [96]. For a long time, apathy has lacked a clear definition, and it has not been distinguished from other clinical phenomena. For instance, the symptoms of apathy have been considered exclusively as depressive symptoms even in the traditional DSM diagnostic nomenclature and the main psychiatric rating scales, such as the HDRS [97–103]. However, many authors have underlined that apathy is a distinct construct from depression. On the one hand, it can occur as a separate dimension in the absence of a mood disorder; on the other hand, in several depressed patients, other dimensions are more prominent than apathy [104]. Indeed, the literature suggests that not all cases of apathy are caused by depression and that this is true across a number of diseases [105].

For this reason, psychometric instruments have been designed that concentrate exclusively on this dimension, such as the Apathy Evaluation Scale [96], the Apathy Inventory [106], and the Lille Apathy Rating Scale [107]. These instruments cover the main clinical manifestations of apathy, such as flattening of emotional response, lack of interest, reduced initiative, low effort, lack of concern, decreased productivity, reduction in novelty seeking or curiosity, and poor social engagement. Moreover, there are some scales that specifically include an "apathy" item, such as the Neuropsychiatric Inventory [108]. In other scales, such as the Brief Psychiatric Rating Scale (BPRS), the apathy dimension is investigated through a series of related items. For instance, apathy is mapped in the BPRS by the "blunted affect", "emotional withdrawal", and "motor retardation" items. Furthermore, most rating scales for depression include apathy-related items, which assess symptoms of lassitude, lack of motivation, loss of interest, and anhedonia—such as the "work and activities", "psychomotor retardation", and "insight" items of the HDRS [109].

Apathy, stemming from the Greek word pathos ("passion"), modified by an alpha privative to express negation or absence, is customarily defined as passivity accompanied by decreased motivation, flattened affect, and diminished interest and concern. Some authors emphasise lack of motivation as the main clinical feature of apathy, while others underscore the lack of voluntary, goal-directed behaviour as the key disturbance in apathetic patients [110]. Actually, the two approaches to the definition of apathy tend to converge, as the recognition of reduced motivation requires examining a number of goal-related aspects of behaviour, such as overt behaviour (e.g. diminished productivity, effort, and initiative), goal-related thought content (e.g. decreased interests, plans, or goals for the future), and diminished emotional responses to goal-related events (e.g. flattened affect, emotional indifference). Typically, patients with apathy show impairment in four behavioural domains, i.e. intellectual curiosity, self-awareness, emotion, and action initiation. While some symptoms (such as decreased interest, psychomotor retardation, diminished energy, and lack of insight) may be shared between apathy and depression, research in this area suggests that certain symptoms (such as flattened affect, indifference, low social engagement, diminished initiation, and persistence of activities) are specific to apathy, and other symptoms (such as depressed mood, suicidal ideation, self-devaluation, guilt, pessimism, hopelessness, and sleep and appetite abnormalities) are unique to depression [97, 103, 111].

The construct of apathy also shows a considerable overlap with the negative symptom domain of schizophrenia. There is an emerging consensus, supported by factor analysis [112], that negative symptoms comprise two separate yet interrelated subdomains. The first subdomain, designated as diminished emotional expression, incorporates blunted affect and poverty of speech; the second, designated as avolition, encompasses decreased motivation, social withdrawal, and anhedonia. Both the blunted affect component of the diminished emotional expression subdomain, which reflects reduced intensity and range of emotional expression, and the whole avolition subdomain (deficits in the initiation and maintenance of goal-directed behaviours; deficits in desire to undertake activities; reduced ability to experience or anticipate pleasure; diminished interest in, motivation towards, and appreciation of social interactions with others) show a strong overlap with most of the basic clinical features listed in the classical definitions of apathy. It is worth noting that the kind of anhedonia subsumed under the negative symptoms domain does not reflect the inability to experience pleasure ("consummatory anhedonia") which characterises depressive states but rather conforms to deficient anticipation of pleasure ("anticipatory anhedonia"), as reflected in a lack of coordinated behaviour and effort devoted to its attainment [113].

In 2008, a task force including members of the Association Française de Psychiatrie Biologique, the European Psychiatric Association, and the European Alzheimer's Disease Consortium, and experts from Europe, Australia, and North America, proposed diagnostic criteria for apathy in Alzheimer's disease and other neuropsychiatric disorders [114]. These criteria constituted a modified version of previously suggested criteria [96, 115]. A recent study of 306 patients with Alzheimer's disease, mixed dementia, mild cognitive impairment, Parkinson's disease, schizophrenia, or major depressive episode provided evidence of inter-rater reliability and concurrent validity for these criteria [116]. According to the criteria, a diagnosis of apathy requires (1) loss of or diminished motivation in comparison to the patient's previous level of functioning, which is not consistent with his age or culture and is present for at least 4 weeks; (2) presence of symptoms in at least two of the three domains of apathy (loss of, or diminished, goal-directed behaviour; loss of, or diminished, goal-directed cognitive activity; loss of, or diminished, emotion) for at least 4 weeks; and (3) clinically significant impairment in personal, social, occupational, or other important areas of functioning attributable to the apathetic symptoms. In addition, the symptoms of apathy must not be exclusively explained or due to sensorimotor disabilities, diminished level of consciousness, or the direct physiological effects of a substance.

6.2.2 Neurobiology

Apathy may arise from dysfunctions occurring at any of the steps that are necessary to achieve goal-directed behaviour, such as the processing of external and internal determinants that affect the intention to act, elaboration of the plan of action, or initiation, execution, and feedback control of behaviour. Therefore, it is likely that apathy might result from dysfunction in several different components of decision-making mechanisms and that the pathophysiology of apathy may vary depending on which specific process is altered. Elaborating on a previous distinction between three forms of apathy, labelled as emotional-affective, cognitive, and behavioural [117], three subtypes of disrupted processing were proposed by Levy and Dubois as functioning in the apathetic condition: deficits in emotional-affective, cognitive, or auto-activation function [110]. The concept of "auto-activation deficit" refers to a fundamental deficit of activation of behaviour that is not primarily due to an emotional or a cognitive deficit and can be reversed by external stimulation or "hetero-activation".

Apathy is a common feature of dysfunctions or lesions of the frontal-subcortical circuits involving the prefrontal cortex (PFC) and the basal ganglia. These circuits share some common features. They originate in specific zones of the prefrontal cortex, project to the striatum, connect to the globus pallidus and substantia nigra, and from there connect to the thalamus. Each circuit forms a closed loop, as there is a final link back to the frontal cortex. On the one hand, the anatomical positions of the circuit structures remain segregated as they pass through the caudate and putamen, globus pallidus, substantia nigra, and thalamus. On the other hand, in addition to these closed frontal-subcortical loops, there are open connections involving projections to and from other cortical and subcortical structures functionally related to each circuit, which mediate coordination between functionally similar areas of the brain and the frontal–subcortical circuits [118]. As the PFC is functionally and anatomically heterogeneous, different subtypes of apathy depend on which PFC region is affected (Fig. 6.7).

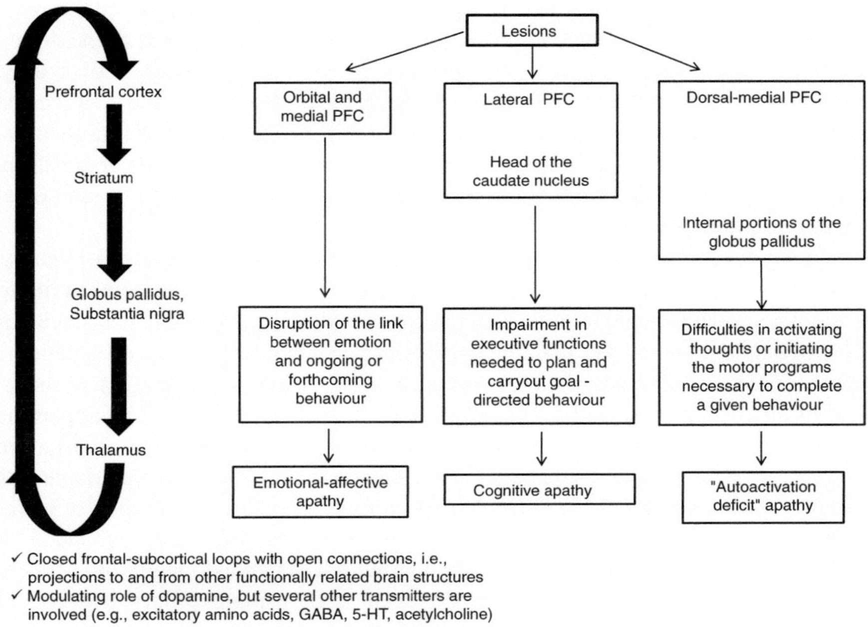

Fig. 6.7 Neuroanatomy of the different forms of apathy

Apathy related to disruption of emotional processing is postulated to be due to an inability to associate emotional signals with ongoing and forthcoming behaviours. Emotions are necessary to decode the context of a given ongoing or forthcoming behaviour, to provide its motivational value, and to orient decision-making. Any disruption of the link between emotion and behaviour may lead to apathy, either by reducing the willingness to perform actions and bring them to completion or by diminishing one's ability to assess the consequences of future actions. This form of apathy is believed to result from lesions in the frontal-subcortical circuits involving the orbital and medial PFC [110].

Apathy related to disruption of cognitive processing is posited to be due to impairments in a number of executive functions that are needed to plan and carry out goal-directed behaviour, such as deficits in working memory and planning (maintenance and mental manipulation of goals and subgoals), difficulty in generating new rules or strategies, and difficulty in shifting from one mental and behavioural set to another. This form of apathy is ascribed to lesions of the lateral PFC and of the dorsal (associative) areas of the basal ganglia, in particular the dorsal portion of the head of the caudate nucleus [110].

Apathy related to difficulties in activating thoughts or initiating the motor programme necessary to complete a given behaviour ("auto-activation deficit") is the most severe form of apathy, characterised by difficulties in self-initiating actions or thoughts ("mental emptiness"), contrasting with externally driven responses that remain relatively normal. This form of apathy is observed after lesions of the dorsal-medial PFC, and after restricted and specific lesions in the basal ganglia, in most cases affecting, bilaterally, the internal portion of the pallidum [110].

The findings of functional and structural neuroimaging studies in patients with Alzheimer's disease [119] and late-life depression [120] lend support to the notion that frontal-subcortical networks are involved in the pathophysiology of apathy. Imaging studies in patients with schizophrenia have also consistently suggested anomalies in processes related to the anticipation of reward and to the motivation, effort, approach behaviour, and goal-directed actions required for its acquisition, which can primarily be attributed to abnormal operation of the ventral and dorsal striatum in interaction with the PFC [113].

Among neurotransmitters possibly implicated in the pathophysiology of apathy, dopamine is the one that received the most research attention, as the opportunity to manipulate dopaminergic treatment in Parkinson's disease (PD) provides researchers with a model that allows insight into the neural substrates of apathy. Indeed, the frequent presence of apathy in a disease such as PD, where there is no gross structural brain damage, but rather an array of dysfunctions secondary to dopamine depletion in brain regions implicated in motivation, such as the ventral striatum, ventral tegmental area, and substantia nigra pars compacta, suggests that dopamine may play an important role in the pathophysiology of apathy. The finding of a significant difference in the severity of apathy between the "off" and "on" states in fluctuating PD patients suggests that apathy is at least in part a dopamine-dependent syndrome [121]. In line with this notion, a recent study using pupillary measures of reward sensitivity reported that reduced pupillary modulation by incentives was predictive of apathy severity and that reward sensitivity was modulated by a dopaminergic state, with blunted sensitivity when patients were "off" dopaminergic drugs [122]. Further supporting a role of dopamine, convergent clinical and preclinical evidence suggests the importance of aberrant striatal and prefrontal dopaminergic signalling in the motivational deficits of schizophrenia [113].

However, apart from dopamine, a variety of neurotransmitters, including excitatory amino acids, gamma-amino-butyric acid, serotonin, and acetylcholine, are involved in frontal-subcortical circuits [118], and still others are involved in the open connections involving projections to and from other cortical and subcortical structures functionally related to each frontal-subcortical circuit. In schizophrenia, studies using magnetic resonance spectroscopy have suggested that a disruption of GABA-glutamatergic dialogue in the PFC and striatum is correlated with negative symptoms. Also, decreased oxytocinergic transmission appears to be implicated in the genesis of negative symptoms, including loss of social motivation and interaction [113].

6.2.3 Clinical and Therapeutic Aspects

In comparison with Sadness/Demoralisation, Apathy has a less "cross-nosological" character. Its presence is less ubiquitous, as it mainly characterises the patients with a diagnosis of mood disorder or psychotic disorder. In our outpatient sample, the highest mean scores were observed in patients with schizophrenia and with unipolar or bipolar major depressive episode, who showed similar average levels of this dimension. Other mood disorders, such as dysthymic disorder, depressive disorder NOS, and other psychotic disorders, such as psychotic disorder NOS and delusional disorder, showed lower scores. Of all the other diagnostic groups, only borderline personality disorder and obsessive-compulsive disorder showed a non-negligible average level of Apathy (Fig. 6.8). Similar findings were observed in our inpatient sample. The highest mean scores were observed in patients with unipolar major depressive episode, obsessive-compulsive disorder, schizophrenia, and bipolar major depressive episode; other disorders, such as depressive disorder NOS, psychotic disorder NOS, schizoaffective disorder, mixed mood episode, and borderline personality disorder showed lower scores (Fig. 6.9). In both outpatients and inpatients, the findings observed in the whole group of patients with a given primary diagnosis were similar to those observed in the subgroup of patients who were free from other Axis I comorbidity.

In outpatients, the Apathy dimension displayed several significant ($p < 0.001$) correlations, such as a moderate to strong correlation with Sadness/Demoralisation (rho = 0.48), small correlations with Reality Distortion (rho = 0.18) and Thought Disorganisation (rho = 0.15), and a small negative correlation with Activation (rho = −0.11). In the inpatient sample, significant ($p < 0.001$) correlations included a moderate to strong correlation with Sadness/Demoralisation (rho = 0.39), small correlations with Apprehension/Fear (rho = 0.15) and Obsessiveness (rho = 0.15), and small to moderate negative correlations with Impulsivity (rho = −0.16), Anger/Aggressiveness (rho = −0.21), and Activation (rho = −0.38).

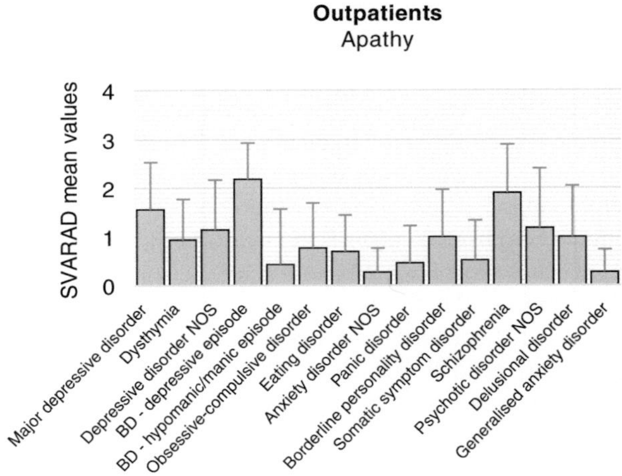

Fig. 6.8 SVARAD Apathy dimension across outpatients' diagnostic categories: mean scores and standard deviations

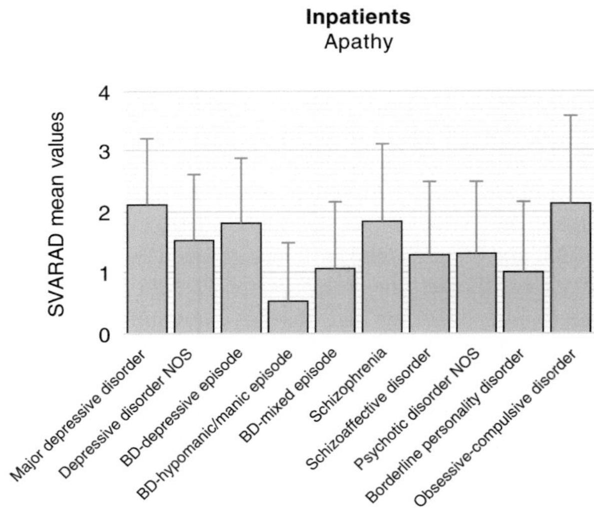

Fig. 6.9 SVARAD Apathy dimension across inpatients' diagnostic categories: mean values and standard deviations

Despite being present to a considerable degree in depressive and psychotic disorders, the prominence of Apathy varies substantially from patient to patient within these diagnostic groups. For instance, in unipolar major depression, a proportion as high as 44% of our outpatients displayed absent or very low levels of Apathy, as attested by a score of 0 (17%) or 1 (27%) on the corresponding SVARAD item. Even in the inpatient sample, absent (9%) or low (16%) levels of Apathy as measured by the SVARAD were observed in one-quarter of patients. In schizophrenia, too, there appears to be a sizable proportion of patients with absent or low levels of Apathy, as it was rated as absent in 10% of outpatients and 19% of inpatients and as mild in 20% of outpatients and 20% of inpatients. As illustrated in detail below, the relative prominence of Apathy in the clinical picture is a major factor to consider in treatment selection, both in patients with depression and in those with psychosis.

The pharmacological treatment of apathy is still an evolving field, with many unresolved questions. The specific impact on apathy symptoms of several different medications has been investigated, though only a few reports have used well-controlled study designs. The neurotransmitter systems implicated by the treatment studies include dopamine, acetylcholine, serotonin, glutamate, and norepinephrine, which is consistent with the involvement of frontal-subcortical-limbic systems in the pathophysiology of apathy. Dopaminergic agents have received only preliminary study, with most evidence coming from open-label trials. In patients with various neurological diseases or traumatic brain injury, the dopamine agonists bromocriptine and amantadine were reported to decrease apathy symptoms and improve motivation, participation, or spontaneity. Also, psychostimulants, such as methylphenidate and dextroamphetamine, were reported to reduce apathy symptoms and increase motivation, socialisation, and participation. In Alzheimer's disease, there is evidence from randomised controlled trials that acetylcholinesterase inhibitors, such as metrifonate, tacrine, and donepezil, reduce apathy symptoms, even though participants in these studies were not selected on the basis of the presence of apathy [105, 123]. In apathetic and depressed stroke patients, a randomised controlled trial suggested that nefiracetam, a pyrrolidone-type nootropic agent, may ameliorate apathy [124].

As far as depression is concerned, the presence of substantial levels of apathy, as in the patient with the SVARAD profile illustrated in Fig. 6.4, poses significant treatment challenges. First, there are hints that the presence of apathy in patients with depression is correlated with worse treatment outcome. In patients with major depression, the presence of apathy symptoms at baseline was found to be related to poor outcome [125]. Similarly, severe apathy at baseline predicted poor outcome after treatment in drug-resistant patients with major depression treated with deep transcranial magnetic stimulation over the PFC [126]. Second, in recent years, evidence has accumulated that the commonly used SSRI antidepressants may actually cause or exacerbate apathy when used in the treatment of depression [127, 128].

Case reports about SSRI-induced apathy began to appear in the literature in the 1990s [129, 130] and continued to appear in subsequent years [131–135]. A phenomenological description of this adverse effect was provided by a qualitative study based on semi-structured individual interviews of 38 patients with depressive or anxiety disorders who received SSRIs. Patients described varying degrees of emotional detachment, which ranged from feeling as "just not caring" about things previously considered as important, to complete emotional numbing. Some patients reported financial and working problems because of "just not caring". Although this detachment was experienced as a beneficial effect by some patients, others experienced it as a decrease in normal emotional responsiveness [136].

While no systematic, large-scale data are available regarding the incidence of an apathetic syndrome in patients receiving SSRIs, clinical experience and the available data suggest that this adverse effect is dose-dependent and reversible after drug dose reduction or discontinuation and is likely to be under-recognised, as it has an insidious and delayed onset and is often characterised by low insight concerning loss of motivation in those afflicted [127, 128]. In a study performed on patients who completed a trial of antidepressants, nearly one fifth of 161 subjects who received SSRIs reported apathy and loss of creativity [137]. A cross-sectional study of 117 patients with major depressive disorder found that about 30% of patients on SSRIs had some form of apathy [138]. In a study of 15 patients who were being treated with SSRIs for major depression and drug-induced sexual dysfunction, 80% reported a clinically significant blunting of several emotions [139]. Although it is not established that the incidence of this adverse effect is greater with SSRIs than with other antidepressants, anecdotal observations and epidemiological studies suggest that SSRIs might be particularly prone to inducing apathy. In some case reports, it has been noted that patients who developed apathy with SSRIs did not have a similar reaction with monoamine oxidase inhibitors, tricyclic antidepressants, or clomipramine [129, 130]. In a retrospective case-control study on elderly depressed patients on antidepressants, either SSRIs or non-SSRIs, it was found that apathy at discharge was associated with SSRI use [140]. A recent study suggested that the phenomenon of emotional blunting is not restricted to SSRIs but may require a serotonergic effect [141].

The exact mechanism by which altered serotonergic function may cause apathy is unknown. Experimental studies in animals and humans indicate that serotonergic pathways have an inhibitory influence over neural systems, mediating not only negative but also positive affective processes [142]. When serotonergic transmission is enhanced, at the same time, there is a dampening of the activity of noradrenergic and dopaminergic neurons through inhibitory 5-HT2A and 5-HT2C receptors, respectively [143]. Therefore, increases in serotonin function may induce a state of flattened affect in which the salience of both rewarding and aversive stimuli is reduced. Results from a double-blind, parallel group design study of 45 healthy participants randomly allocated to receive citalopram, reboxetine, or placebo for 7 days were consistent with this hypothesis. Citalopram was found to decrease the neural processing of both aversive and rewarding stimuli, whereas reboxetine decreased neural responses to the aversive stimuli conditions, but did not suppress ventral striatal activity, and even increased neural responses within the medial orbitofrontal cortex to reward [144].

While it might still be debated whether serotonergic antidepressants actually induce apathy, there does appear to be a cluster of symptoms—including loss of pleasure, loss of interest, fatigue, and loss of energy—that are not satisfactorily addressed by these drugs. Preliminary evidence suggests that antidepressants that enhance catecholaminergic activity may offer advantages over serotonergic antidepressants in the treatment of symptoms associated with a reduction in positive affect [145]. Other evidence suggests that they may also be effective in increasing social motivation [146]. However, a recent multicentre, double-blind, randomised study on patients who were no longer depressed but continued to suffer from apathy after SSRI treatment suggests that switching to serotonin-norepinephrine reuptake inhibitors with insufficient norepinephrine reuptake inhibitory potency has no significant beneficial effect on apathy symptoms [147]. In light of the previous considerations, when facing a depressed patient with prominent apathy symptoms, the clinician may prefer to choose an antidepressant with much greater noradrenergic, rather than serotonergic, effects, such as reboxetine, desipramine, or nortriptyline [148]. In cases with poor response, the pathophysiology of apathy suggests the use of drugs that increase dopaminergic activity, such as bupropion, which has noradrenergic effects, inhibiting norepinephrine and dopamine reuptake, but is devoid of clinically significant serotonergic effects or direct effects on postsynaptic receptors [149]. Alternatively, other possible options to increase dopaminergic activity are psychostimulants, which can be effectively used in combination with antidepressants [150, 151], or even as monotherapy [152], and possibly the dopamine agonists bromocriptine and amantadine. Modafinil may also be an option, as it is usually better tolerated and seems to potentiate both dopaminergic and noradrenergic neurotransmission, while also leading to significantly elevated extracellular serotonin, glutamate, and histamine levels, while decreasing GABA levels. The effects on dopamine and norepinephrine appear to be primary, while the effects on serotonin, GABA, glutamate, and histamine may be secondary to catecholamine effects [153]. In the case of nonresponse, or if drugs enhancing dopaminergic function are contraindicated or not tolerated, acetylcholinesterase inhibitors may also be tried.

Apart from medication, there is a place for psychotherapeutic interventions in the treatment of apathy in depression. Behavioural interventions are particularly relevant in this regard. Although different versions of cognitive behavioural therapy (CBT) place varying emphasis on the cognitive and behavioural components of treatment, in the early stages, all of them usually focus on symptom relief, emphasise behavioural change, and aim at re-engaging patients in their daily activities and restoring psychosocial functioning. Essentially, the behavioural aspect of CBT involves monitoring behaviour and using the resulting data to help motivate patients to make positive behavioural changes by replacing behaviours that may be contributing to depression with healthier ones. This treatment component borrows greatly from behavioural therapy, particularly from behavioural activation, which is an effective treatment for depression in its own right. Behavioural activation, which entails the assignment and scheduling of weekly activities, helps patients return to activities they have ceased and engage in new activities. Patients are instructed to monitor their daily activities and rate their level of enjoyment of each experience by keeping an activity log. In this way, they learn to recognise the relationship between their behaviour and their mood, and they gather information about activities that enhance their mood, as opposed to those that impair it. The therapist will then develop strategies for helping patients to increase the number of pleasant activities and will also carefully look for any skill deficit that might play a role in maintaining not only depressed mood but also inaction and passivity [48].

In patients with schizophrenia, the presence of considerable levels of apathy, as in the patient with the SVARAD profile illustrated in Fig. 6.10, constitutes a severe challenge for the clinician. In this patient population, on a practical level, the treatment of apathy overlaps with the treatment of negative symptoms. This is due to the fact that the whole avolition subdomain and the blunted affect component of the diminished emotional expression subdomain display substantial similarity to the clinical features described in the classical definitions of apathy. While there is no information specifically concerning the effects of treatments on apathy, ample literature describes the effect of a variety of treatments on negative symptoms in

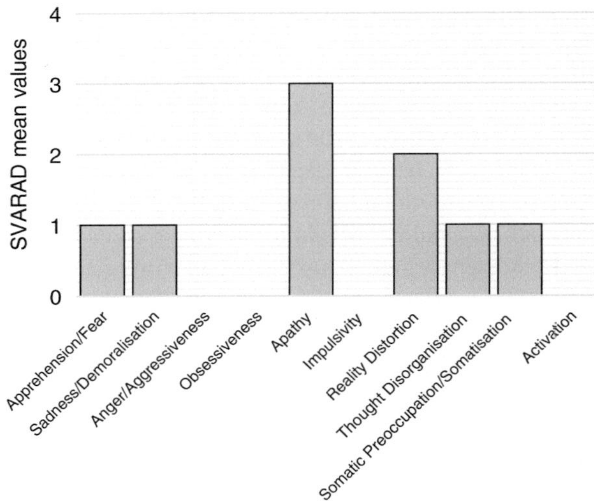

Fig. 6.10 A patient with prominent Apathy and Reality Distortion

schizophrenia. Regrettably, a number of comprehensive reviews of this literature [113, 154, 155] suggest that it is quite difficult to bring about substantial improvement in this dimension. Despite great interest in the topic, and a growing number of studies addressing negative symptoms as the identified primary outcome, currently there is insufficient evidence to support a specific treatment for negative symptoms.

Both typical and atypical antipsychotic drugs have shown modest efficacy, with no evidence of superiority of the newer agents over their conventional counterparts. This is hardly surprising, given that, according to current models of psychosis and antipsychotic action, the emergence of apathy and lack of initiative can be seen as an unwanted consequence of the same mechanism of attenuation of motivational salience that relieves psychotic symptoms [156]. In addition, the newer agents' preferential antagonist activity at the 5-HT2a receptor relative to that at the D2 receptor would not, on a mechanistic basis, be anticipated to reinforce the control of negative symptoms [113]. Conflicting results were obtained with the use of antidepressants, psychostimulants, modafinil, anticonvulsants, drugs enhancing NMDA receptor function (e.g. d-serine, sarcosine, N-acetyl-cysteine, D-cycloserine), NMDA receptor antagonists (e.g. amantadine, memantine), acetylcholinesterase inhibitors (e.g. rivastigmine, donepezil, galantamine), and antibiotics with neuroprotective properties (e.g. minocycline). Preliminary work with selective 5-HT2 (e.g. ritanserin) and 5-HT3 (e.g. ondansetron, tropisetron, granisetron) antagonists has shown promising results [157–160]. As a matter of fact, a recent comprehensive meta-analysis of randomised controlled trials of interventions for negative symptoms in schizophrenia reported that atypical antipsychotics, antidepressants, and glutamatergic medications all significantly reduced negative symptoms as compared with placebo, but the effect was not large enough to be clinically meaningful [161]. Positive effects may possibly be obtained by removing or reducing, rather than adding, drugs. Indeed, the sensible recommendation for the management of apathy to eliminate or reduce doses of psychotropics that aggravate motivational loss, such as serotonergic antidepressants and dopamine antagonists [162], likely also holds true for patients with schizophrenia.

High-frequency repetitive transcranial magnetic stimulation of the PFC has been reported in several studies to be effective in the relief of negative symptoms independent of any change in depressive symptoms. Interestingly, there is some evidence that its actions involve modulation of NMDA receptors and striatal DA release. However, not all studies have yielded positive results and a recent comprehensive meta-analysis of randomised controlled trials found that brain stimulation techniques did not have a significant effect on negative symptoms as compared with placebo [161].

A different approach to restore the functionality of cortico-subcortical circuits involves the application of psychosocial therapies. Cognitive remediation therapy aims to endurably rekindle, through behavioural practice, some key skills related to neurocognition and social cognition. This treatment has been shown to recruit a distributed network of frontocortical structures implicated in apathy. Cognitive behavioural therapy (CBT) more specifically targets poor motivation and anhedonia, as well as negative and pessimistic beliefs, and promotes the active engagement of patients to achieve defined aims. The significance of higher-level operations in the mediation of CBT is suggested by a study indicating a role for functional changes in the striatum in transducing its effects on emotional processing, though the relationship of such changes to apathy is still to be clarified [163]. However, not all studies of psychosocial interventions have been successful and their effect is modest in size. In fact, a recent comprehensive meta-analysis of randomised controlled trials found that although psychological interventions displayed a statistically significant effect on negative symptoms, this effect was not large enough to be clinically meaningful [161]. The treatment of apathy in psychotic patients therefore remains a difficult problem. In the absence of clear information about how to prioritise the currently available treatments in a rational and optimal manner, some degree of trial and error and careful consideration of patient preferences and issues of compliance are required to find the best treatment options for each individual patient, in line with the suggestion that integrated and personalised programmes should be provided as standard treatment to people with schizophrenia [164].

References

1. Burton R. The anatomy of melancholy. New York, NY: The New York Review of Books; 2001.
2. Horwitz AV, Wakefield JC, Lorenzo-Luaces L. History of depression. In: Derubeis RJ, Strunk DR, editors. The Oxford handbook of mood disorders. New York: Oxford University Press; 2017.
3. Kendler KS. The phenomenology of major depression and the representativeness and nature of DSM criteria. Am J Psychiatry. 2016;173:771–80.
4. Pancheri P, Picardi A, Pasquini M, Gaetano P, Biondi M. Psychopathological dimensions of depression: a factor study of the 17-item Hamilton depression rating scale in unipolar depressed outpatients. J Affect Disord. 2002;68:41–7.
5. Shafer AB. Meta-analysis of the factor structures of four depression questionnaires: Beck, CES-D, Hamilton, and Zung. J Clin Psychol. 2006;62:123–46.

6. Biondi M, Picardi A, Pasquini M, Gaetano P, Pancheri P. Dimensional psychopathology of depression: detection of an 'activation' dimension in unipolar depressed outpatients. J Affect Disord. 2005;84:133–9.
7. Clark LA, Watson D. Tripartite model of anxiety and depression: psychometric evidence and taxonomic implications. J Abnorm Psychol. 1991;100:316–36.
8. Anderson ER, Hope DA. A review of the tripartite model for understanding the link between anxiety and depression in youth. Clin Psychol Rev. 2008;28:275–87.
9. Watson D, Clark LA, Weber K, Assenheimer JS, Strauss ME, McCormick RA. Testing a tripartite model: II. Exploring the symptom structure of anxiety and depression in student, adult, and patient samples. J Abnorm Psychol. 1995;104:15–25.
10. Bech P, Gram LF, Dein E, Jacobsen O, Vitger J, Bolwig TG. Quantitative rating of depressive states. Acta Psychiatr Scand. 1975;51:161–70.
11. Bech P, Cialdella P, Haugh M, Birkett MA, Hours A, Boissel JP, Tollefson GD. Meta-analysis of randomised controlled trials of fluoxetine versus placebo and tricyclic antidepressants in the short-term treatment of major depression. Br J Psychiatry. 2000;176:421–8.
12. Bech P, Boyer P, Germain JM, Padmanabhan K, Haudiquet V, Pitrosky B, Tourian KA. HAM-D17 and HAM-D6 sensitivity to change in relation to desvenlafaxine dose and baseline depression severity in major depressive disorder. Pharmacopsychiatry. 2010;43:271–6.
13. Binswanger L. Melancholie und Manie. Phänomenologische Studien Pfullingen. Germany: Verlag Günther Neske; 1960.
14. Le ME. temps vécu. Etudes phénoménologiques et psychopathologiques. Paris: D'Artrey; 1933.
15. Tatossian A. Phénoménologie des psychoses. Paris: Masson; 1979.
16. Ratcliffe M. Experiences of depression: a study in phenomenology. New York: Oxford University Press; 2015.
17. Picardi A, Pallagrosi M, Fonzi L, Biondi M. Psychopathological dimensions and the clinician's subjective experience. Psychiatry Res. 2017;258:407–14.
18. Frank JD. Psychotherapy. The restoration of morale. Am J Psychiatr. 1974;131:271–4.
19. Clarke DM, Kissane DW. Demoralization: its phenomenology and importance. Aust N Z J Psychiatry. 2002;36:733–42.
20. De Figueiredo JM. Distress, demoralization and psychopathology: diagnostic boundaries. Eur J Psychiatry. 2013;27:61–73.
21. Tecuta L, Tomba E, Grandi S, Fava GA. Demoralization: a systematic review on its clinical characterization. Psychol Med. 2015;45:673–91.
22. De Figueiredo JM. Depression and demoralization: phenomenological differences and research perspectives. Compr Psychiatry. 1993;34:308–11.
23. Willner P, Scheel-Krüger J, Belzung C. The neurobiology of depression and antidepressant action. Neurosci Biobehav Rev. 2013;37:2331–71.
24. Savitz J, Drevets WC. Bipolar and major depressive disorder: neuroimaging the developmental-degenerative divide. Neurosci Biobehav Rev. 2009;33:699–771.
25. Phillips ML, Chase HW, Sheline YI, Etkin A, Almeida JR, Deckersbach T, Trivedi MH. Identifying predictors, moderators, and mediators of antidepressant response in major depressive disorder: neuroimaging approaches. Am J Psychiatry. 2015;172:124–38.
26. Pizzagalli DA. Depression, stress, and anhedonia: toward a synthesis and integrated model. Annu Rev Clin Psychol. 2014;10:393–423.
27. Hirschfeld RM. History and evolution of the monoamine hypothesis of depression. J Clin Psychiatry. 2000;61(Suppl 6):4–6.
28. Krishnan V, Nestler EJ. The molecular neurobiology of depression. Nature. 2008;455:894–902.
29. Nestler EJ, Barrot M, DiLeone RJ, Eisch AJ, Gold SJ, Monteggia LM. Neurobiology of depression. Neuron. 2002;34:13–25.
30. Mahar I, Rodriguez Bambico F, Mechawar N, Nobrega JN. Stress, serotonin, and hippocampal neurogenesis in relation to depression and antidepressant effects. Neurosci Biobehav Rev. 2014;38:173–92.

31. Felger JC, Lotrich FE. Inflammatory cytokines in depression: neurobiological mechanisms and therapeutic implications. Neuroscience. 2013;246:199–229.
32. Miller AH, Maletic V, Raison CL. Inflammation and its discontents: the role of cytokines in the pathophysiology of major depression. Biol Psychiatry. 2009;65:732–41.
33. Duman RS, Aghajanian GK, Sanacora G, Krystal JH. Synaptic plasticity and depression: new insights from stress and rapid-acting antidepressants. Nat Med. 2016;22:238–49.
34. Kohler O, Benros ME, Nordentoft M, Farkouh ME, Iyengar RL, Mors O, Krogh J. Effect of anti-inflammatory treatment on depression, depressive symptoms, and adverse effects: a systematic review and meta-analysis of randomized clinical trials. JAMA Psychiat. 2014;71:1381–91.
35. Lavebratt C, Herring MP, Liu JJ, Wei YB, Bossoli D, Hallgren M, Forsell Y. Interleukin-6 and depressive symptom severity in response to physical exercise. Psychiatry Res. 2017;252:270–6.
36. Lieverse R, Van Someren EJ, Nielen MM, Uitdehaag BM, Smit JH, Hoogendijk WJ. Bright light treatment in elderly patients with nonseasonal major depressive disorder: a randomized placebo-controlled trial. Arch Gen Psychiatry. 2011;68:61–70.
37. Perera S, Eisen R, Bhatt M, Bhatnagar N, de Souza R, Thabane L, Samaan Z. Light therapy for non-seasonal depression: systematic review and meta-analysis. BJPsych Open. 2016;2:116–26.
38. Gaynes BN, Lloyd SW, Lux L, Gartlehner G, Hansen RA, Brode S, Jonas DE, Swinson Evans T, Viswanathan M, Lohr KN. Repetitive transcranial magnetic stimulation for treatment-resistant depression: a systematic review and meta-analysis. J Clin Psychiatry. 2014;75:477–89.
39. Noda Y, Silverstein WK, Barr MS, Vila-Rodriguez F, Downar J, Rajji TK, Fitzgerald PB, Mulsant BH, Vigod SN, Daskalakis ZJ, Blumberger DM. Neurobiological mechanisms of repetitive transcranial magnetic stimulation of the dorsolateral prefrontal cortex in depression: a systematic review. Psychol Med. 2015;45:3411–32.
40. Baeken C, De Raedt R. Neurobiological mechanisms of repetitive transcranial magnetic stimulation on the underlying neurocircuitry in unipolar depression. Dialogues Clin Neurosci. 2011;13:139–45.
41. UK ECT Review Group. Efficacy and safety of electroconvulsive therapy in depressive disorders: a systematic review and meta-analysis. Lancet. 2003;361:799–808.
42. Loo C, Katalinic N, Mitchell PB, Greenberg B. Physical treatments for bipolar disorder: a review of electroconvulsive therapy, stereotactic surgery and other brain stimulation techniques. J Affect Disord. 2011;132:1–13.
43. Leroy A, Naudet F, Vaiva G, Francis A, Thomas P, Amad A. Is electroconvulsive therapy an evidence-based treatment for catatonia? A systematic review and meta-analysis. Eur Arch Psychiatry Clin Neurosci. 2017. [Epub ahead of print]. https://doi.org/10.1007/s00406-017-0819-5.
44. Grover S, Kumar Mattoo S, Gupta N. Theories on mechanism of action of electroconvulsive therapy. German. J Psychiatry. 2005;8:70–84.
45. Fosse R, Read J. Electroconvulsive treatment: hypotheses about mechanisms of action. Front Psych. 2013;4:94.
46. Merkl A, Heuser I, Bajbouj M. Antidepressant electroconvulsive therapy: mechanism of action, recent advances and limitations. Exp Neurol. 2009;219:20–6.
47. Scott A. Mode of action of electroconvulsive therapy: an update. Adv Psychiatr Treat. 2011;17:15–22.
48. Picardi A, Gaetano P. Psychotherapy of mood disorders. Clin Prac Epidemiol Ment Health. 2014;10:140–58.
49. Freud S. Project for a scientific psychology. In: Strachey J, editor. Standard edition of the complete psychological works of Sigmund Freud, vol. 1. London: Hogarth Press; 1895/1966. p. 283–397.
50. Kandel ER. A new intellectual framework for psychiatry. Am J Psychiatry. 1998;155:457–69.

51. Linden DE. How psychotherapy changes the brain–the contribution of functional neuroimaging. Mol Psychiatry. 2006;11:528–38.
52. Beauregard M. Functional neuroimaging studies of the effects of psychotherapy. Dialogues Clin Neurosci. 2014;16:75–81.
53. Barsaglini A, Sartori G, Benetti S, Pettersson-Yeo W, Mechelli A. The effects of psychotherapy on brain function: a systematic and critical review. Prog Neurobiol. 2014;114:1–14.
54. Weingarten CP, Strauman TJ. Neuroimaging for psychotherapy research: current trends. Psychother Res. 2015;25:185–213.
55. Dichter GS, Felder JN, Petty C, Bizzell J, Ernst M, Smoski MJ. The effects of psychotherapy on neural responses to rewards in major depression. Biol Psychiatry. 2009;66:886–97.
56. Fournier JC, Price RB. Psychotherapy and neuroimaging. Focus. 2014;12:290–8.
57. Quidé Y, Witteveen AB, El-Hage W, Veltman DJ, Olff M. Differences between effects of psychological versus pharmacological treatments on functional and morphological brain alterations in anxiety disorders and major depressive disorder: a systematic review. Neurosci Biobehav Rev. 2012;36:626–44.
58. Boccia M, Piccardi L, Guariglia P. How treatment affects the brain: meta-analysis evidence of neural substrates underpinning drug therapy and psychotherapy in major depression. Brain Imaging Behav. 2016;10:619–27.
59. Busch FN, Sandberg LS. Combined treatment of depression. Psychiatr Clin North Am. 2012;35:165–79.
60. Cuijpers P, Dekker J, Hollon SD, Andersson G. Adding psychotherapy to pharmacotherapy in the treatment of depressive disorders in adults: a meta-analysis. J Clin Psychiatry. 2009;70:1219–29.
61. Cuijpers P, van Straten A, Warmerdam L, Andersson G. Psychotherapy versus the combination of psychotherapy and pharmacotherapy in the treatment of depression: a meta-analysis. Depress Anxiety. 2009;26:279–88.
62. Cuijpers P, van Straten A, Hollon SD, Andersson G. The contribution of active medication to combined treatments of psychotherapy and pharmacotherapy for adult depression: a meta-analysis. Acta Psychiatr Scand. 2010;121:415–23.
63. Guidi J, Fava GA, Fava M, Papakostas GI. Efficacy of the sequential integration of psychotherapy and pharmacotherapy in majordepressive disorder: a preliminary meta-analysis. Psychol Med. 2011;41:321–31.
64. Biondi M. Beyond the brain-mind dichotomy and toward a common organizing principle of pharmacological and psychological treatments. Psychother Psychosom. 1995;64:1–8.
65. Picardi A, Biondi M. Towards an integrated treatment of depression. Riv Psichiatr. 2002;37:284–9.
66. Pasquini M, Picardi A, Biondi M, Scarciglia P, Pancheri P. Analisi Psicopatologica Dimensionale del Disturbo Ossessivo-Compulsivo: appartiene realmente ai Disturbi d'Ansia? Riv Psichiatr. 2003;38(2):86–91.
67. Fava M, Rush AJ, Alpert JE, Balasubramani GK, Wisniewski SR, Carmin CN, Biggs MM, Zisook S, Leuchter A, Howland R, Warden D, Trivedi MH. Difference in treatment outcome in outpatients with anxious versus nonanxious depression: a STAR*D report. Am J Psychiatry. 2008;165:342–51.
68. Wiethoff K, Bauer M, Baghai TC, Möller HJ, Fisher R, Hollinde D, Kiermeir J, Hauth I, Laux G, Cordes J, Brieger P, Kronmüller KT, Zeiler J, Adli M. Prevalence and treatment outcome in anxious versus nonanxious depression: results from the German algorithm project. J Clin Psychiatry. 2010;71:1047–54.
69. Martin EI, Ressler KJ, Binder E, Nemeroff CB. The neurobiology of anxiety disorders: brain imaging, genetics, and Psychoneuroendocrinology. Psychiatr Clin North Am. 2009;32:549–75.
70. Ionescu DF, Niciu MJ, Richards EM, Zarate CA. Pharmacologic treatment of dimensional anxious depression: a review. Prim Care Companion CNS Disord. 2014;16(3). https://doi.org/10.4088/PCC.13r01621.

71. De la Mora MP, Gallegos-Cari A, Arizmendi-García Y, Marcellino D, Fuxe K. Role of dopamine receptor mechanisms in the amygdaloid modulation of fear and anxiety: structural and functional analysis. Prog Neurobiol. 2010;90:198–216.
72. Markowitz JC, Lipsitz J, Milrod BL. Critical review of outcome research on interpersonal psychotherapy for anxiety disorders. Depress Anxiety. 2014;31:316–25.
73. Smits JA, Hofmann SG. A meta-analytic review of the effects of psychotherapy control conditions for anxiety disorders. Psychol Med. 2009;39:229–39.
74. Roshanaei-Moghaddam B, Pauly MC, Atkins DC, Baldwin SA, Stein MB, Roy-Byrne P. Relative effects of CBT and pharmacotherapy in depression versus anxiety: is medication somewhat better for depression, and CBT somewhat better for anxiety? Depress Anxiety. 2011;28:560–7.
75. Carey TA. Exposure and reorganization: the what and how of effective psychotherapy. Clin Psychol Rev. 2011;31:236–48.
76. Picardi A, Morosini P, Gaetano P, Pasquini M, Biondi M. Higher levels of anger and aggressiveness in unipolar depression as compared with anxiety and somatoform disorders. J Clin Psychiatry. 2004;65:442–3.
77. Piacentino D, Pasquini M, Tarsitani L, Berardelli I, Roselli V, Maraone A, Biondi M. The association of anger with symptom subtypes in severe obsessive-compulsive disorder outpatients. Psychopathology. 2016;49:40–6.
78. Olivier B. Serotonin and aggression. Ann N Y Acad Sci. 2004;1036:382–92.
79. Yanowitch R, Coccaro EF. The neurochemistry of human aggression. Adv Genet. 2011;75:151–69.
80. Duke AA, Bègue L, Bell R, Eisenlohr-Moul T. Revisiting the serotonin-aggression relation in humans: a meta-analysis. Psychol Bull. 2013;139:1148–72.
81. Coccaro EF, Fanning JR, Phan KL, Lee R. Serotonin and impulsive aggression. CNS Spectr. 2015;20:295–302.
82. Malhi GS, Tanious M, Das P, Coulston CM, Berk M. Potential mechanisms of action of lithium in bipolar disorder. Current understanding. CNS Drugs. 2013;27:135–53.
83. Lewitzka U, Severus E, Bauer R, Ritter P, Müller-Oerlinghausen B, Bauer M. The suicide prevention effect of lithium: more than 20 years of evidence-a narrative review. Int J Bipolar Disord. 2015;3:32.
84. Pasquini M, Picardi A, Speca A, et al. Combining an SSRI with an anticonvulsant in depressed patients with dysphoric mood: open study. Clin Pract Epidemol Ment Health. 2007;3:3–4.
85. Paris J. Clinical trials of treatment for personality disorders. Psychiatr Clin North Am. 2008;31:517–26.
86. Bateman A, Fonagy P. Mentalization based treatment for borderline personality disorder. World Psychiatry. 2010;9:11–5.
87. Bear RA, Nietzel MT. Cognitive and behavioral treatment of impulsivity in children: a meta-analytic review of the outcome literature. J Clin Child Psychol. 1991;20:400–12.
88. Lee AH, DiGiuseppe R. Anger and aggression treatments: a review of meta-analyses. Curr Opin Psychol. 2017;19:65–74.
89. Segal ZV, Williams JMG, Teasdale JD. Mindfulness-based cognitive therapy for depression: a new approach to preventing relapse. New York: Guilford; 2002.
90. Amutio A, Franco C, Pérez-Fuentes Mde C, Gázquez JJ, Mercader I. Mindfulness training for reducing anger, anxiety, and depression in fibromyalgia patients. Front Psychol. 2014;5:1572.
91. Tang YY, Hölzel BK, Posner MI. The neuroscience of mindfulness meditation. Nat Rev Neurosci. 2015;16:213–25.
92. Linehan MM, Wilks CR. The course and evolution of dialectical behavior therapy. Am J Psychother. 2015;69:97–110.
93. Guidano VF. The self in process. New York: Guilford Press; 1991.
94. Gaetano P, Maselli P, Meldolesi GN, Picardi A. Experience-centred cognitive psychotherapy: towards a phenomenologically oriented therapy. Riv Psichiatr. 2015;50:51–60.

95. Enright RD, Fitzgibbons RP. Helping clients forgive: an empirical guide for resolving anger and restoring hope. Washington, DC: American Psychological Association Books; 2000.
96. Marin RS, Biedrzycki RC, Firinciogullari S. Reliability and validity of the apathy evaluation scale. Psychiatry Res. 1991;38:143–62.
97. Marin RS, Firinciogullari S, Biedrzycki RC. The sources of convergence between measures of apathy and depression. J Affect Disord. 1993;28:117–24.
98. Levy ML, Cummings JL, Fairbanks LA, Masterman D, Miller BL, Craig AH, Paulsen JS, Litvan I. Apathy is not depression. J Neuropsychiatry Clin Neurosci. 1998;10:314–9.
99. Kirsch Darrow L, Fernandez HF, Marsiske M, Okun MS, Bowers D. Dissociating apathy and depression in Parkinson disease. Neurology. 2006;67:33–8.
100. Pedersen KF, Larsen JP, Alves G, Aarslan D. Prevalence and clinical correlates of apathy in Parkinson's disease: a community-based study. Parkinsonism Relat Disord. 2009;5:95–9.
101. Tagariello P, Girardi P, Amore M. Depression and apathy in dementia: same syndrome or different construct? A critical review. Arch Gerontol Geriatr. 2009;49:246–9.
102. Starkstein SE, Leentjens AFG. The nosological position of apathy in clinical practice. J Neurol Neurosurg Psychiatry. 2008;79:1088–92.
103. Marin RS. Differential diagnosis and classification of apathy. Am J Psychiatry. 1990;147:22–30.
104. Marin RS, Frinciogullari S, Biedrzycki RC. Group differences in the relationship between apathy and depression. J Nerv Ment Dis. 1994;182:235–9.
105. Van Reekum R, Stuss DT, Ostrander L. Apathy: why care? J Neuropsychiatry Clin Neurosci. 2005;17:7–19.
106. Robert PH, Clairet S, Benoit M, Koutaich J, Bertogliati C, Tible O, Caci H, Borg M, Brocker P, Bedoucha P. The apathy inventory: assessment of apathy and awareness in Alzheimer's disease, Parkinson's disease and mild cognitive impairment. Int J Geriatr Psychiatry. 2002;17:1099–105.
107. Sockeel P, Dujardin K, Devos D, Deneve C, Desèe A, Defebvre L. The Lille apathy rating scale (LARS), a new instrument for detecting and quantifying apathy: validation in Parkinson's disease. J Neurol Neurosurg Psychiatry. 2006;77:579–84.
108. Clarke DE, Ko JY, Kuhl EA, Van Reekum R, Salvador R, Marin RS. Are the available apathy measures reliable and valid? A review of the psychometric evidence. J Psychosom Res. 2011;70:73–97.
109. Weiss HD, Marsh L. Apathy in Parkinson's disease. In: Pfeiffer RF, Wszolek ZK, Ebadi M, editors. Parkinson's disease. 2nd ed. Boca Raton: CRC Press; 2012.
110. Levy R, Dubois B. Apathy and the functional anatomy of the prefrontal cortex-basal ganglia circuits. Cereb Cortex. 2006;16:916–28.
111. Landes AM, Sperry SD, Strauss ME, Geldmacher DS. Apathy in Alzheimer's disease. J Am Geriatr Soc. 2001;49:1700–7.
112. Blanchard JJ, Cohen AS. The structure of negative symptoms within schizophrenia: implications for assessment. Schizophr Bull. 2006;32:238–45.
113. Millan MJ, Fone K, Steckler T, Horan WP. Negative symptoms of schizophrenia: clinical characteristics, pathophysiological substrates, experimental models and prospects for improved treatment. Eur Neuropsychopharmacol. 2014;24:645–92.
114. Robert P, Onyike CU, Leentjens AF, Dujardin K, Aalten P, Starkstein S, Verhey FR, Yessavage J, Clement JP, Drapier D, Bayle F, Benoit M, Boyer P, Lorca PM, Thibaut F, Gauthier S, Grossberg G, Vellas B, Byrne J. Proposed diagnostic criteria for apathy in Alzheimer's disease and other neuropsychiatric disorders. Eur Psychiatry. 2009 Mar;24(2):98–104.
115. Starkstein SE, Petracca G, Chemerinski E, Kremer J. Syndromic validity of apathy in Alzheimer's disease. Am J Psychiatry. 2001;158(6):872–7.
116. Mulin E, Leone E, Dujardin K, Delliaux M, Leentjens A, Nobili F, Dessi B, Tible O, Agüera-Ortiz L, Osorio RS, Yessavage J, Dachevsky D, Verhey FR, Cruz Jentoft AJ, Blanc O, Llorca PM, Robert PH. Diagnostic criteria for apathy in clinical practice. Int J Geriatr Psychiatry. 2011 Feb;26(2):158–65.

117. Stuss DT, Van Reekum R, Murphy KJ. Differentiation of states and causes of apathy. In: Borod JC, editor. The neuropsychology of emotion. New York: Oxford University Press; 2000. p. 340–63.
118. Tekin S, Cummings JL. Frontal-subcortical neuronal circuits and clinical neuropsychiatry: an update. J Psychosom Res. 2002;53:647–54.
119. Theleritis C, Politis A, Siarkos K, Lyketsos CG. A review of neuroimaging findings of apathy in Alzheimer's disease. Int Psychogeriatr. 2014;26:195–207.
120. Alexopoulos GS, Hoptman MJ, Yuen G, Kanellopoulos D, Seirup JK, Lim KO, Gunning FM. Functional connectivity in apathy of late-life depression: a preliminary study. J Affect Disord. 2013 Jul;149(1–3):398–405.
121. Czernecki V, Pillon B, Houeto JL, Pochon JB, Levy R, Doubois B. Motivation, reward, and Parkinson's disease: influence of dopatherapy. Neuropsychologia. 2002;40:2257–67.
122. Muhammed K, Manohar S, Lennox G, Ben Yehuda M, Chong TT, Tofaris G, et al. Reward sensitivity deficits modulated by dopamine are associated with apathy in Parkinson's disease. Brain. 2016;139:2706–21.
123. Roth RM, Flashman LA, McAllister TW. Apathy and its treatment. Curr Treat Options Neurol. 2007 Sep;9(5):363–70.
124. Robinson RG, Jorge RE, Clarence-Smith K, et al. Double-blind treatment of apathy in patients with poststroke depression using nefiracetam. J Neuropsychiatry Clin Neurosci. 2009;21(2):144–51.
125. Chaturvedi SK, Sarmukaddam SB. Prediction of outcome in depression by negative symptoms. Acta Psychiatr Scand. 1986;74:183–6.
126. Levkovitz Y, Sheer A, Harel EV, Katz LN, Most D, Zangen A, Isserles M. Differential effects of deep TMS of the prefrontal cortex on apathy and depression. Brain Stimul. 2011 Oct;4(4):266–74.
127. Barnhart WJ, Makela EH, Latocha MJ. SSRI-induced apathy syndrome: a clinical review. J Psychiatr Pract. 2004;10:196–9.
128. Sansone RA, Sansone LA. SSRI-induced indifference. Psychiatry (Edgmont). 2010;7(10):14–8.
129. Hoehn-Saric R, Lipsey JR, McLeod DR. Apathy and indifference in patients on fluvoxamine andfluoxetine. J Clin Psychopharmacol. 1990;10:343–5.
130. Hoehn-Saric R, Harris GJ, Pearlson GD, et al. A fluoxetine-induced frontal lobe syndrome in an obsessive compulsive patient. J Clin Psychiatry. 1991;52:131–3.
131. Reinblatt SP, Riddle MA. Selective sero- tonin reuptake inhibitor-induced apathy: a pediatric case series. J Child Adolesc Psychopharmacol. 2006;16:227–33.
132. Garland EJ, Baerg EA. Amotivational syndrome associated with selective serotonin reuptake inhibitors in children and adolescents. J Child Adolesc Psychopharmacol. 2001;11:181–6.
133. Sato S, Asada T. Sertraline-induced apathy syndrome. J Neuropsychiatry Clin Neurosci. 2011;23:E19.
134. Kodela S, Venkata PD. Antidepressant induced apathy responsive to dose reduction. Psychopharmacol Bull. 2010;43:76–9.
135. Padala PR, Padala KP, Monga V, Ramirez DA, Sullivan DH. Reversal of SSRI-associated apathy syndrome by discontinuation of therapy. Ann Pharmacother. 2012;46(3):e8.
136. Price J, Cole V, Goodwin GM. Emotional side-effects of selective serotoninreuptake inhibitors: qualitative study. Br J Psychiatry. 2009;195:211–7.
137. Bolling MY, Kohlenberg RJ. Reasons for quitting serotonin reuptake inhibitor therapy: paradoxical psychological side effects and patient satisfaction. Psychother Psychosom. 2004;73:380–5.
138. Fava M, Graves LM, Benazzi F, Scalia MJ, Iosifescu DV, Alpert JE, Papakostas GI. A cross-sectional study of the prevalence of cognitive and physical symptoms during long-term antidepressant treatment. J Clin Psychiatry. 2006;67(11):1754–9.
139. Opbroek A, Delgado PL, Laukes C, et al. Emotional blunting associated with SSRI-induced sexual dysfunction. Do SSRIs inhibit emotional responses? Int J Neuropsychopharmacol. 2002;5:147–51.

140. Wongpakaran N, van Reekum R, Wongpakaran T, Clarke D. Selective serotonin reuptake inhibitor use associates with apathy among depressed elderly: a case-control study. Ann General Psychiatry. 2007;6:7.
141. Goodwin GM, Price J, De Bodinat C, Laredo J. Emotional blunting with antidepressant treatments: a survey among depressed patients. J Affect Disord. 2017;221:31–5.
142. Zald DH, Depue RA. Serotonergic functioning correlates with positive and negative affect in psychiatrically healthy males. Pers Individ Dif. 2001;30:71–86.
143. Blier P, Briley M. The noradrenergic symptom cluster: clinical expression and neuropharmacology. Neuropsychiatr Dis Treat. 2011;7(1):15–20.
144. McCabe C, Mishor Z, Cowen PJ, Harmer CJ. Diminished neural processing of aversive and rewarding stimuli during selective serotonin reuptake inhibitor treatment. Biol Psychiatry. 2010;67(5):439–45.
145. Nutt D, Demyttenaere K, Janka Z, Aarre T, Bourin M, Canonico PL, Carrasco JL, Stahl S. The other face of depression, reduced positive affect: the role of catecholamines in causation and cure. J Psychopharmacol. 2007;21:461–71.
146. Dubini A, Bosc M, Polin V. Do noradrenaline and serotonin differentially affect social motivation and behaviour? Eur Neuropsychopharmacol. 1997;7(1):S49–55.
147. Raskin J, George T, Granger RE, Hussain N, Zhao GW, Marangell LB. Apathy in currently nondepressed patients treated with a SSRI for a major depressive episode: outcomes following randomized switch to either duloxetine or escitalopram. J Psychiatr Res. 2012;46:667–74.
148. Gillman PK. Tricyclic antidepressant pharmacology and therapeutic drug interactions updated. Br J Pharmacol. 2007;151:737–48.
149. Stahl SM, Pradko JF, Haight BR, Modell JG, Rockett CB, Learned-Coughlin S. A review of the neuropharmacology of bupropion, a dual norepinephrine and dopamine reuptake inhibitor. Prim Care Companion J Clin Psychiatry. 2004;6:159–66.
150. Masand PS, Anand VS, Tanquary JF. Psychostimulant augmentation of second generation antidepressants: a case series. Depress Anxiety. 1998;7:89–91.
151. Lavretsky H, Reinlieb M, St Cyr N, Siddarth P, Ercoli LM, Senturk D. Citalopram, methylphenidate, or their combination in geriatric depression: a randomized, double-blind, placebo-controlled trial. Am J Psychiatry. 2015;172:561–9.
152. Ciraulo DA, Shader RI, Greenblatt DJ. Clinical pharmacology and therapeutics of antidepressants. In: Ciraulo DA, Shader RI, editors. Pharmacotherapy of depression. 2nd ed. New York: Humana Press; 2011. p. 33–124.
153. Minzenberg MJ, Carter CS. Modafinil: a review of neurochemical actions and effects on cognition. Neuropsychopharmacology. 2008;33:1477–502.
154. Tsapakis EM, Dimopoulou T, Tarazi FI. Clinical management of negative symptoms of schizophrenia: an update. Pharmacol Ther. 2015;153:135–47.
155. Remington G, Foussias G, Fervaha G, Agid O, Takeuchi H, Lee J, Hahn M. Treating negative symptoms in schizophrenia: an update. Curr Treat Options Psychiatry. 2016;3:133–50.
156. Kapur S. How antipsychotics become anti-"psychotic" – from dopamine to salience to psychosis. Trends Pharmacol Sci. 2004;25(8):402–6.
157. Akhondzadeh S, Malek-Hosseini M, Ghoreishi A, Raznahan M, Rezazadeh SA. Effect of ritanserin, a 5HT2A/2C antagonist, on negative symptoms of schizophrenia: a double-blind randomized placebocontrolled study. Prog Neuro-Psychopharmacol Biol Psychiatry. 2008;32:1879–83.
158. Duinkerke SJ, Botter PA, Jansen AA, van Dongen PA, van Haaften AJ, Boom AJ, van Laarhoven JH, Busard HL. Ritanserin a selective 5- HT2/1C antagonist, and negative symptoms in schizophrenia. A placebo-controlled double-blind trial. Br J Psychiatry. 1993;163:451–5.
159. Khodaie-Ardakani MR, Seddighi S, Modabbernia A, Rezaei F, Salehi B, Ashrafi M, Shams-Alizadeh N, Mohammad-Karimi M, Esfandiari GR, Hajiaghaee R, Akhondzadeh S. Granisetron as an add-on to risperidone for treatment of negative symptoms in patients with stable schizophrenia: randomized double-blind placebo-controlled study. J Psychiatr Res. 2013;47:472–8.

160. Noroozian M, Ghasemi S, Hosseini SM, Modabbernia A, Khodaie-Ardakani MR, Mirshafiee O, Farokhnia M, Tajdini M, Rezaei F, Salehi B, Ashrafi M, Yekehtaz H, Tabrizi M, Akhondzadeh S. A placebo-controlled study of tropisetron added to risperidone for the treatment of negative symptoms in chronic and stable schizophrenia. Psychopharmacol (Berl). 2013;228:595–602.
161. Fusar-Poli P, Papanastasiou E, Stahl D, et al. Treatments of negative symptoms in schizophrenia: meta-analysis of 168 randomized placebo-controlled trials. Schizophr Bull. 2015;41(4):892–9.
162. Marin RS, Wilkosz PA. Disorders of diminished motivation. J Head Trauma Rehabil. 2005;20:377–88.
163. Kumari V, Fannon D, Peters ER, Ffytche DH, Sumich AL, Premkumar P, Anilkumar AP, Andrew C, Phillips ML, Williams SC, Kuipers E. Neural changes following cognitive behaviour therapy for psychosis: a longitudinal study. Brain. 2011 Aug;134(Pt 8):2396–407.
164. Galderisi S, Rossi A, Rocca P, Bertolino A, Mucci A, Bucci P, Rucci P, Gibertoni D, Aguglia E, Amore M, Bellomo A, Biondi M, Brugnoli R, Dell'Osso L, De Ronchi D, Di Emidio G, Di Giannantonio M, Fagiolini A, Marchesi C, Monteleone P, Oldani L, Pinna F, Roncone R, Sacchetti E, Santonastaso P, Siracusano A, Vita A, Zeppegno P, Maj M. Italian network for research on psychoses. The influence of illness-related variables, personal resources and context-related factors on real-life functioning of people with schizophrenia. World Psychiatry. 2014;13:275–87.

The "Outer Dimensions": Impulsivity, Anger/Aggressiveness, Activation

7

Francesco Saverio Bersani and Massimo Pasquini

7.1 Introduction

Across time and cultures, anger has been recognised as among the most salient of emotions for human beings. "*Menin*" (anger) is the first word used in the Iliad by Homer, while the Odyssey begins with the word "*andra*" (man). Anger is one of the "three poisons of the mind" in Buddhist teachings. Today, psychopathologists consider symptoms of anger and activation in the context of several clinical syndromes. Dysphoric states are often linked to manic-depressive illness, but several European schools of thought consider dysphoria a third independent polarity, different from depression and mania. Further, anger and impulsivity, not necessarily associated with each other, are present among several personality and non-affective disorders. In this chapter we discuss the relationships between impulsivity, anger/aggressiveness, and activation and the burden of each dimension in psychiatric syndromes.

The fifth edition of the Diagnostic and Statistical Manual of Mental Disorders (DSM-5) formally recognises the category of *disruptive, impulse-control, and conduct disorders*, which includes a range of disorders characterised by problems in emotional and behavioural self-control, such as oppositional defiant disorder, intermittent explosive disorder, conduct disorder, pyromania, and kleptomania [1]. However, impaired impulsivity is a core characteristic of a much wider range of psychiatric disturbances, and it can be found in virtually all psychiatric diseases [2]. Thus, it represents a transdiagnostic dimensional symptom that spans traditional diagnostic boundaries.

Impulsivity has been variously defined. From a biopsychosocial perspective, impulsivity can be defined as "the tendency to suddenly act in ways that are improper or potentially harmful to oneself or others, without adequate reflection on the causes or the consequences of one's own actions" [3, 4]. Moeller and colleagues defined it

F. S. Bersani (✉) · M. Pasquini
Department of Human Neurosciences, Sapienza University of Rome, Rome, Italy
e-mail: francescosaverio.bersani@uniroma1.it; massimo.pasquini@uniroma1.it

M. Biondi et al. (eds.), *Dimensional Psychopathology*,
https://doi.org/10.1007/978-3-319-78202-7_7

as "a predisposition toward rapid, unplanned reactions to internal or external stimuli without regard to the negative consequences of these reactions to the impulsive individual or to others" [2].

Although impaired impulse control can be present in any individual with or without a DSM-defined diagnosis, it is often a core psychopathological feature in several disorders—not only in the so-called impulse-control disorders but also in other conditions, such as borderline, antisocial, histrionic, and narcissistic personality disorders, substance use disorders, gambling disorder, bipolar disorder (especially during manic, hypomanic, or mixed episodes), attention deficit hyperactivity disorder (ADHD), paraphilic disorders, trichotillomania, and suicidal behaviour disorder [1, 2, 5, 6]. The recently developed Research Domain Criteria (RDoC), aimed at explicating fundamental biobehavioural dimensions that cut across disorder categories, include impulsive behaviours in the *cognitive control* construct [7], thus further substantiating the cross-cutting nature of impulsive behaviours.

Hollander and colleagues have argued that impulsivity and compulsivity are opposite ends of a spectrum [6, 8, 9]. While, in fact, the concept of impulsivity covers a wide range of "actions that are poorly conceived, prematurely expressed, unduly risky, or inappropriate to the situation and that often result in undesirable outcomes", compulsivity refers to "repetitive behaviours that are performed according to certain rules or in a stereotypical fashion, i.e. a tendency to repeat the same, often purposeless acts, which are sometimes associated with undesirable consequences" [6, 8, 9]. Impulsivity and compulsivity can therefore be conceptualized as diametrically opposed or, from a different perspective, as similar, in that both imply an impairment of impulse control [6, 8, 9]. According to Hollander's model, impulsive behaviours initially present an element of pleasure, although they tend to lose their pleasurable quality over time. Some patients with impulse control disturbances may engage in the behaviour to increase arousal, but there might be a compulsive component to their behaviour in that they continue to engage in the behaviour to decrease dysphoria. So, in general, while compulsivity may be driven by an attempt to alleviate anxiety or discomfort, impulsivity may be driven by the desire to obtain pleasure, arousal, or gratification [6, 8, 9]. Hollander and colleagues have also highlighted how both compulsivity and impulsivity, despite their intrinsic differences, share an impairment in inhibiting or suppressing repetitive behaviours and can change over time: impulsive behaviours can become compulsive and vice versa [6, 8, 9].

According to Charles Spielberger, anger can be defined as "an emotional state that may range in intensity from mild irritability to intense fury and rage" [10]. It is considered pathological when it does not realistically reflect the actual circumstances of the individual. Anger, considered one of the "basic emotions" by Paul Ekman [11], has also been defined by Berkowitz and Harmon-Jones as "a syndrome of relatively specific feelings, cognitions, and physiological reactions linked associatively with an urge to injure some target" [12]. The expressions of anger vary widely in different individuals and may be considered functional under certain controlled conditions. While a robust relationship between anger and violent behaviour has consistently been observed, it is true that anger does not always lead to aggressiveness nor is anger a necessary cause of aggressiveness [13].

Aggressiveness has been defined by Vitiello and Stoff as "a behaviour deliberately aimed at inflicting physical damage to persons or property" [14]. As reviewed by Vitiello and Stoff, in adult psychiatry two different areas of investigation can be identified which are relevant to human aggressiveness: one is related to impulsive forms of aggression, while the other is related to non-impulsive aggression in a context of antisocial tendencies [14]. The first type (i.e. impulsive aggressiveness) is usually explosive, and it is often characterized by disinhibition, anger, affective instability, and high levels of arousal, but not necessarily by antisocial personality traits; it can be associated with specific neurological dysfunctions (e.g. in the temporal or frontal cortices) or with the use of certain substances; it can, at times, be self-directed [14]. The second type (i.e. non-impulsive aggression) usually occurs in individuals who are less likely to have affective instability, and their aggressive behaviour is usually goal-oriented: it is initiated in order to achieve a specific goal other than physical harm of the victim [14]. As summarized by Vitiello and Stoff, the dichotomies *overt* vs. *covert*, *reactive* vs. *proactive*, *affective* vs. *predatory*, and *hostile* vs. *instrumental* have been identified to differentiate the qualitatively different forms of aggression; overall, in an evolutionary perspective, aggression could be conceptualised as defensive aggression (i.e. an impulsive-reactive-hostile-affective subtype) or offensive aggression (i.e. a controlled-proactive-instrumental-predatory subtype) [14].

The concept of activation (or psychomotor activation) summarises the psychopathological symptoms related to increased motor activity, agitation, acceleration of ideas, disinhibition, increased energy and self-confidence, euphoria, or irritability [15]. Psychomotor activation is a disturbance of movement, cognition, and behaviour associated with psychiatric or physical conditions [15, 16]. According to Carrol, it could be defined as "the rate of thought process, speech, and non-verbal communication, as well as the more obvious aspects of posture, speed of movement, and whole-body motility" [16, 17]. As reviewed by Day, the motor manifestations of activation are generally described as restless, repetitive to the extent of being almost stereotypic, aimless, non-purposeful, unproductive, and goalless; together with fixed, repetitive, accelerated, and incessant thoughts, activation can progress to an attitude of irritability, uncooperativeness, hostility, belligerence, or assaultiveness [16]. This condition has been described in a wide range of psychiatric syndromes including neurocognitive, depressive, substance use-related, psychotic, bipolar, and anxiety disorders [15, 16], and its management often raises important clinical and therapeutic challenges.

7.2 Neurobiological Aspects: Brain Circuitry

In terms of brain circuitry, impulsive and aggressive behaviours can be conceptualised as the consequence of decreased "top-down" stimuli provided by the frontal lobe (mainly from the orbital frontal cortex and the anterior cingulate cortex) and increased "bottom-up" stimuli provided by limbic regions (mainly from the insula and the amygdala) [18, 19]. In fact, while frontal "top-down" stimuli are involved in the calibration of behaviour to social cues and in the prediction of expectancies of reward and punishment, limbic "bottom-up" stimuli mainly mediate more primordial cognitive processes [19, 20]. An imbalance of these neurobiological triggers, with decreased frontal activity or increased limbic activity, may result in increased impulsivity and impulsive aggressiveness.

As summarised by Siever [18], it is possible to conceptualize a three-step cognitive and neurobiological pathway which leads emotionally provocative or challenging stimuli to determine impulsive or impulsive aggressive reactions. In the first step, the stimuli are processed by sensory processing centres; sensory deficits or disturbances may result in incomplete or distorted perception, thus increasing the perception of the stimuli as threatening or provocative [18]. Sensory distortions can contribute to mediate the impact of certain substances (e.g. alcohol or cocaine) in increasing violent behaviours [21, 22]. In the second step proposed by Siever, the cognitive appraisal and evaluation of the stimuli take place in a widespread cerebral network (anatomically located mainly in the prefrontal, temporal, and parietal cortices) involving social information processing regions and higher-association regions [18]. At this stage the appraisal can be affected or biased by cultural, social, and neuropsychological factors, as well as by negative cognitive or behavioural schemata, often derived from early or prolonged negative experiences; consistently, there is a large amount of evidence suggesting associations between the exposure to adversities or traumas in childhood and an increased risk for aggressiveness in adulthood [18, 23, 24]. In the last step of Siever's model, the processing of perceptive and cognitive stimuli in the amygdala and related limbic regions can stimulate the bottom-up "drive" to an aggressive or impulsive action, while the frontal cortices provide "top-down" modulation and suppression of those emotional responses and behavioural outcomes with the potential of leading to negative consequences [18]. Such key modulatory role of the frontal cortex has been confirmed by studies involving brain modulation techniques (mainly transcranial magnetic stimulation [TMS] or transcranial direct current stimulation [TDCS]) in which (1) reduced cortical inhibition was found in frontal cortices of violent offenders and (2) increases in neural activity of dorsolateral prefrontal cortex reduced certain aspects of aggressiveness [25–27]. A summary of the pathway described by Siever [18] is given in Fig. 7.1.

The critical role of the prefrontal cortex in the regulation of impulsivity or impulsive aggressiveness was first recognised in the context of prefrontal cortical lesions resulting in disinhibited behaviour. A famous example of disinhibition resulting from the disruption of cortical-subcortical networks is found in the case of neuroscience's most famous patient, Phineas Gage [28]. As reviewed by Thiebaut de Schotten, when he was 25 years old, Phineas Gage made a mistake at his workplace that resulted in an iron bar passing through the left side of his skull [28]. Despite extensive damage to his forehead, he survived the accident, but not without consequences: according to John Harlow, the local doctor who followed Gage throughout his recovery, he became "fitful, irreverent, indulging at times in the grossest profanities (which was not previously his custom), manifesting little deference for his fellows, impatient of restraint or advice when it conflicts with his desires" [28, 29].

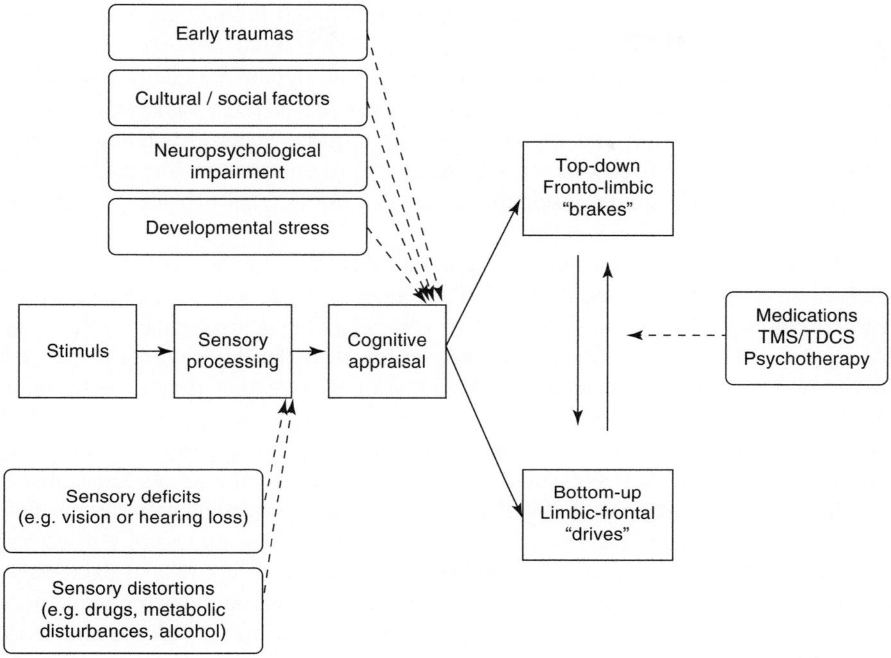

Fig. 7.1 Schematic representation of initiation/modulation of impulsive aggression. *TMS* transcranial magnetic stimulation, *TDCS* transcranial direct current stimulation

Table 7.1 Some of the psychiatric manifestations observed in subjects with acquired prefrontal or cerebellar disturbances (e.g. stroke, tumour, acute inflammation, trauma, neurodegeneration)

Mood	Cognition	Behaviour
Irritability	Impaired attention	Impulsivity
Rage	Impaired memory	Aggressiveness
Anger	Impaired executive function	Hyperactivity
Dysphoria	Distractibility	
	Conceptual disorganisation	

At the clinical level, it is known that individuals with congenital and acquired frontal lobe disturbances may present with a range of psychiatric manifestations in the cognitive, emotional, affective, perceptive, delusional, behavioural, and social areas, with impulsivity, anger, aggressiveness, impaired judgement, and increased psychomotor activation being especially represented (Table 7.1) [30].

In addition to empirical observations, the overall model of an impaired cortical-subcortical regulation underlying impulsive behaviours has been substantiated through neuroimaging findings. Significant volume reductions have been demonstrated in the left orbital frontal cortex and right anterior cingulate cortex in patients with borderline personality disorder and antisocial personality disorder, i.e. two clinical conditions characterised by impaired impulse control [31, 32]. Consistent findings have been obtained in studies using functional brain imaging that have provided evidence for concomitant reduced glucose metabolism/blood flow in frontal areas and enhanced activity in the amygdala and hypothalamus in association with aggressiveness or impulsive personality traits [33–37]. Catani et al. proposed that impaired inhibition and executive functioning, key neuropsychological elements related to impulsivity, may result from impairments in the white matter tracts of the superior longitudinal fasciculus, uncinate, and internal capsule (i.e. fronto-striatal projections) [38].

The prefrontal cortex also has an important role in the regulation of the degree of activation. Barkley proposed a model of executive dysfunctions located in the prefrontal cortex that explains the cognitive and behavioural disturbances related to increased activation [39, 40]. As reviewed by Spencer, Barkley's model comprises five major executive functions that enable individuals to recognise and control their actions to achieve a goal: response inhibition, nonverbal working memory, verbal working memory, self-regulation of emotion and motivation, and reconstitution [39, 40]. Response inhibition delays and interrupts responses and controls interference to allow individuals to control verbal and motor impulses; working memory enables a person have a sense of the past and future and a cognitive awareness of self; verbal working memory gives people the ability to internalise receptive and expressive language for self-questioning, self-description, and establishing rules for behaviour; self-regulation of emotion and motivation provides individuals with the ability to control their emotions and the motivation and persistence necessary to meet their goals; reconstitution is a form of play that allows people to analyse the experiences in their working memories to synthesise new responses, which they accept or reject based on the likelihood that the response can help them to achieve their goals [39, 40]. As reviewed by Spencer (in a paper mainly focusing on subjects with ADHD), Barkley has proposed that, of these five executive functions, response inhibition is deficient in highly activated patients and that this deficit may lead to the impairments observed in the psychological and social abilities associated with the other four executive functions [39, 40].

Findings from structural and functional neuroimaging have confirmed the hypothesis of a key role of prefrontal cortex in the regulation of activation [39]. However, an increasing amount of studies suggest that the pathophysiology of such a complex psychopathological domain reflects abnormal interplay among large-scale brain circuits. In relation to this point, Catani et al. have proposed that a condition of increased behavioural activation may result from impairments in the white matter tracts of the uncinate, inferior fronto-occipital fasciculus, and internal capsule (i.e. fronto-striatal projections) [38].

Intriguing recent findings also point out a possible role of the cerebellum to explain the conditions of increased impulsivity, aggressiveness, and activation. The physiology of the cerebellum has traditionally been limited to coordination of voluntary movement, gait, posture, speech, and motor function. However, recent anatomical studies demonstrate that the output of the cerebellum targets multiple non-motor areas in the prefrontal and posterior parietal cortex, and evidence from studies of patients with overt cerebellar diseases as well as from healthy subjects suggests a possible role for the cerebellum in higher cognitive functions and behavioural changes [41, 42]. In relation to impulsivity, aggressiveness, and activation, (1) patients with borderline personality disorder and those with ADHD have been reported to have increased reactivity of specific cerebellar regions [18, 43], (2) a positive correlation has been found between motor impulsivity and grey matter in the right cerebellum in a sample of psychiatric patients characterised by self-control problems [44], and (3) changes in impulsive behaviour in rats have been found to be associated with gene expression changes in cerebellar nuclei [45]. Similarly to what was observed in relation to frontal lobe disturbances, individuals with congenital (e.g. agenesis, dysplasia, and hypoplasia) and acquired (e.g. stroke, tumour, acute inflammation, trauma, and neurodegeneration) cerebellar disorders may present with a range of psychiatric manifestations in the cognitive, emotional, affective, perceptive, delusional, behavioural, and social areas, with impulsivity, anger, aggressiveness, impaired judgement, and increased psychomotor activation being especially represented [46, 47] (Table 7.1). While the exact mechanisms are still not known, these findings led Schmahmann et al. to express the concept of *cerebellar cognitive affective syndrome (CCAS)* [46, 47].

7.3 Neurobiological Aspects: Molecular Pathways

Twin and family studies suggest that anger, impulsivity, and hyperactivity/hyperactivation have substantial heritability. Men homozygous for a specific gene variant of *DAT1* (the 10-repeat allele) have higher levels of hyperactivity and impulsivity than males from all other groups [48]. A meta-analysis performed in 1997 on data from 24 twin studies found that the genetic effect accounted for up to 50% of the variance in aggressiveness-related outcomes [49]. The strongest link between genetic variation and aggressiveness exists for the monoamine oxidase A (MAOA) gene, located on the X chromosome and coding for a key enzyme in the catabolism of monoamines [50]. Mice knockout for the *MAOA gene* have increased levels of serotonin, norepinephrine, and dopamine in the brain and show enhanced amygdala-dependent emotional, but not motor, learning, and males exhibit highly increased aggressive behaviour [50–53]. In humans, a Dutch family with a missense mutation in the *MAOA* gene has been identified: hemizygous males, representing functional gene knockouts, exhibited a pattern of impulsively violent behaviour for generations [50, 51].

Overall, the majority of candidate genes explored in association with impulsivity, aggressiveness, and hyperactivity/hyperactivation are related to neurotransmitters [18, 48, 54, 55]. However, as traumatic-, sociocultural-, parenting-, and peer-related environmental factors are also independently conducive to increased anger, activation, and aggressiveness per se [18, 48, 54, 55], gene-environment interactions are thought to play a crucial role. Individuals with a biological risk for aggression, in fact, may be particularly vulnerable to the effect of psychosocial adversity [56]. Genes for the serotonin transporter (5-HTT) and the MAOA enzyme can interact with early traumas and psychological adversity to predispose to impulsivity and violence [18, 57, 58].

From a neurochemical perspective, the first evidence indicating that reduction of serotonergic activity could lead to impulsive, violent, or self-destructive behaviour in humans comes from research conducted over two decades ago [59]. Since then, consistent data from clinical and preclinical studies have evidenced the hypofunction of the serotonergic system as primarily involved in impulsive aggressiveness and other impulsivity-related behaviours (e.g. drug addiction and violent suicide). Animal models of psychopathology have showed that knockout mice for the gene encoding the 5-HT1B receptor show increased maternal aggression, territorial aggression, and drug addiction behaviour [60]. The 5-HT2A TYR 452 allele and certain polymorphisms of *5-HTT* have been associated with aggression [61, 62]. A functional polymorphism in the promoter region of the *5-HTT* gene (*5-HTTLPR*), leading to lower levels of expression of the gene, has been associated with violent suicidal behaviour [63]. Further, serotonin has been found to be mechanistically associated with so-called cognitive impulsivity, i.e. a distorted judgement of alternative outcomes which results in a loss of reward in the long term and may underlie many impulsivity-related phenomena [60].

The dopaminergic system also plays an active role in the modulation of impulsive behaviours. In fact, (1) in animal studies, hyperactivity in the dopamine system is associated with increases in impulsive aggressiveness; (2) studies on aggressive behaviours in rodents have shown that elevated dopamine levels can be observed before, during, and after aggressive fights; (3) there is evidence that impulsive behaviour may be enhanced by elevated dopaminergic function; and (4) medications acting as dopamine agonists (e.g. pramipexole, ropinirole) have been associated with pathological gambling and excessive or problematic behaviours in other impulsivity-related domains (relating to sex, eating, and shopping) in individuals with Parkinson's disease (Case Vignette 1) (for reviews, see [64, 65]).

Case Vignette 1: Impulsivity in Patient with Parkinson's Disease Receiving Dopamine Replacement Therapy

Mrs. F, a 54-year-old married woman working as a housewife and diagnosed with idiopathic Parkinson's disease 12 years before, presented with impulse control disturbances. She was receiving dopamine replacement therapy via a number of medications, including ropinirole, levodopa, and amantadine, for about 6 years. After the beginning of the therapy with dopaminergic medications, a range of impulsive behaviours emerged, among which gambling was the most problematic. She reported that she had lost large sums of money through casino games playing, racetrack bets, and online gambling. She also experienced symptoms of compulsive shopping and hypersexuality. In an effort to attenuate the symptoms, ropinirole was discontinued; this resulted in a significant reduction in her symptoms of gambling, while compulsive shopping and hypersexuality remained unchanged. Symptoms further improved when gabapentin was included as additional therapy.

As the serotonergic and dopaminergic system have strong anatomical and functional mutual relationships [66], it is possible that the interaction between the two systems represents a relevant *locus of interest* to understand the molecular mechanisms underlying impulsivity and impulsive aggressiveness. Considering the functional regulation of serotonin over the dopamine system, in fact, deficient serotonergic function may result in hyperactivity of the dopamine system, promoting impulsive behaviours [65]. This relationship may account for co-occurring serotonin and dopamine dysfunctions in individuals with impulsive aggression. In support of this, several studies have found that prefrontal serotonin levels in rats significantly decrease during and after fights, whereas prefrontal dopamine levels significantly increase [65, 67, 68].

An imbalance in glutamatergic/GABAergic activity (i.e. reduced activity at GABA receptors and glutamatergic enhancement) may contribute to aggression through hyperactivity of subcortical limbic regions [18].

At the peripheral level, the relations between endocrine factors and aggression have been assessed extensively through studies on (1) the effects of early hormonal "programming" of adult aggressiveness, (2) direct effects (presumably via an action on the central nervous system) of endocrine manipulations on fighting and threat, (3) indirect effects (presumably via changed social signals, etc.) of endocrine manipulations on fighting and threat, (4) hormone-aggression correlations, and (5) influences of fighting on endocrine function (for a review, see [69]). Overall, although it seems unlikely that casual relationships exist between aggression and a single hormone, testosterone has been consistently found to be crucially involved with aggressive behaviour in different experimental approaches [18, 69]. Exposure to testosterone in childhood has been found to increase aggressiveness in adulthood in both animal and human studies; high concentrations of testosterone have been reported in populations characterised by high levels of aggressiveness (e.g. criminals with personality disorders, violent offenders, and abusers); at the neurobiological level, testosterone and steroids may enhance responsiveness of brain circuitry related to social aggression (for a review, see [18, 69]). On the other hand, cortisol concentrations have generally been found to be low in individuals with high aggression, and aggressive behaviour has been linked to corticotropin releasing factor reactive autoantibodies [18, 69, 70].

It is also known that increased anger, aggressiveness, and activation can influence certain patterns of autonomic nervous system arousal through adrenergic and noradrenergic pathways, as indicated by heart rate, skin conductance, and blood pressure. Autonomic nervous system arousal can therefore be used as a biological marker or predictor of physically aggressive or hyperactive behaviour, although research in the field is still preliminary [71].

Intriguing recent findings also point out the possible role of inflammation in impulsivity, hyperactivity/hyperactivation, anger, and impulsive aggressiveness. A modulating role for cytokines in aggressive behaviour in mammals is suggested by direct experimental manipulation in lower mammals, the effect of inflammatory cytokines on increased anger and aggression in patients treated with cytokine immunotherapy, and correlative studies of plasma inflammatory cytokines in otherwise healthy humans [72]. Recently, Coccaro et al. reported significant positive correlations between aggression and levels of inflammatory cytokines in both plasma (C-reactive protein and interleukin 6) and cerebrospinal fluid (soluble interleukin 1 receptor II protein) in impulsively aggressive subjects and controls [73, 74]. Further, preclinical studies recently showed that cytokines are present in brain regions, such as the hypothalamus and midbrain periaqueductal gray, that are known to play a key role in aggression and rage behaviour [72]. As the immune and central nervous systems communicate in a bidirectional manner, and this molecular interplay has a prominent role in mediating behaviour and psychiatric conditions, this field of exploration may be of particular interest in further revealing the biological underpinnings of impulsivity, activation, and related phenomena.

7.4 Clinical Aspects: Impulsivity, Anger/Aggressiveness, and Activation Across Psychiatric Conditions

From a clinical perspective, it is known that the combination of clusters of symptoms differs from patient to patient and gives rise to a wide variety of clinical pictures, even among subjects with the same diagnosis. Subsequently, it is important that the clinical features related to impulsivity, anger/aggressiveness, and activation are specifically and *multiparametrically* investigated and treated independently of the concomitant DSM-defined condition. For example, in 2015, our research group conducted a prospective study on two separate cohorts of patients presenting to hospital emergency departments. We found that high levels of impulsivity among the patients were among the strongest predictors of subsequent hospitalisation in an acute inpatient psychiatric unit, independently of the concomitant DSM-defined diagnosis [75]. This approach is also an avenue for bringing together a more specific pathophysiology with a more heuristic approach to clinical symptoms.

Although several assessments have been used as "measures" of impulsivity, there are primarily three main classes of instruments that appear to measure key aspects of impulsivity: clinical scales, behavioural laboratory measures, and event-related potentials (ERPs) (for a review, see [2]). Among the clinical scales, the Momentary Impulsivity Scale, the Eysenck Impulsiveness Questionnaire, and the Barratt Impulsiveness Scale have been used extensively [2, 76–78]. Among laboratory paradigms, punishment and/or extinction paradigms, reward-choice paradigms, and response disinhibition/attentional paradigms have been used to measure impulsivity-related phenomena [2]. Among the ERPs, reduced amplitude of the P300 ERP, recorded in response to target stimuli during the performance of "oddball" tasks, has been related to impulsivity and impulse-control disorders [2, 79, 80]. The assessment of aggressiveness is of particular interest in the field of forensic psychiatry. The Violence Risk Appraisal Guide (VRAG) is based on four components: (1) identifying empirically valid risk factors, (2) determining a method for measuring (or "scoring") these risk factors, (3) establishing a procedure for combining scores on the risk factors, and (4) producing an estimate of violence risk [81]. The VRAG has reliable predictive and incremental validity and it is considered among the best instruments related to the aggressiveness risk assessment [81].

A range of clinical scales have been used to assess activation, including the *Brief Psychiatric Rating Scale* (BPRS), the *Conners Rating Scale* (CRS), the *Vanderbilt Rating Scale*, and the *ADHD Self-Report System*, with some of these rating scales being originally conceptualised to capture hyperactivity as a core symptom of ADHD [82, 83]. As increased physical activity is considered a meaningful behavioural correlate of the condition of increased activation, several instruments have been recently validated to objectively measure it in a clinical context. These include accelerometers, actigraphs, and infrared motion tracking [82]. The SVARAD is one of the few instruments that can be used to rate the severity of the three "outer" dimensions (impulsivity, anger/aggressiveness, and activation) at the same time. While it lacks the ability to measure biological correlates of the explored psychopathological conditions, it presents several strengths, among which are short administration duration, low cost, and good prediction of clinical outcomes [4, 15, 75, 78, 84–86].

As written above, high levels of impulsivity, anger/aggression, and activation can be present in any individual with or without a DSM-defined diagnosis. Traditionally, these psychopathological dimensions have been considered a core characteristic of certain psychiatric disturbances (listed above), but they can also play a relevant role in the clinical picture of an even wider spectrum of conditions. Over the last few years, our research team has used the SVARAD to evaluate 846 inpatients and 1124 outpatients consecutively recruited at the Department of Neurology and Psychiatry of Policlinico Umberto I Hospital, Sapienza University of Rome. We found that (1) in inpatients, the degree/severity of impulsivity and anger/aggressiveness was similar across patients with different psychiatric diagnoses (mainly major depressive, bipolar, borderline personality, and psychotic disorders). However, in outpatients, these dimensions were markedly higher in patients with borderline personality disorder and with eating disorders, in comparison with patients with other diagnoses (Figs. 7.2 and 7.3). We further found that (2) the degree/severity of activation was

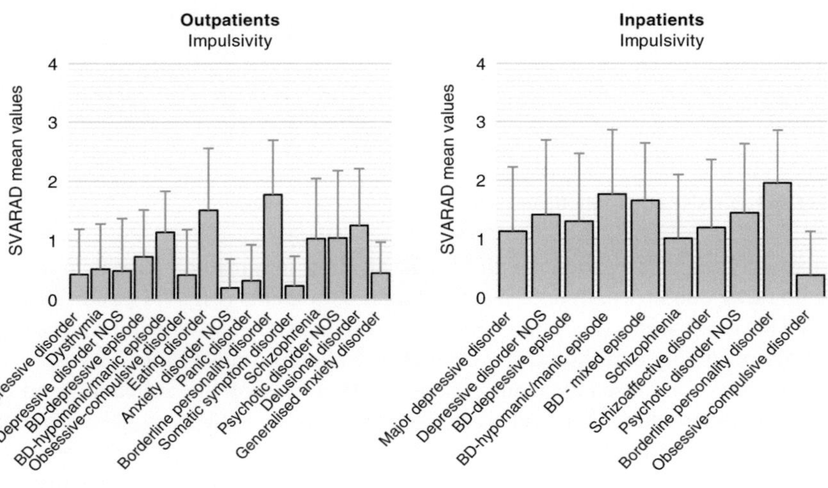

Fig. 7.2 SVARAD impulsivity dimension across diagnostic categories

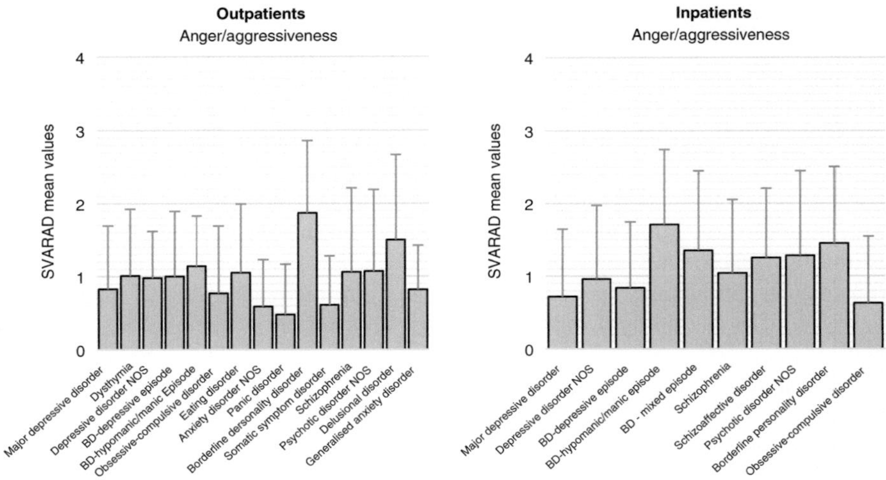

Fig. 7.3 SVARAD anger/aggressiveness dimension across diagnostic categories

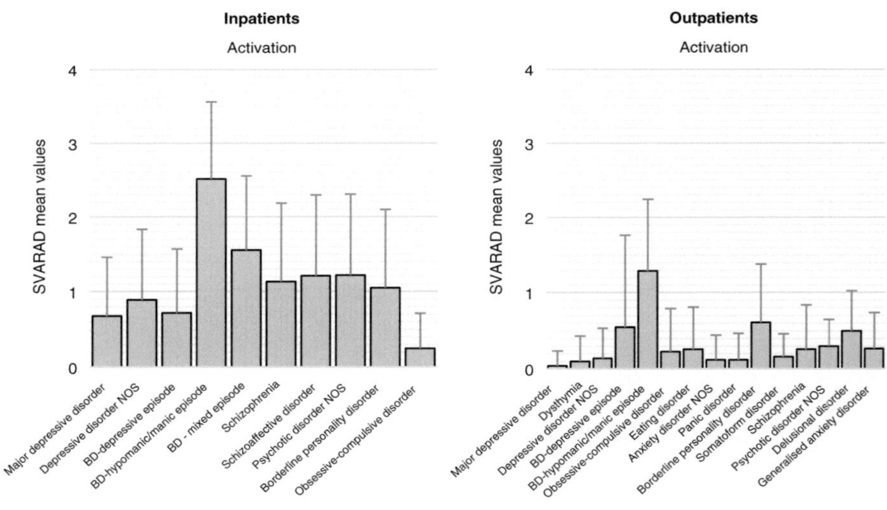

Fig. 7.4 SVARAD activation dimension across diagnostic categories

markedly higher in patients with bipolar disorder in a manic phase than in patients with other diagnoses, in both inpatients and outpatients (Fig. 7.4), and (3) the three SVARAD subscales (i.e. impulsivity, anger/aggression, and activation) were significantly positively correlated with each other in both groups (Table 7.2).

In 2005, our research team detected high levels of the activation and anger/aggressiveness dimensions, characterised by anger, irritability, aggressiveness, hostility, and psychomotor activation, in a sample of 380 patients with depressive disorders [15, 85, 87]. Consistently, in the present sample, we found that, across

Table 7.2 Two-tailed Spearman correlation between the three explored SVARAD subscales (i.e. anger/aggressiveness, impulsivity, and activation) across all subjects in both cohorts (i.e. inpatients and outpatients)

	Inpatients		Outpatients	
	r	*p*	*r*	*p*
Anger/aggressiveness				
Anger/aggressiveness	–	–	–	–
Impulsivity	0.57	<0.01	0.59	<0.01
Activation	0.49	<0.01	0.33	<0.01
Impulsivity				
Anger/aggressiveness	0.57	<0.01	0.59	<0.01
Impulsivity	–	–	–	–
Activation	0.41	<0.01	0.35	<0.01
Activation				
Anger/aggressiveness	0.49	<0.01	0.34	<0.01
Impulsivity	0.41	<0.01	0.34	<0.01
Activation	–	–	–	–

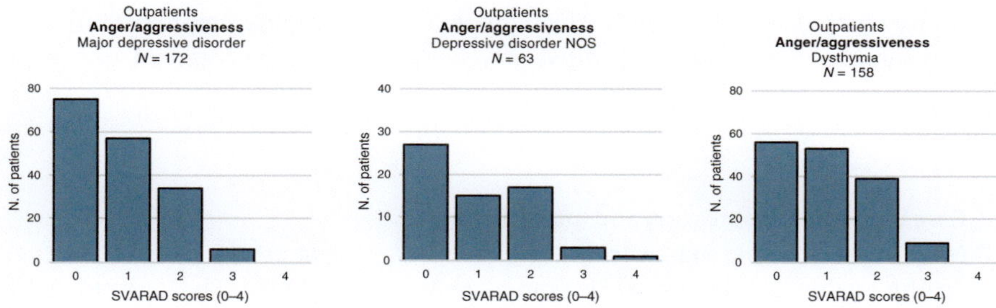

Fig. 7.5 SVARAD scores for the anger/aggressiveness dimension across outpatients with major depressive disorder ($n = 172$), depressive disorder NOS ($n = 63$), and dysthymia ($n = 158$)

outpatients with major depressive disorder, depressive disorder NOS, and dysthymia, a substantial number of subjects had moderate or high levels of anger or aggressiveness (Fig. 7.5).

Historically, the importance of anger and aggressive impulses in depressed patients has been underlined extensively by scholars in the fields of psychoanalysis [88, 89] and cognitivism [90]. Also, many intellectuals and artists, among them the Italian poet Giacomo Leopardi [91], have pointed out that, for people who do not love and esteem themselves, it is difficult to be good and kind with others. It has also been underscored that non-melancholic depressions often entail anger and that even during melancholic phases anger can emerge as an antidote against anhedonia or affective anaesthesia (i.e. the disturbing state of having no feelings) [92]. Consistently, a close inspection of the literature in clinical psychiatry suggests that symptoms of anger, irritability, aggressiveness, and hostility have often been observed in patients with depression when the clinical assessment was performed with instruments aimed at exploring such symptoms (Case Vignettes 2 and 3)

[93–97]. Of clinical relevance, when depression is associated with high levels of anger and impulsivity, the risk of suicide is markedly high [2]. When assessment instruments covering anger and irritability are used, these symptoms are often detected in many psychiatric disorders, particularly in depressed outpatients [98–100]. In fact, anger and aggression are prominent in depressed outpatients to a degree similar to that of depression per se or anxiety [98], and the presence of anger is more frequent among depressed patients than among patients with anxiety or somatoform disorders [100].

Case Vignette 2: Depression and Anger

Ms. G, a 32-year-old woman working as a designer, presented with symptoms of sadness, fatigue, sleep disturbances, and irritability related to conflicts with her boyfriend. She felt he was not emotionally responsive enough to meet her needs; he was often critical of her intelligence, her appearance, or her life-style. Ms. G felt deeply hurt by these comments, which always had a long-lasting negative impact on her. She reported to concomitantly experience sadness, feeling of inadequacy, and anger; she was worried about the risk of overreacting to these slights. Her symptoms significantly improved after 2–3 months of individual cognitive psychotherapy and concomitant use of escitalopram and valproic acid.

Case Vignette 3: Depression and Impulsivity

Mr. A, a 31-year-old man working as a labourer, presented with symptoms of apathy, sadness, anhedonia, insomnia, and guilt feelings. The symptoms began after his wife discovered that he had lost a large amount of money in gambling, and then left him. Three weeks before the first psychiatric visit, he had a car accident during which he bled profusely; he tasted his own blood and liked its smell and taste. After that event, he started to often self-cut and then suck his own blood. The acts of self-injury were preceded by a mounting tension or arousal and were followed by a feeling of relief or pleasure. He was admitted to our inpatient unit and treated with group therapy, venlafaxine, and valproic acid. His symptoms started to improve after 1 month of treatment.

Although high levels of anger/aggressiveness were not observed in patients with obsessive-compulsive disorder (OCD) in our sample (Fig. 7.3), in a recent study of 57 OCD patients, we found a significant direct correlation between levels of anger and a number of obsession subtypes, i.e. aggressive, contamination, and sexual obsessions [101]. This is consistent with OCD cognitive explanation models for which anger and disgust are important components of moral judgement and moral violation, and aggressive and sexual obsessions or thoughts may represent elicitors of anger [102].

The finding of high levels of impulsivity among patients with eating disorders (Fig. 7.2) is consistent with previous clinical and theoretical observations. McElroy et al. suggested in 1994 that the various eating disorders (i.e. anorexia nervosa, bulimia nervosa, and binge eating disorder) could be situated on a spectrum with varying degrees of obsessive-compulsive and impulsive traits, with a more obsessional nature in anorexia nervosa versus a more impulsive nature in bulimia nervosa and binge eating disorder [103].

Taken together, these data confirm that (1) high levels of impulsivity, anger/aggressiveness, and activation play a critical role in the psychopathology of a variety of psychiatric disturbances, (2) their severity changes from patient to patient at the individual level largely independently of the concomitant DSM-defined diagnosis, and (3) these clinical features should be specifically investigated and treated in order to achieve the best possible treatment outcome.

7.5 Clinical Aspects: Implications for Treatment

Both psychotherapy and pharmacotherapy have shown some effectiveness in treating impulsivity, anger, aggressiveness, activation, and related symptoms.

As reviewed by Moeller, examples of psychoanalytic therapies for impulsivity are found primarily in the clinical literature on treatment of borderline personality disorder, with some authors emphasising a supportive, ego-building approach, and others emphasising the use of more intensive, expressive, and regressive techniques aimed at achieving a more fundamental personality change [2]. Cognitive and behavioural psychotherapies (CBTs) have been used to reduce impulsivity in chronic psychiatric patients [104], impulsive preschool children [105], and drug-dependent patients [106, 107]. In a meta-analysis of 36 outcome studies of CBT in children, Baer and Nietzel reported that improvements in impulsivity were significantly stronger in subjects receiving CBT than in untreated or placebo-treated subjects [108]. Contingency management treatments for impulsive-related disorders, involving the use of predetermined positive or negative consequences to reward or punish (and thus deter) the occurrence of a target behaviour, have also received clinical and research attention [2].

In a recent review of meta-analyses, Lee and DiGiuseppe highlighted that anger does not always lead to aggressiveness nor is anger a necessary cause of aggressiveness, and therefore the psychological treatments should address the dimensions of anger and aggression independently [13]. The large majority of studies on psychological interventions specifically targeting anger have been performed using CBT and its classical constructs, showing an overall good clinical effectiveness [13]. However, two meta-analyses comparing relaxation, social skills, cognitive, and relaxation treatments for anger found that relaxation treatments yielded the highest effect size ($d > 0.8$ in both) [109, 110]. One meta-analysis comparing nine types of psychological treatments (i.e. cognitive, cognitive behaviour therapy, exposure, psychodynamic, psychoeducational, relaxation-based, skills-based, stress inoculation, and multicomponent) concluded that psychodynamic therapy yielded the largest effect size (*d* = 1.40) [111]. Regarding aggression, almost all studies have employed CBT, and the mean effect sizes reported by the meta-analyses on the topic vary greatly from small (four studies) to medium (three studies) to large (two studies), ranging from 0.10 to 1.14 [13]. Overall, although the body of literature suggests that treatments for anger and aggression are moderately effective, much remains to be done to inform future treatment of these clinically significant and impairing problems, and, as reported by Lee and DiGiuseppe, "we are a long way from answering Gordon Paul's classic questions: what types of psychotherapy (for anger) work best for which types of problems?" [13, 112].

Evidence supports the idea that increased impulsivity, impulsive aggressiveness, and activation are associated with (and at least partially due to) an imbalance between excessive "bottom-up drives" triggered or signalled by limbic regions (such as the amygdala and insula) and reduced "top-down" control or "brakes" provided by the frontal cortex (which is involved in calibration of behaviour to social cues, prediction of expectancies of reward and punishment, and modulation or suppression of aggressive behaviour with negative consequences) [18, 19]. Given this evidence, it is possible that certain psychotherapeutic approaches show effectiveness in ameliorating these conditions because they can specifically modulate the fronto-limbic circuitry. Preliminary data from functional neuroimaging studies suggest that treatment with CBT, mindfulness-based cognitive therapy, and, to a lesser extent, psychodynamic psychotherapy can lead to increased activity of the dorsolateral prefrontal cortex and decreased activity of the amygdala [113, 114].

In relation to psychopharmacotherapy, there is no standardised treatment for complex disorders involving impulsivity, anger, aggression, or activation, although a range of different medication classes have been investigated. As reviewed by Hollander and colleagues, pharmacological treatments may reduce impulsivity and normalise arousal through a range of pathways including (1) decrease of dopaminergic activity, (2) enhancement of serotonergic activity, (3) shift in the neurotransmitter balance from excitatory (glutamatergic) toward inhibitory (GABAergic) transmission, and (4) reduction or stabilisation of noradrenergic effects [6, 8, 9].

Overall, the efficacy of anticonvulsants and lithium for decreasing impulsivity, aggression, and suicidality has consistently been reported [2, 115–119]. On the other hand, contrasting findings exist in relation to selective serotonin reuptake inhibitors (SSRIs): while fluoxetine, fluvoxamine, and citalopram have been shown to significantly decrease impulsive and aggressive behaviours compared with placebo in certain controlled studies [120–122], other studies reported this class of medications to lead to increased impulsive-related phenomena such as suicidality [123].

In 2007, our research team showed that the concomitant use of one medication of the SSRI class and one anticonvulsant (valproate, carbamazepine, or gabapentin) was highly effective in the treatment of depressed subjects with a substantial level of anger/aggressiveness (assessed using the SVARAD) [86]. The idea underlying this approach was to concomitantly use the two different classes of medications (i.e. antidepressants and anticonvulsants) in order to simultaneously treat the two major aspects of the clinical picture of the patients (i.e. depression *per se* and impulsivity/anger/aggressiveness/activation) [86], in a sort of "dimensional psychopharmacology" in which the treatment was function-oriented rather than exclusively diagnosis-oriented. We also used a similar therapeutic strategy in a population of subjects with psychiatric symptoms reactive to the condition of having cancer (Case Vignette 4), and we obtained significant reductions in the severity of both depression and anger symptoms over time [84].

Case Vignette 4: Depression, Cancer, and Anger
Ms. G is a 54-year-old woman, married with three children. She has breast cancer with metastases at the bone, uterus, bone marrow, and liver. She has a life expectancy of a few months. She expressed her feelings thus: "Is this a kind of life worth living? I don't laugh anymore. I don't cook anymore. I don't eat anymore. Why did I have to be castigated? It is not right". Her daughter reported that, after an initial phase of apathy, she was becoming more and more aggressive, angry, and dictatorial: "It seems that she envies healthy people".

Accumulating evidence suggests a specific antiaggressive effect for certain atypical antipsychotics: placebo-controlled trials of risperidone in adults with dementia [124, 125], adults with autism spectrum disorders [126], and children with conduct disorder [127] reported a significant decrease in aggression without significant sedation. The β-adrenergic antagonists constitute another class of medications that have been used to treat impulsive aggression, although data are preliminary [2].

Overall, this evidence is consistent with the model of a "dimensional psychopathology", in which clinicians *multiparametrically* identify and comprehensively treat impulsivity, anger, aggressiveness, and activation based on the individual symptom profile of each patient and the underlying neurobiology of these phenomena partially independently of the DSM-defined diagnosis.

References

1. American Psychiatric Association. Diagnostic and statistical manual of mental disorders. 5th ed. Washington, DC: American Psychiatric Association; 2013.
2. Moeller FG, et al. Psychiatric aspects of impulsivity. Am J Psychiatry. 2001;158(11):1783–93.
3. Biondi M, Pasquini M. Dimensional psychopharmacology of the cancer patient. In: Biondi M, Costantini A, Wise TN, editors. Psycho-oncology. Arlington, VA: American Psychiatric Publishing; 2013.
4. Pancheri P, et al. Costruzione della "SVARAD", una scala per la valutazione rapida dimensionale. Rivista di Psichiatria. 1999;34:72–83.
5. Turecki G. Dissecting the suicide phenotype: the role of impulsive-aggressive behaviours. J Psychiatry Neurosci. 2005;30(6):398–408.
6. Hollander E, Rosen J. Impulsivity. J Psychopharmacol. 2000;14(2 Suppl 1):S39–44.
7. Insel T, et al. Research domain criteria (RDoC): toward a new classification framework for research on mental disorders. Am J Psychiatry. 2010;167(7):748–51.
8. Oldham JM, Hollander E, Skodol AE. Impulsivity and compulsivity. Arlington, VA: American Psychiatric Publishing; 1996.

9. Berlin HA, Hollander E. Understanding the differences between impulsivity and compulsivity. Psychiatric Times, 2008;25(8). http://www.psychiatrictimes.com/impulse-control-disorders/understanding-differences-betweenimpulsivity-and-compulsivity.
10. Spielberger CD, Sarason IG. Stress and emotion. Didcot: Taylor & Francis; 1995.
11. Ekman P. An argument for basic emotions. Cognit Emot. 1992;6:169–200.
12. Berkowitz L, Harmon-Jones E. Toward an understanding of the determinants of anger. Emotion. 2004;4(2):107–30.
13. Lee AH, DiGiuseppe R. Anger and aggression treatments: a review of meta-analyses. Curr Opin Psychol. 2018;19:65–74.
14. Vitiello B, Stoff DM. Subtypes of aggression and their relevance to child psychiatry. J Am Acad Child Adolesc Psychiatry. 1997;36(3):307–15.
15. Biondi M, et al. Dimensional psychopathology of depression: detection of an 'activation' dimension in unipolar depressed outpatients. J Affect Disord. 2005;84(2–3):133–9.
16. Day RK. Psychomotor agitation: poorly defined and badly measured. J Affect Disord. 1999;55(2–3):89–98.
17. Carroll BJ. Psychopathology and neurobiology of manic-depressive disorders. In: Carroll BJ, Barrett JE, editors. Psychopathology and the Brain. New York: Raven Press; 1991.
18. Siever LJ. Neurobiology of aggression and violence. Am J Psychiatry. 2008;165(4):429–42.
19. Blair RJ. The roles of orbital frontal cortex in the modulation of antisocial behavior. Brain Cogn. 2004;55(1):198–208.
20. Frith C, Dolan R. The role of the prefrontal cortex in higher cognitive functions. Brain Res Cogn Brain Res. 1996;5(1–2):175–81.
21. Beck A, Heinz A. Alcohol-related aggression-social and neurobiological factors. Dtsch Arztebl Int. 2013;110(42):711–5.
22. Yudofsky SC, Silver JM, Hales RE. Cocaine and aggressive behavior: neurobiological and clinical perspectives. Bull Menn Clin. 1993;57(2):218–26.
23. Frazzetto G, et al. Early trauma and increased risk for physical aggression during adulthood: the moderating role of MAOA genotype. PLoS One. 2007;2(5):e486.
24. Sansone RA, Leung JS, Wiederman MW. Five forms of childhood trauma: relationships with aggressive behavior in adulthood. Prim Care Companion CNS Disord. 2012;14(5). https://doi.org/10.4088/PCC.12m01353.
25. Dambacher F, et al. Reducing proactive aggression through non-invasive brain stimulation. Soc Cogn Affect Neurosci. 2015;10(10):1303–9.
26. Philipp-Wiegmann F, et al. Reduced cortical inhibition in violent offenders: a study with transcranial magnetic stimulation. Neuropsychobiology. 2011;64(2):86–92.
27. Riva P, et al. Reducing aggressive responses to social exclusion using transcranial direct current stimulation. Soc Cogn Affect Neurosci. 2015;10(3):352–6.
28. Thiebaut de Schotten M, et al. From Phineas Gage and Monsieur Leborgne to H.M.: Revisiting disconnection syndromes. Cereb Cortex. 2015;25(12):4812–27.
29. Harlow J. Recovery from the passage of an iron bar through the head. Publ Mass Med Soc. 1868;2:327–47.
30. Nicholl J, LaFrance WC Jr. Neuropsychiatric sequelae of traumatic brain injury. Semin Neurol. 2009;29(3):247–55.
31. Hazlett EA, et al. Reduced anterior and posterior cingulate gray matter in borderline personality disorder. Biol Psychiatry. 2005;58(8):614–23.
32. Raine A, et al. Hippocampal structural asymmetry in unsuccessful psychopaths. Biol Psychiatry. 2004;55(2):185–91.
33. Donegan NH, et al. Amygdala hyperreactivity in borderline personality disorder: implications for emotional dysregulation. Biol Psychiatry. 2003;54(11):1284–93.
34. Gobrogge KL, et al. Anterior hypothalamic neural activation and neurochemical associations with aggression in pair-bonded male prairie voles. J Comp Neurol. 2007;502(6):1109–22.
35. Goyer PF, et al. Positron-emission tomography and personality disorders. Neuropsychopharmacology. 1994;10(1):21–8.

36. Herpertz SC, et al. Evidence of abnormal amygdala functioning in borderline personality disorder: a functional MRI study. Biol Psychiatry. 2001;50(4):292–8.
37. Koenigsberg HW, et al. Neural correlates of emotion processing in borderline personality disorder. Psychiatry Res. 2009;172(3):192–9.
38. Catani M, et al. Beyond cortical localization in clinico-anatomical correlation. Cortex. 2012;48(10):1262–87.
39. Spencer TJ, et al. Overview and neurobiology of attention-deficit/hyperactivity disorder. J Clin Psychiatry. 2002;63(Suppl 12):3–9.
40. Barkley RA. ADHD and the nature of self-control. New York: Guilford Press; 1997.
41. Minichino A, et al. The role of cerebellum in unipolar and bipolar depression: a review of the main neurobiological findings. Riv Psichiatr. 2014;49(3):124–31.
42. Strick PL, Dum RP, Fiez JA. Cerebellum and nonmotor function. Annu Rev Neurosci. 2009;32:413–34.
43. King JA, et al. Neural substrates underlying impulsivity. Ann N Y Acad Sci. 2003;1008:160–9.
44. Lee AK, et al. Neural correlates of impulsivity factors in psychiatric patients and healthy volunteers: a voxel-based morphometry study. Brain Imaging Behav. 2011;5(1):52–64.
45. Moers-Hornikx VM, et al. Cerebellar nuclei are involved in impulsive behaviour. Behav Brain Res. 2009;203(2):256–63.
46. Schmahmann JD, Sherman JC. The cerebellar cognitive affective syndrome. Brain. 1998;121(Pt 4):561–79.
47. Schmahmann JD, Weilburg JB, Sherman JC. The neuropsychiatry of the cerebellum - insights from the clinic. Cerebellum. 2007;6(3):254–67.
48. Curatolo P, D'Agati E, Moavero R. The neurobiological basis of ADHD. Ital J Pediatr. 2010;36(1):79.
49. Miles DR, Carey G. Genetic and environmental architecture of human aggression. J Pers Soc Psychol. 1997;72(1):207–17.
50. Meyer-Lindenberg A, et al. Neural mechanisms of genetic risk for impulsivity and violence in humans. Proc Natl Acad Sci U S A. 2006;103(16):6269–74.
51. Brunner HG, et al. Abnormal behavior associated with a point mutation in the structural gene for monoamine oxidase A. Science. 1993;262(5133):578–80.
52. Cases O, et al. Aggressive behavior and altered amounts of brain serotonin and norepinephrine in mice lacking MAOA. Science. 1995;268(5218):1763–6.
53. Kim JJ, et al. Selective enhancement of emotional, but not motor, learning in monoamine oxidase A-deficient mice. Proc Natl Acad Sci U S A. 1997;94(11):5929–33.
54. Deater-Deckard K, et al. Multiple risk factors in the development of externalizing behavior problems: group and individual differences. Dev Psychopathol. 1998;10(3):469–93.
55. Fergusson DM, Lynskey MT. Physical punishment/maltreatment during childhood and adjustment in young adulthood. Child Abuse Negl. 1997;21(7):617–30.
56. Marks DJ, et al. The interaction of psychosocial adversity and biological risk in childhood aggression. Psychiatry Res. 2007;151(3):221–30.
57. Kim-Cohen J, et al. MAOA, maltreatment, and gene-environment interaction predicting children's mental health: new evidence and a meta-analysis. Mol Psychiatry. 2006;11(10):903–13.
58. Reif A, et al. Nature and nurture predispose to violent behavior: serotonergic genes and adverse childhood environment. Neuropsychopharmacology. 2007;32(11):2375–83.
59. Linnoila M, et al. Low cerebrospinal fluid 5-hydroxyindoleacetic acid concentration differentiates impulsive from nonimpulsive violent behavior. Life Sci. 1983;33(26):2609–14.
60. Brunner D, Hen R. Insights into the neurobiology of impulsive behavior from serotonin receptor knockout mice. Ann N Y Acad Sci. 1997;836:81–105.
61. Davidge KM, et al. Association of the serotonin transporter and 5HT1Dbeta receptor genes with extreme, persistent and pervasive aggressive behaviour in children. Psychiatr Genet. 2004;14(3):143–6.
62. Mik HM, et al. Serotonin system genes and childhood-onset aggression. Psychiatr Genet. 2007;17(1):11.

63. Courtet P, et al. Association between violent suicidal behavior and the low activity allele of the serotonin transporter gene. Mol Psychiatry. 2001;6(3):338–41.
64. Potenza MN. How central is dopamine to pathological gambling or gambling disorder? Front Behav Neurosci. 2013;7:206.
65. Seo D, Patrick CJ, Kennealy PJ. Role of serotonin and dopamine system interactions in the neurobiology of impulsive aggression and its comorbidity with other clinical disorders. Aggress Violent Behav. 2008;13(5):383–95.
66. Daw ND, Kakade S, Dayan P. Opponent interactions between serotonin and dopamine. Neural Netw. 2002;15(4–6):603–16.
67. Ferrari PF, et al. Accumbal dopamine and serotonin in anticipation of the next aggressive episode in rats. Eur J Neurosci. 2003;17(2):371–8.
68. van Erp AM, Miczek KA. Aggressive behavior, increased accumbal dopamine, and decreased cortical serotonin in rats. J Neurosci. 2000;20(24):9320–5.
69. Brain PF. Hormonal aspects of aggression and violence. In: N.R. Council, editor. Understanding and preventing violence, Biobehavioral influences, vol. 2. Washington, DC: The National Academies Press; 1994.
70. Fetissov SO, et al. Aggressive behavior linked to corticotropin-reactive autoantibodies. Biol Psychiatry. 2006;60(8):799–802.
71. Gower AL, Crick NR. Baseline autonomic nervous system arousal and physical and relational aggression in preschool: the moderating role of effortful control. Int J Psychophysiol. 2011;81(3):142–51.
72. Zalcman SS, Siegel A. The neurobiology of aggression and rage: role of cytokines. Brain Behav Immun. 2006;20(6):507–14.
73. Coccaro EF, Lee R, Coussons-Read M. Cerebrospinal fluid inflammatory cytokines and aggression in personality disordered subjects. Int J Neuropsychopharmacol. 2015;18(7):pyv001.
74. Coccaro EF, Lee R, Coussons-Read M. Elevated plasma inflammatory markers in individuals with intermittent explosive disorder and correlation with aggression in humans. JAMA Psychiat. 2014;71(2):158–65.
75. Dazzi F, et al. Predictors of inpatient psychiatric admission in patients presenting to the emergency department: the role of dimensional assessment. Gen Hosp Psychiatry. 2015;37(6):587–94.
76. Tomko RL, et al. Measuring impulsivity in daily life: the momentary impulsivity scale. Psychol Assess. 2014;26(2):339–49.
77. Patton JH, Stanford MS, Barratt ES. Factor structure of the Barratt impulsiveness scale. J Clin Psychol. 1995;51(6):768–74.
78. Pancheri P, et al. Validazione della scala per la valutazione rapida dimensionale "SVARAD". Riv Psichiatr. 1999;34:84–93.
79. Sunohara GA, et al. Effect of methylphenidate on attention in children with attention deficit hyperactivity disorder (ADHD): ERP evidence. Neuropsychopharmacology. 1999;21(2):218–28.
80. Harmon-Jones E, Barratt ES, Wigg C. Impulsiveness, aggression, reading, and the P300 of the event-related potential. Personal Individ Differ. 1997;22(4):439–45.
81. Skeem JL, Monahan J. Current directions in violence risk assessment. Curr Dir Psychol Sci. 2011;20(1):38–42.
82. Wehrmann T, Muller JM. An objective measure of hyperactivity aspects with compressed webcam video. Child Adolesc Psychiatry Ment Health. 2015;9:45.
83. Collett BR, Ohan JL, Myers KM. Ten-year review of rating scales. V: scales assessing attention-deficit/hyperactivity disorder. J Am Acad Child Adolesc Psychiatry. 2003;42(9):1015–37.
84. Pasquini M, et al. Detection and treatment of depressive and anxiety disorders among cancer patients: feasibility and preliminary findings from a liaison service in an oncology division. Depress Anxiety. 2006;23(7):441–8.
85. Pasquini M, et al. Relevance of anger and irritability in outpatients with major depressive disorder. Psychopathology. 2004;37(4):155–60.

86. Pasquini M, et al. Combining an SSRI with an anticonvulsant in depressed patients with dysphoric mood: an open study. Clin Pract Epidemiol Ment Health. 2007;3:3.
87. Picardi A, et al. Higher levels of anger and aggressiveness in major depressive disorder than in anxiety and somatoform disorders. J Clin Psychiatry. 2004;65(3):442–3.
88. Abraham K. Notes on the psychoanalytical investigation of manic-depressive insanity and allied conditions. In: Abraham K, editor. Selected papers of Karl Abraham. London: Hogarth Press; 1948.
89. Freud S. Mourning and melancholia. In: Strachey J, editor. Standard edition of the complete psychological works of Sigmund Freud, vol. 14. London: Hogarth Press; 1975.
90. Guidano VF. Complexity of the self. New York: Guilford Press; 1987.
91. Leopardi G. Zibaldone. Milan: Mondadori Editore; 1996.
92. Stanghellini G. Dysphoria, vulnerability and identity. An eulogy for anger. Psychopathology. 2000;33(4):198–203.
93. Fava GA, et al. Hostility and recovery from melancholia. J Nerv Ment Dis. 1986;174(7):414–7.
94. Fava M, Rosenbaum JF. Anger attacks in depression. Depress Anxiety. 1998;8(Suppl 1):59–63.
95. Pull CB, et al. The principal dimensions of manifest depression. A factor analysis of manifest depressive symptomatology. Neuropsychobiology. 1979;5(4):207–12.
96. Overall JE, Hollister LE. Phenomenological classification of depressive disorders. J Clin Psychol. 1980;36(2):372–7.
97. Riley WT, Treiber FA, Woods MG. Anger and hostility in depression. J Nerv Ment Dis. 1989;177(11):668–74.
98. Posternak MA, Zimmerman M. Anger and aggression in psychiatric outpatients. J Clin Psychiatry. 2002;63(8):665–72.
99. Sato T, et al. Frequency of manic symptoms during a depressive episode and unipolar 'depressive mixed state' as bipolar spectrum. Acta Psychiatr Scand. 2003;107(4):268–74.
100. Koh KB, Kim CH, Park JK. Predominance of anger in depressive disorders compared with anxiety disorders and somatoform disorders. J Clin Psychiatry. 2002;63(6):486–92.
101. Piacentino D, et al. The Association of anger with symptom subtypes in severe obsessive-compulsive disorder outpatients. Psychopathology. 2016;49(1):40–6.
102. Carni S, et al. Intrapsychic and interpersonal guilt: a critical review of the recent literature. Cogn Process. 2013;14(4):333–46.
103. McElroy SL, Phillips KA, Keck PE Jr. Obsessive compulsive spectrum disorder. J Clin Psychiatry. 1994;55(Suppl):33–51. discussion 52-3
104. Hansen DJ, St Lawrence JS, Christoff KA. Effects of interpersonal problem-solving training with chronic aftercare patients on problem-solving component skills and effectiveness of solutions. J Consult Clin Psychol. 1985;53:167–74.
105. Shure MB, Spivack G. Interpersonal problem solving in young children: a cognitive approach to prevention. Am J Community Psychol. 1982;10:341–56.
106. Intagliata J. Increasing the interpersonal problem-solving skills of an alcoholic population. J Consult Clin Psychol. 1978;46:489–98.
107. Platt JJ, McKim P, Husband SD. TIPS: training in interpersonal problem solving: enhanced health promotion and AIDS prevention program. Trainer's Manual: Camden, University of Medicine and Dentistry of New Jersey; 1990.
108. Baer RA, Nietzel MT. Cognitive and behavioral treatment of impulsivity in children: a meta-analytic review of the outcome literature. J Clin Child Psychol. 1991;20:400–12.
109. Edmondson CB, Conger JC. A review of treatment efficacy for individuals with anger problems: conceptual, assessment, and methodological issues. Clin Psychol Rev. 1996;16(3):251–75.
110. Del Vecchio T, O'Leary KD. Effectiveness of anger treatments for specific anger problems: a meta-analytic review. Clin Psychol Rev. 2004;24(1):15–34.
111. Saini M. A meta-analysis of the psychological treatment of anger: developing guidelines for evidence-based practice. J Am Acad Psychiatry Law. 2009;37(4):473–88.

112. Paul GL. Strategy of outcome research in psychotherapy. J Consult Psychol. 1967;31(2):109–18.
113. Marchand WR. Neural mechanisms of mindfulness and meditation: evidence from neuroimaging studies. World J Radiol. 2014;6(7):471–9.
114. Fournier JC, Price RB. Psychotherapy and neuroimaging. Focus (Am Psychiatr Publ). 2014;12(3):290–8.
115. Malone RP, et al. A double-blind placebo-controlled study of lithium in hospitalized aggressive children and adolescents with conduct disorder. Arch Gen Psychiatry. 2000;57(7):649–54.
116. Campbell M, et al. Lithium in hospitalized aggressive children with conduct disorder: a double-blind and placebo-controlled study. J Am Acad Child Adolesc Psychiatry. 1995;34(4):445–53.
117. Tariot PN, et al. Efficacy and tolerability of carbamazepine for agitation and aggression in dementia. Am J Psychiatry. 1998;155(1):54–61.
118. Donovan SJ, et al. Divalproex treatment for youth with explosive temper and mood lability: a double-blind, placebo-controlled crossover design. Am J Psychiatry. 2000;157(5):818–20.
119. Goldstein MR, Mascitelli L. Is violence in part a lithium deficiency state? Med Hypotheses. 2016;89:40–2.
120. Coccaro EF, Kavoussi RJ. Fluoxetine and impulsive aggressive behavior in personality-disordered subjects. Arch Gen Psychiatry. 1997;54(12):1081–8.
121. McDougle CJ, et al. A double-blind, placebo-controlled study of fluvoxamine in adults with autistic disorder. Arch Gen Psychiatry. 1996;53(11):1001–8.
122. Vartiainen H, et al. Citalopram, a selective serotonin reuptake inhibitor, in the treatment of aggression in schizophrenia. Acta Psychiatr Scand. 1995;91(5):348–51.
123. Nischal A, Tripathi A, Trivedi JK. Suicide and antidepressants: what current evidence indicates. Mens Sana Monogr. 2012;10(1):33–44.
124. Katz IR, et al. Comparison of risperidone and placebo for psychosis and behavioral disturbances associated with dementia: a randomized, double-blind trial. Risperidone Study Group. J Clin Psychiatry. 1999;60(2):107–15.
125. De Deyn PP, et al. A randomized trial of risperidone, placebo, and haloperidol for behavioral symptoms of dementia. Neurology. 1999;53(5):946–55.
126. McDougle CJ, et al. A double-blind, placebo-controlled study of risperidone in adults with autistic disorder and other pervasive developmental disorders. Arch Gen Psychiatry. 1998;55(7):633–41.
127. Findling RL, et al. A double-blind pilot study of risperidone in the treatment of conduct disorder. J Am Acad Child Adolesc Psychiatry. 2000;39(4):509–16.

8 The Obsessive-Compulsive Dimension

Ines Taddei, Martina Valentini, and Massimo Pasquini

8.1 Introduction

Obsessive-compulsive disorder (OCD) has been an ancient companion of humans. The tendency to seek psychopathological concordance between OCD and disorders that demonstrate peculiar affinity with it can be traced back to comments by classical authors. Janet described the obsessive character using a large number of psychopathological phenomena including the "*obsession de la honte du corps*" (obsession of the shame of their body), due to dysmorphophobia; motor symptoms, which he defined as "forced agitations" not unlike the expression of a Tourette disorder; and experiences of depersonalisation. Dysmorphophobia was described by Kraepelin [1] as "compulsive neurosis", emphasising the iterative and ego-dystonic nature of the symptoms.

Jaspers [2] highlighted a continuity of obsessive-compulsive symptoms and "impulsive actions," suggesting common psychopathological roots for OCD and impulse control disorders. Krafft-Ebing [3] in Germany used the term Zwangsvorstellung to define invasive and irresistible thoughts that are oppressive (Zwang) representations (Vorstellung). Westphal [4] introduced the concept of compulsions as secondary to obsessive ideas, and obsessive-compulsive suffering in France was described as *folie lucide*, because the patient was quite aware of the wrong and anomalous recurring ideas besieging his mind [5]. Falret [6] introduced the word obsession from the Latin *obsidere*. Early in the nineteenth century, the term appeared in British and American medical literature [7].

In more recent years, the renewed interest in OCD has led researchers to "drill down" in an attempt to define its subtypes and explore other psychiatric and neuropsychiatric disorders that might have clinical and/or aetiopathological links with OCD. Disorders that are posited to be linked to OCD, based on their similarities

I. Taddei (✉) · M. Valentini · M. Pasquini
Department of Human Neurosciences, Sapienza University of Rome, Rome, Italy
e-mail: massimo.pasquini@uniroma1.it

M. Biondi et al. (eds.), *Dimensional Psychopathology*,
https://doi.org/10.1007/978-3-319-78202-7_8

with OCD in a variety of domains, are referred to as "OC spectrum disorders" or "OC continuum" [8, 9]. On the other hand, OCD has undergone significant changes in its classification within categorical systems of mental disorders. The Diagnostic and Statistical Manual (DSM) III and DSM IV included OCD within the category of anxiety disorders, conceiving it as within the anxiety disorders group because of the often very high levels of anxiety in clinical OCD patients, along with the common symptoms of distressing obsessive thoughts and compulsive behaviours [10]. Growing evidence has finally led to a separation of OCD from the other anxiety disorders and to the decision by the DSM-5 task force to classify it within its own separate category: "obsessive-compulsive and related disorders" [11]. Today, the DSM-5 [11] recognises obsessive-compulsive and related disorders (OCRDs) as a new diagnostic category. The category includes OCD, body dysmorphic disorder (BDD), trichotillomania (TTM, hair-pulling disorder), excoriation disorder (skin picking), and hoarding disorder. OCD involves the experience of obsessions (recurrent, distressing intrusive thoughts, images, or urges) and compulsions (ritualised behaviours completed to reduce distress from obsessions). BDD involves an excessive, distressing, and time-consuming preoccupation with an imagined appearance flaw and repetitive rituals performed in response to this preoccupation. TTM is characterised by recurrent hair pulling, resulting in hair loss. Excoriation disorder involves recurrent skin picking, resulting in lesions. Finally, hoarding disorder involves persistent difficulty parting with possessions and a perception that items must be saved, resulting in clutter in one's active living space that reduces the space's usability [11].

The grouping of these conditions is based on their phenomenological similarities to OCD (i.e. obsessive thinking and/or compulsive behaviours), as well as similarity to OCD in course of illness, comorbidity, family history patterns, biological abnormalities, and treatment responses. The research planning agenda for DSM-5 examined possible similarities in phenomenology, comorbidity, familial and genetic features, brain circuitry, and treatment response between OCD and several related disorders that are characterised by repetitive thoughts or behaviours. Certain disorders, such as BDD, obsessive-compulsive personality disorder (OCPD), Tourette syndrome (TS), and TTM, share many commonalities with OCD in phenomenology, comorbidity, familial and genetic features, brain circuitry, and treatment response. Other disorders, such as the impulse control disorders (ICDs), share some common features with OCD but also differ in many ways as well.

According to most epidemiological studies, OCD is currently considered a relatively rare disorder, with a weighted 1-month prevalence of 1.1% in the British National Comorbidity Survey [12]. Similar results were reported by Crino et al. [13] who estimated the 12-month prevalence of DSM-IV OCD to be 0.6%.

Although many advances have been made with regard to the aetiopathogenesis and treatment of OCD, there are still no correct estimates of the prevalence of the disorder in the community. The variation in the different studies is mostly due to limitations in methodology and inconsistencies between lay and clinical diagnosis [14]. Further regarding prevalence in the general population, data obtained from a prospective longitudinal study of an unselected birth cohort has shown that 21–25%

of individuals from the general population exhibit obsessions and/or compulsions, as defined in the DSM-IV, although only 2–3% meet full diagnostic criteria for the disorder [15].

The aim of this chapter is to disentangle obsessive-compulsive manifestations in psychiatric subjects not affected by OCD. In fact, obsessions and compulsions may range from physiologic/transient or normal presentation, to a pathological degree of suffering [16].

The continuum hypothesis, proposed by Clark and Rhyno [17], has been challenged by recent literature and regarded as an oversimplified interpretation of a more complex phenomenon of obsessions and compulsions, characterised not simply by quantitative but by qualitative differences as well, related in particular to the severity of the content of obsessions, triggers, appraisals, and responses. Berry and Laskey [18] argue for a revised continuum model of intrusive thoughts that incorporates the above-mentioned differences in obsessions, with particular reference to content and its severity, with clinically obsessive individuals reporting more bizarre and aggressive thoughts than individuals with symptoms too mild to be considered clinically obsessive.

Over the last 20 years, the concept of spectrum has received strong impetus in international psychiatry literature. The "spectrum" attributed to a psychopathological phenomenon may have various meanings. Strictly speaking, it should refer to a set of disturbances that, beyond possible heterogeneity of symptomatic manifestations, share the same aetiological determinants or similar pathogenic mechanisms. There are a number of reasons for this interest in "spectrum". First, the identification of a psychopathological spectrum can guide the search for the common aetiopathogenic mechanisms underlying it. Second, this could be the first step in a common therapeutic approach. The finding of concordance between disorders, in various nosographic environments, does not relate only to psychopathological criteria but extends to common aetiopathogenic hypotheses, thus constituting a clinical picture with expected response to specific therapeutic treatments.

OCD is in a favourable position from this point of view. In fact, among psychopathological phenomena, obsessions and compulsions have a high level of consensus regarding their definition criteria. Moreover, this is one of the few cases in which the dominant presence of a single symptom allows the identification of a "syndrome" disorder. Despite this, the concept of the obsessive-compulsive (OC) spectrum has been discussed in the literature and increasingly studied, with contrasting evidence depending on the starting hypothesis (5HT system involvement and/or basal ganglia-thalamus-cortex-frontal circuitry).

Finally, the term "spectrum" has been used to mean many issues, as well as many disorders, that have a degree of similarity, at least in terms of symptoms, to OCD. The list includes BDD, hypochondriasis, trichotillomania, the eating disorders, autism spectrum disorders, and several other impulse control disorders, such as pathologic gambling and kleptomania. Other psychiatric disorders, such as depersonalisation disorder, borderline personality disorder, sexual compulsions, and paraphilias, and some neurological disorders such as Tourette's syndrome, Sydenham's chorea, and parkinsonism have been included in this spectrum.

8.2 Neurobiology of Obsessions and Compulsions

Initially, OCD was considered a primary psychiatric disorder. Subsequently, clinical observation increasingly showed that this disorder has a neurodevelopmental basis.

The first observation of the existence of brain alterations in obsessive patients resulted from the finding that patients with neurological dysfunctions, such as streptococcal infection, head trauma, or encephalitis, as well as comorbid tic disorders such as Tourette's syndrome, often develop obsessive phenomena [19–21]. Several studies have reported an abnormally high prevalence of neurological soft signs (NSS) in patients with obsessive-compulsive disorder as compared to healthy people [8, 22].

The classical cortico-striato-thalamo-cortical (CSTC) model of OCD was established a few decades ago. Thanks to modern brain imaging methods, neurocircuitry models of OCD and related disorders have been refined to a higher level of complexity, with possible promising implications.

In addition to cortico-striatal circuitry, which is now regarded as interconnected rather than segregated, the lateral orbitofrontal cortex (OFC) might mediate obsessions, and the dorsal anterior cingulated cortex (dACC) might be implicated in fear expression and conditioning, as well as in aberrant error monitoring, in OCD. On the basis of such a conceptual framework, Milad and Rauch [23] proposed a testable hypothesis on how dysfunctions in these areas might be linked to fear inhibition and severity of symptoms. This information, in turn, could be used to predict treatment response.

Subsequent studies using functional magnetic resonance imaging (fMRI), positron emission tomography (PET), and single photon emission computed tomography (SPECT), as well as other more sophisticated techniques, such as voxel-based morphometry (VBM), diffusion tensor imaging (DTI), and proton magnetic resonance spectroscopy (1H-MRS), have compared OCD patients with healthy controls. These studies have tended to confirm the key role of the CSTC and its connections with the limbic system (amygdalo-cortical circuitry) in the pathogenesis of OCD [24, 25].

The above-mentioned functional imaging studies generally showed that patients with OCD have reduced grey matter density in the dorsolateral prefrontal cortex (DLPFC) and OFC and reduced volume of the anterior cingulate cortex (ACC) and OFC [26, 27]. Further, patients with OCD showed hyperactivity in the area of the OFC, caudate nucleus, and thalamus, compared with healthy controls [28–30]. PET studies of OCD patients have supported these findings, showing changes in brain metabolism, more or less in the same areas as previously described (head of the caudate nucleus and the orbital gyrus, OFC, and prefrontal cortex, which are a part of the CSTC circuit) [31]. Furthermore, the increased glucose metabolism along orbitofrontal-basal ganglia-thalamo-cortical circuits measured by PET generally decreases after treatment with antidepressants [32–35] and/or psychotherapy [36–38].

The empirical demonstration of the effectiveness of antidepressants (especially those blocking serotonin reuptake) for OCD dates back to more than 30 years ago. The new neuroimaging approaches, allowing the examination of regional cerebral neurochemistry and permitting in vivo quantification of specific neurochemicals in various brain regions, have allowed greater clarification of the role of serotonin in OCD.

Neurochemistry techniques have been used in the study of the transport of neurotransmitters, in order to better explain the role of serotonin in OCD. These studies have provided an increased understanding of the biology underlying the well-known empirical evidence that antidepressant drugs blocking serotonin reuptake improve obsessive symptoms over time. Also, they indicated not only the involvement of the serotonin system but also, albeit to a lesser extent, the involvement of dopamine [39].

These studies, mainly conducted using SPECT, allowed not only the confirmation of the role of serotonin and other neurotransmitters in OCD but also the examination of changes following treatment through cerebral metabolic monitoring [37, 38]. They also facilitated the development of hypotheses about both pharmacological and psychotherapeutic models that are potentially valid and useful in clinical practice [40].

As evidenced by early studies using PET and SPECT, hyperactivity of the head of the caudate nucleus and the orbital gyrus may be the source of complex obsessive symptomatology. Modern functional neuroimaging methods permit the establishment of a relationship between cerebral activity and a particular symptom. This allows comparison of cerebral activation in the presence of the symptoms of interest with corresponding activation in the same subject when these symptoms are resolved [41].

Several functional neuroimaging studies report that the cortico-striatal-thalamo-cortical (CSTC) circuit is dysfunctional in obsessive patients. This dysfunctional CSTC circuit consists mainly of altered communication between the lateral OFC and the ventral striatum [23, 42]. According to some authors, the "dysfunction" of this circuit would be hyperactivity during the resting or neutral state, which would increase during the onset of the symptoms, subsequently attenuating later pharmacological and behavioural treatment [32, 43, 44].

Studies by fMRI in OCD patients exploring brain metabolism alterations during a brain activation task report the involvement of the same areas that make up the CSTC circuit [30, 45, 46]. Most studies have examined these various areas, each of which, stimulated with specific tasks, produces a different result of increased, decreased, or stable glucose metabolism [47, 48].

Other studies have investigated possible changes in brain activation before and after drug therapy and/or psychotherapy [49, 50]. These neuroimaging studies have also, more sophisticatedly, sought to correlate the therapeutic response with neuropsychological tests, with the aim of finding predictive response indices. The resulting neuropsychological and treatment findings further support neuromorphological data, obtained through several different methods, implicating the role of CSTC circuitry in OCD pathophysiology. It has been mentioned several times before that OCD symptoms are manifested through a hyperactivity of the regions involved in the CSTC circuitry (coronary orbital cortex, cortical cortege, thalamus, and head of the caudate nucleus), causing a "malfunction" of this circuitry. Such a malfunction has been proposed to be the origin of the intrusive symptoms and neuropsychological dysfunctions [22, 49].

Regarding molecular imaging studies that have considered the role of neurotransmitters in brain sites involved in OCD obsessive phenomena, the results obtained, though encouraging, are limited due to small sample sizes and mixed diagnoses.

In approximately one third of patients with OCD, standard treatment with selective serotonin reuptake inhibitors (SSRIs) fails to bring satisfactory relief of obsessive-compulsive symptoms. Atypical antipsychotics can augment the effectiveness of SSRIs in such patients, but the mechanism underlying this synergetic effect still needs to be better clarified [51–53].

Another important aspect of OCD and related disorders, likely to be a common trait across specific diagnostic categories, has been recently considered and studied: deficits in goal-directed control. Deficits of this type have been observed across the spectrum of OCD and related disorders, for example, OCD [54–56], drug addiction [57, 58], and binge eating disorder [59]. These deficits are associated with dysfunction of the caudate and medial orbitofrontal cortex. Goal-directed control permits deliberate behaviour regulation. Its deficits lead to lack of flexibility; more rigid habits are induced; and repetition of thoughts and actions is enhanced. Gillan et al. looked at this trait as a clue to a transdiagnostic compulsivity dimension and, in two large general population samples, found it to be associated with (1) OCD symptoms and (2) symptoms of other DSM diagnostic categories. Exactly these kinds of studies reflect the concept of dimensional psychopathology as suggested by Pancheri [60].

Obsessive phenomena may appear in other psychiatric disorders, in comorbidity, and occasionally even in healthy subjects. Therefore, it is conceivable that these obsessive phenomena share at least part of the same pathogenic mechanism.

The empirical demonstration of the effectiveness of antidepressants, especially serotonin reuptake blockers, on OCD and, consequently, also on obsessive phenomena can be traced back to 30 years ago.

8.3 Dimensional Psychopathology of Obsessions

A psychopathological dimension is defined as an alteration of psychic function phenomenologically expressed by symptoms (referred to by the patient) or by signs (observed) that are indicative of and specific to the observed function [61].

Obsessions and compulsions, like any other psychiatric symptoms, can appear as psychopathologically dominant and exclusive elements, or they can be associated with a psychopathological framework in which one or more symptoms, and possibly a cluster of other symptoms, characterise a specific clinical picture. In the first case, "symptom" and "disorder" coincide and qualify as the OCD. In other conditions, obsessive symptoms occur in association with, or in the presence of, other symptoms within the main disorder (e.g. schizophrenic episode) or in comorbidity.

The dimensional model posits that obsession-compulsion may, however, manifest itself with a certain frequency in association with typical symptoms of a given syndrome (e.g. depressive episode). In this case, the obsessive phenomenon must be considered as a symptom, even if an atypical one, that takes on particular characteristics in relation to the disorder in which it is co-appearing. In other words, the basis of the obsessive symptom probably follows pathogenic mechanisms and has neurobiological, genetic, neuroanatomical, neurochemical, or psychophysiological correlates, stemming from the disturbance with which it is associated. This means that, rather than the obsessive symptom mechanisms being separate from the mechanisms underlying the main disorder, the different mechanisms can influence and trigger each other in the context of a particular clinical presentation. Obsession, therefore, understood as a symptom with a specific pathogenic mechanism with its own psychopathological manifestations, can appear in many psychiatric disorders with sometimes-different phenomenological characteristics. Taking this into account, in the next analysis, transnosographic aspects of this dimension will be fully explored.

8.4 Obsessions and Schizophrenia

The frequency of obsessive symptoms in schizophrenia seems to be higher than in the general population. The percentage of schizophrenic patients having "obsessive-compulsive symptoms" reported in the literature fluctuates between 30% and 59% [62], and the frequency of OCD comorbidity in patients affected by schizophrenia has been estimated to range from approximately 8 to 23% [63]. In contrast, the appearance of schizophrenic symptoms in the course of OCD seems to be relatively rare and only slightly higher than the incidence of schizophrenia in the general population [64]. These data indicate that the schizophrenic disorder could present a vulnerability to obsessions, but not vice-versa. It has even been proposed that the presence of obsessive-compulsive symptoms could be an extreme mechanism of defence that takes place during the initial phase of the psychotic disorder [65].

Data suggest that, because of the involvement of similar functional brain networks (e.g. frontal cortex, basal ganglia) and neurotransmission systems (serotonin and dopamine), in these two disorders, it may be possible to identify a subgroup, designated schizo-obsessive [66]. However, it is not yet clear whether the presence of obsessive symptoms in schizophrenic patients is an indicator of good prognosis [67, 68] or, as has been found in the latest literature, of poor prognosis (in terms of clinical and functional outcomes, including responsiveness to treatment) [69–71]. It should be emphasised that these studies are affected by certain limitations related to sample inequalities and methodologies used.

Some studies have investigated obsessive-compulsive symptoms induced or markedly aggravated by second-generation antipsychotic treatment, particularly with clozapine (most frequent), olanzapine, and risperidone [72]. These effects, however uncomfortable and poorly tolerated by patients, must always be evaluated in the context of the overall intervention. It should be remembered that, especially in the case of clozapine, this is a drug used in resistant schizophrenia and one that is also effective for suicidal ideation [73]. Further, it is often used in monotherapy. Therefore, before suspending treatment, all possible caution should be exercised. In this regard, various strategies for augmentation with anti-obsessive drugs have been proposed, before discontinuing clozapine, and have also been shown to be effective (aripiprazole and amisulpride) [74, 75]. Alternatively, CBT has also proved effective in some schizophrenic patients with obsessive-compulsive symptoms who did not respond well to the various pharmacological protocols [76].

8.5 Obsessions and Mood Disorders

The presence of obsessive-compulsive symptoms in mood disorders has been recognised especially in depression. Regarding bipolar disorder, obsessive-compulsive symptoms manifest themselves more often during depressive episodes and almost never during manic episodes. There is an apparent comorbidity between bipolar disorder and OCD: among OCD patients, 18.3% also have bipolar disorder, and among bipolar patients, 17% also have OCD. However, studies have shown that obsessive-compulsive symptoms occur frequently in depressive patients as well, with an approximate incidence of 20–40% [77].

This suggests that when there is a depressive disorder, obsessions can be, as in the case of schizophrenia, a defensive factor that appears to ease the suffering of the depressed patient. In the case of depression, an additional predisposing factor seems to be represented by a personality with traits or features of "obsessive" type. In contrast to what may happen in schizophrenic disorder, the obsessive-compulsive symptoms in depression keep their fundamental characteristics, especially with respect to awareness of illness.

8.6 Obsessions and Eating Disorders

In light of the new definition of eating disorders that emerged in the DSM-5 and new findings on their aetiopathogenesis, their relationship with OCD has also gained new significance. While earlier emphasis was placed on personality aspects, emphasising obsessive personality traits or comorbid OCD in patients with anorexia nervosa, the concept of spectrum in this group of diseases is currently spreading and becoming increasingly accepted. In line with this, eating disorders can be considered as a spectrum within which, in addition to the more classical anorexia and bulimia nervosa, there are a number of other new disorders or behavioural alterations, including anorexia, "reverse anorexia" or "bigorexia" (i.e. muscle dysmorphia, a phenomenon started among professional athletes and now rising in the general population) [78], "diabulimia" (an eating disorder specific to patients with diabetes, characterised by limiting insulin treatment to lose weight through sustained hyperglycaemia) [79], "drunkorexia" (restricting food intake prior to drinking alcohol to avoid weight) [80], and "pregorexia" (pregnant women who will reduce calories and exercise in excess in an effort to control pregnancy weight gain) [81]. Many of these eating disorders show strong affinity with OCD, in light of the well-known exercise of control. One condition in particular, which is not entirely pathological, is orthorexia nervosa. According to some authors, it appears that, aside from anorexia nervosa, obsessive phenomena seem to be most marked in this disorder [82]. The term orthorexia nervosa describes people whose extreme diets, intended for health reasons, end up leading to malnutrition and/or impairment of daily functioning. Although studies on this are still few and with different methodological limits, it turns out that these patients, apart from poor insight, share many features with OCD and obsessive personality disorder, as anorexia nervosa does [83, 84]. The phenomenological aspects these disorders share include perfectionism, rigid thinking, preoccupation with details and perceived rules, and high levels of anxiety [85]. However, it is important to remember that there is a significant difference between this continuum and OCD, represented by the ego-syntonic experience present in eating disorders and usually absent in OCD. Nevertheless, these data give rise to new questions and further insights [86].

8.7 Treatments

The available research indicates that evidence-based psychotherapy options targeting the Obsessiveness dimension include Exposure with Response Prevention (ERP) and cognitive behavioural therapy (CBT) [87]. An interesting study suggests that basolateral amygdala-ventromedial prefrontal cortex connectivity seems to predict CBT outcome in obsessive-compulsive patients. Although this is reported only by one study, this data suggests that this cerebral structure might be a target of CBT for Obsessiveness [88].

Third-wave cognitive-behavioural therapies for OCD have become more popular in the past few years. These are therapies such as ACT (acceptance and commitment therapy) and MBCT (mindfulness-based cognitive therapy), which encompass cultivating a different relationship with the symptoms, in which the patient is asked to observe the manifestation of the symptoms and accept them as they are. The hypothesised mechanism underlying these therapies is that the maintenance of the symptoms is curtailed as the patient assumes a decentralised perspective towards them. Currently, there is no neuroimaging evidence on the circuits or brain areas modulated by third-wave CBT targeting the Obsessiveness dimension, but it is reasonable to argue that these psychotherapeutic treatments target brain structures and circuits that are involved in the manifestation of obsessive and compulsive behaviour. Despite the good outcome of these therapies in targeting the Obsessiveness dimension, there is currently too little evidence in the literature to support their choice as a first-line therapy. However, they are good candidates for add-on therapies to complement classical CBT and exposure with response prevention [89].

As concerns other available treatments targeting the Obsessiveness dimension, novel neurostimulatory techniques are able to target the cortico-striato-thalamo-cortical loop, involved in obsessive and compulsive behaviour. In particular, these techniques include deep brain stimulation (DBS), repetitive transcranial magnetic stimulation (rTMS), and transcranial direct current stimulation (tDCS), which all can elicit changes in specific brain regions implicated in obsessive and compulsive symptoms. A growing body of literature is showing that brain stimulation techniques can improve symptoms by modulating cortico-striatal circuit activity [90,

91]. DBS targeting the nucleus accumbens (NAcc) reduced anxiety and obsessive symptoms in subjects with OCD and enhanced libido [92]. Even DBS of the ventral caudate led in some cases to an improvement of OCD symptoms [93]. Two studies have shown that the application of inhibitory rTMS to both the left and right lateral orbitofrontal cortex (lOFC) led to a significant improvement of obsessive and compulsive symptoms, and this improvement was associated with a decrease in local metabolism of lOFC [94, 95]. Only two studies investigated tDCS targeting the OFC: Mondino et al. [96] showed a decrease of OCD symptomatology that was maintained after a 1-month follow-up. The protocol consisted of ten twice-daily sessions of tDCS inhibitory stimulation of the left OFC. Bation et al. [97] used the same protocol as Mondino et al. and reported similar results. There is no evidence that electroconvulsive therapy (ECT) improves obsessive or compulsive symptoms [98].

These studies are still in their infancy, but the preliminary results and preclinical trials appear promising in their findings, mostly with regard to resistant symptoms that exist transdiagnostically across traditional categories of psychiatric suffering.

8.8 Findings

The SVARAD describes the Obsessiveness dimension as "Doubtfulness, rigidity, meticulousness, perfectionism; repetitive behaviours aimed at preventing, checking, controlling; presence of obsessions, compulsions". As described in the previous chapters, each item of the SVARAD is scored on a 5-point rating scale (from 0 to 4), where the highest score represents a profound presence of "invasive obsessions and compulsions, present for the vast majority of the day, non-controllable, with impairment of the social and occupational activities". Among the SVARAD dimensions, Obsessiveness is probably the one that is most closely linked to a specific diagnostic category. However, it was introduced into dimensional assessment because clinical experience has suggested that non-OCD disturbances often have an Obsessiveness component or manifestation: for example, body dysmorphic disorder (BDD), trichotillomania, and hoarding disorder. SVARAD studies reported Obsessiveness findings from both the inpatient and outpatient groups of our sample. We present here data concerning our diagnostic groups, exploring how the SVARAD Obsessiveness dimension is present as a transdiagnostic feature (Table 8.1).

This section reviews Obsessiveness findings from both the inpatient and outpatient groups of our sample.

The SVARAD Obsessiveness dimension shows low mean scores in the psychiatric inpatient group as a whole (Fig. 8.1), as well as in the outpatient sample (Fig. 8.2).

High scores on the SVARAD Obsessiveness dimension in the inpatient group were most highly correlated with OCD, followed by schizophrenia, bipolar disorder, depressive episode, major depressive disorder, and psychotic disorder NOS. Figure 8.3 shows the mean (±standard deviation) value of the SVARAD Obsessiveness dimension in the inpatient sample for several DSM-IV-TR diagnostic categories. The highest mean values were observed in OCD (2.72 ± 1.48), followed by schizophrenia

Table 8.1 Descriptive phenomenology of the Obsessiveness SVARAD dimension. Several properties of the SVARAD Obsessiveness dimension share phenomenological aspects in these psychiatric disorders

	Insight	Iterativity	Intrusiveness	Persistence	Ego-dystonia	Tension reduction behaviour	Pleasure-driven behaviour	Guilt	Response to SSRI
OCD	+	+	+	+	+	±	–	+	+
Body dysmorphic disorder	–	+	+	+	–	±	–	–	
Hypochondria	–	+	+	+	–	±	–	–	±
Trichotillomania	+	+	+	+	+	+	+	+	±
Compulsive shopping	±	+	+	+	±	+	+	+	±
Anorexia nervosa	–	+	+	+	–	+	+	–	–
Bulimia nervosa	+	+	+	+	+	+	+	+	+
Tic	+	+	+	+	+	+	–	–	+
Tourette syndrome	±	+	+	+	±	+	–	–	±
Hoarding	–	+	+	+	–	–	±	–	–
Intermittent explosive disorder	–	+	±	+	–	+	–	+/-	–
Gambling	+	+	+	+	–	+	+	+	±

Fig. 8.1 SVARAD profile of the whole psychiatric inpatient sample: Mean scores and standard deviations ($N = 847$)

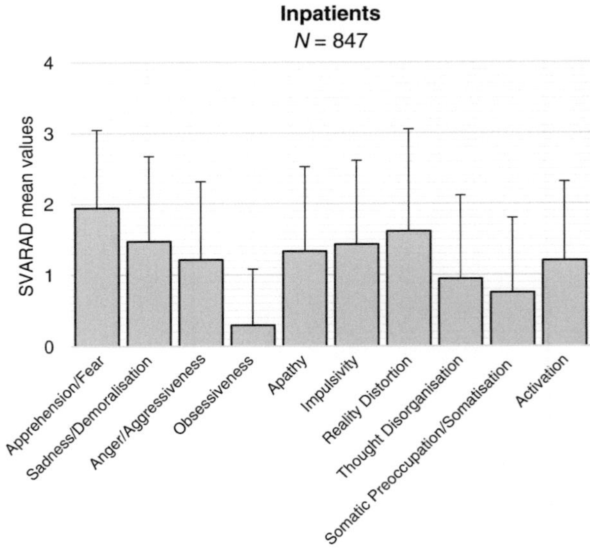

Fig. 8.2 SVARAD profile of the whole psychiatric outpatient sample: Mean scores and standard deviations ($N = 1124$)

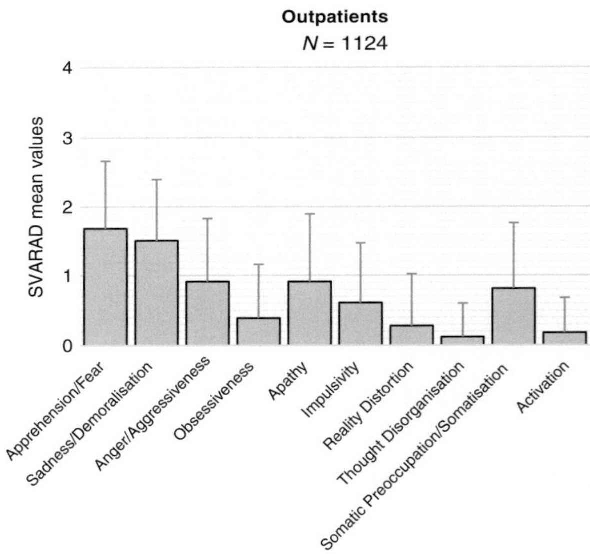

(0.45 ± 0.95), bipolar disorder-depressive episode (0.38 ± 0.87), major depressive disorder (0.36 ± 0.71), and finally, psychotic disorder NOS (0.30 ± 0.86).

It's interesting to point out that, although mean Obsessiveness scores were low throughout the inpatient group, 25.1% of depressive patients (Fig. 8.4), 20% of bipolar-depressive episode patients (Fig. 8.5), and 23.3% of schizophrenic patients (Fig. 8.6) have an Obsessiveness score greater than zero. From a clinical viewpoint, this suggests that taking this dimension into account could lead to better understanding and treatment of one out of four cases.

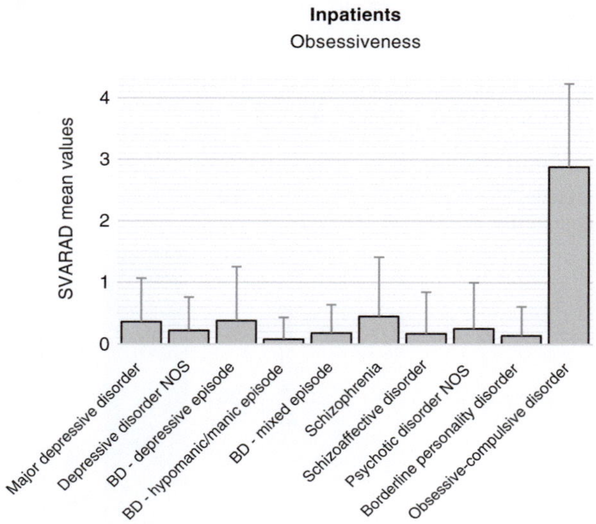

Fig. 8.3 SVARAD Obsessiveness dimension across inpatients' diagnostic categories: Mean values and standard deviations

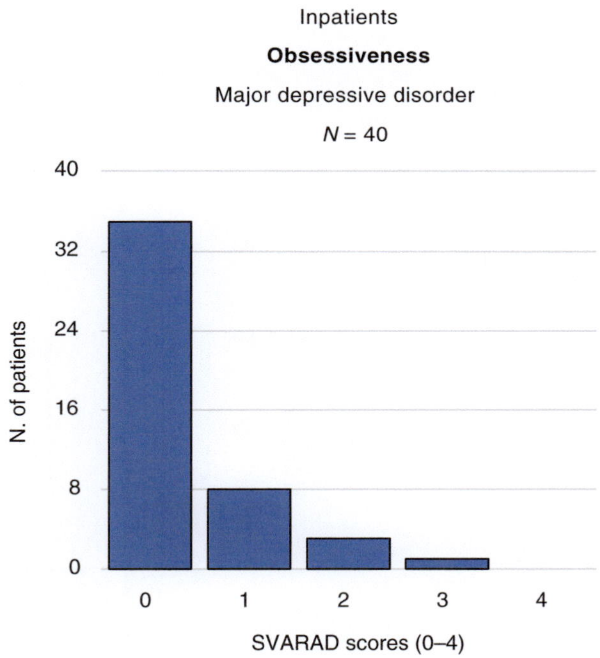

Fig. 8.4 Obsessiveness scores in major depressive disorder patients. Although mean scores are modest, findings suggest that the suffering of 25.1% of the patients includes an obsessive-compulsive component

The mean SVARAD profile for OCD in the inpatient sample displayed a 4-1-2 code type and is shown in Fig. 8.7. Multiparametric analysis showed the highest peak for the Obsessiveness dimension (2.88 ± 1.35), two peaks for the Apprehension/Fear (2.25 ± 1.16) and Sadness/Demoralisation (2.25 ± 1.03) dimensions, and a final peak for the Apathy (2.13 ± 1.45) dimension, all with mean scores above 2. Four out of eight cases had a moderate grade of Apprehension/Fear, and one out of three had a severe/

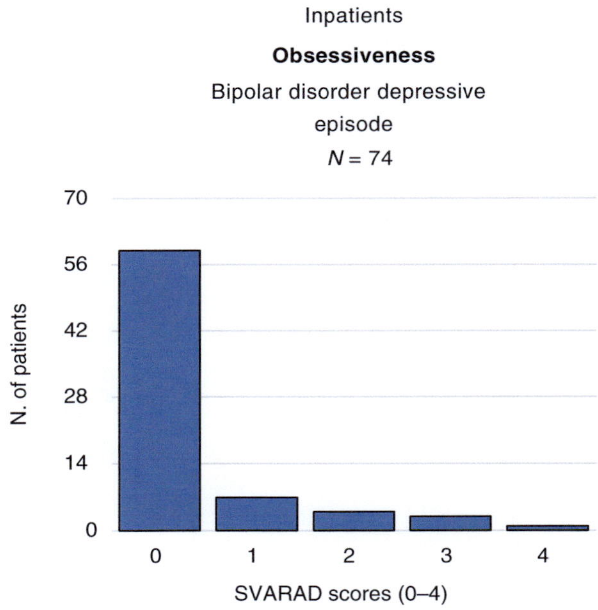

Fig. 8.5 Obsessiveness scores in bipolar disorder-depressive episode patients. Although mean scores are modest, findings suggest that the suffering of 20% of the patients includes an obsessive-compulsive component

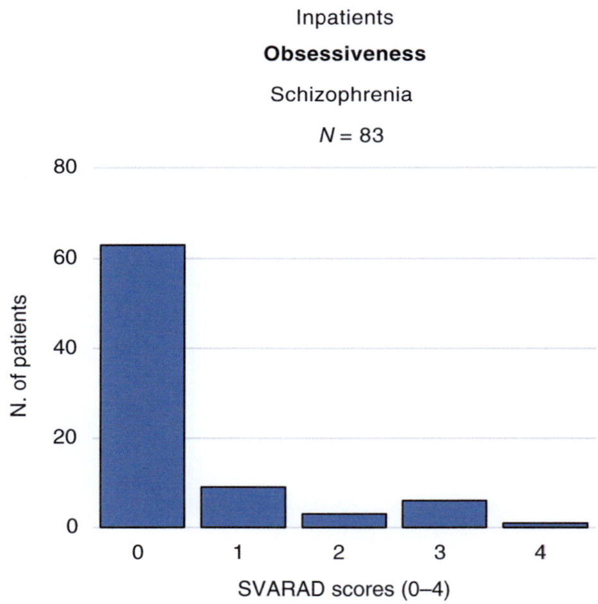

Fig. 8.6 Obsessiveness scores in the schizophrenia inpatient group. Although mean scores are modest, findings suggest that the suffering of 23.3% of the patients includes an obsessive-compulsive component

extreme value for this dimension. One patient out of three had a moderate degree of Sadness/Demoralisation, while one out of three had severe to extreme degrees of this dimension. One half of the group had severe to extreme degrees of Apathy.

Obsessiveness was negatively correlated with Impulsivity (0.636 ± 1.027, Rho = −0.71, $p = 0.038$) and Activation (0.545 ± 1.214, Rho = −0.629, $p = 0.038$).

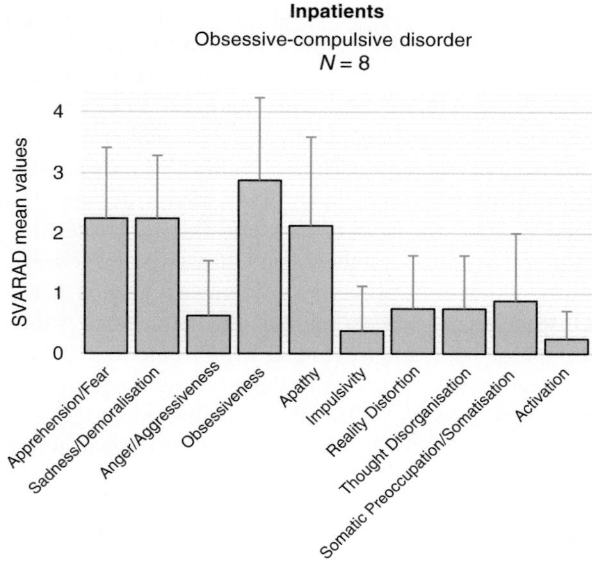

Fig. 8.7 SVARAD profile of inpatients with obsessive-compulsive disorder: Mean values and standard deviations. Code type: 4-1-2 ($N = 8$)

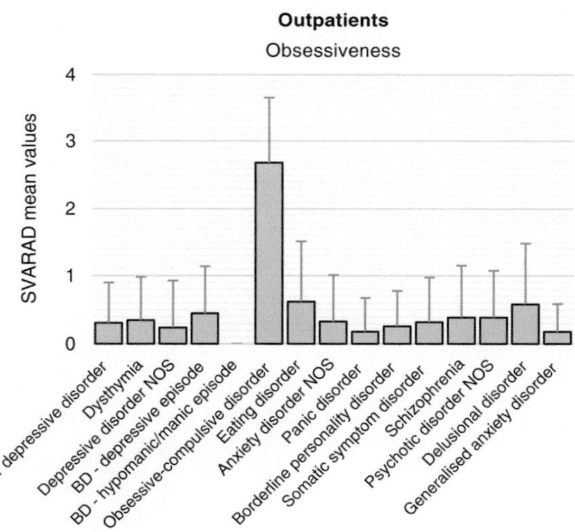

Fig. 8.8 SVARAD Obsessiveness dimension across outpatients' diagnostic categories: Mean scores and standard deviations

Significant positive correlations were also seen between Obsessiveness and Apprehension/Fear (Rho = 0.147, $p < 0.001$), Sadness/Demoralisation (Rho = 0.148, $p < 0.001$), and Apathy (Rho = 0.51, $p < 0.001$).

Regarding the Obsessiveness dimension, values were highest for the OCD DSM-IV diagnostic outpatient group, which displayed mean values just above a score of 2, which is lower than those in the OCD inpatient group, as expected, because of less severe psychopathology in the outpatient group (Fig. 8.8). The highest mean values were observed in OCD (2.68 ± 0.97), followed by eating disorders (0.62 ± 0.89), delusional disorder (0.58 ± 0.90), bipolar disorder-depressive episode (0.45 ± 0.73), and schizophrenia (0.39 ± 0.76).

One patient out of three in the following diagnostic categories had an Obsessiveness score between mild and severe: eating disorders (40%) (Fig. 8.9), bipolar disorder-depressive episode (36%) (Fig. 8.10), delusional disorder (33.8%) (Fig. 8.11), and schizophrenia (30%) (Fig. 8.12). One patient out of four in the following diagnostic categories had an Obsessiveness score between mild and severe: psychotic disorder NOS (27%), dysthymia (27%) (Fig. 8.13), somatoform disorders (26%) (Fig. 8.14), major depressive disorder (25%) (Fig. 8.15), anxiety disorder NOS (24%), and finally, borderline personality disorder (23%) (Fig. 8.16). Although the mean Obsessiveness scores seem low, one quarter to one third of patients of several diagnostic groups had a mild to severe degree of Obsessiveness, indicating that this dimension should be considered for better comprehension and treatment. Mild to moderate Obsessiveness probably predicts to some extent a more problematic relationship with rumination, compulsivity, and doubtfulness, while from a pharmacological perspective, it suggests the choice of 5-HT antidepressants rather than noradrenergic or dopaminergic antidepressants.

The SVARAD dimensional profile of the outpatient OCD group showed a 4-1-2 code type. The Obsessiveness dimension had the highest mean values (2.68 ± 0.98), followed by Apprehension/Fear (2.16 ± 0.86), Sadness/Demoralisation (1.48±0.96), Apathy (0.77 ± 0.92), and Aggressiveness (0.77 ± 0.92) (Fig. 8.17). As Fig. 8.17 shows, more than one half of the group displayed a moderate degree of Apprehension/

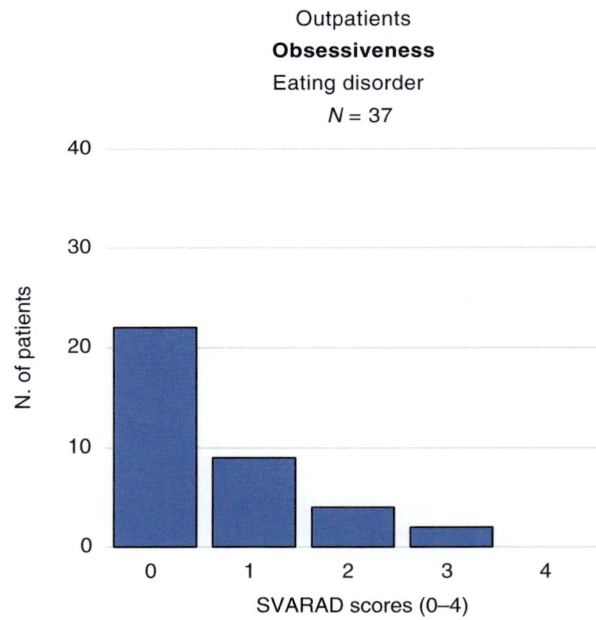

Fig. 8.9 Obsessiveness scores in eating disorders outpatient group. Findings suggest that the suffering of 40% of the patients includes an obsessive-compulsive component

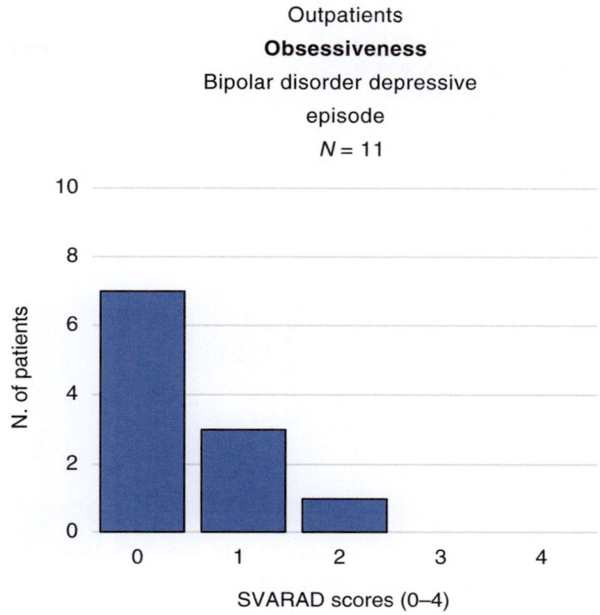

Fig. 8.10 Obsessiveness scores in the bipolar disorder-depressive episode outpatient group. Findings suggest that the suffering of 36% of the patients includes an obsessive-compulsive component of the patients includes an obsessive-compulsive component

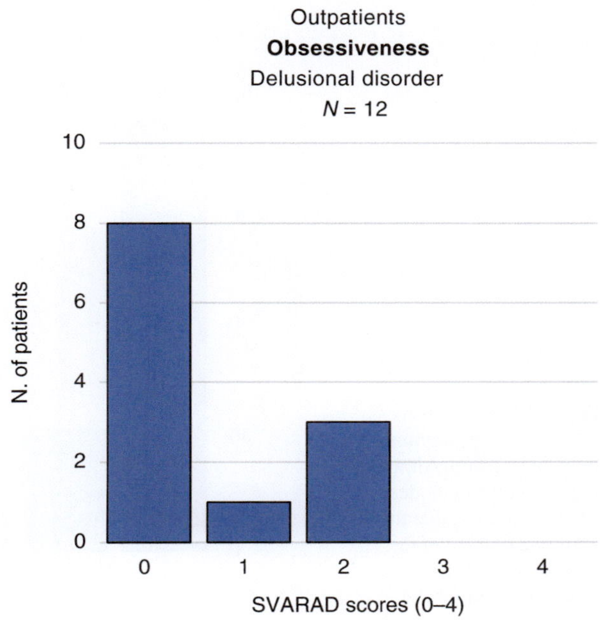

Fig. 8.11 Obsessiveness scores in the delusional disorder outpatient group. Findings suggest that the burden of obsessiveness in the delusional disorder outpatient group account for 33.8%

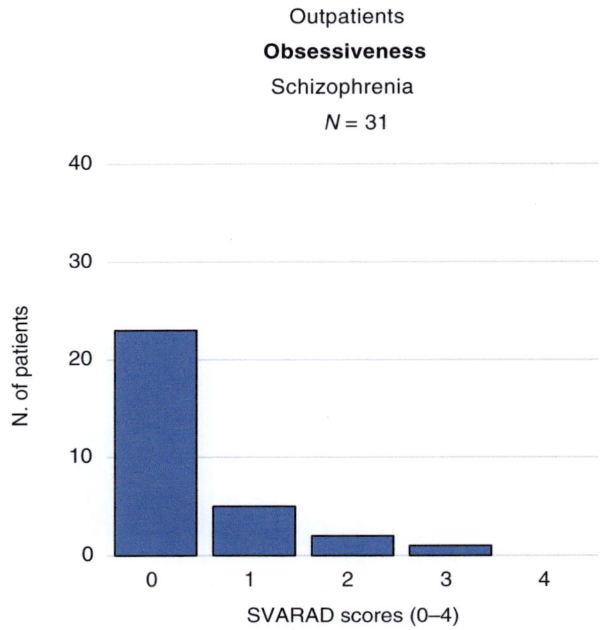

Fig. 8.12 Obsessiveness scores in the schizophrenia outpatient group. Although mean scores are modest, findings suggest that the suffering of 30% of the patients includes an obsessive-compulsive component

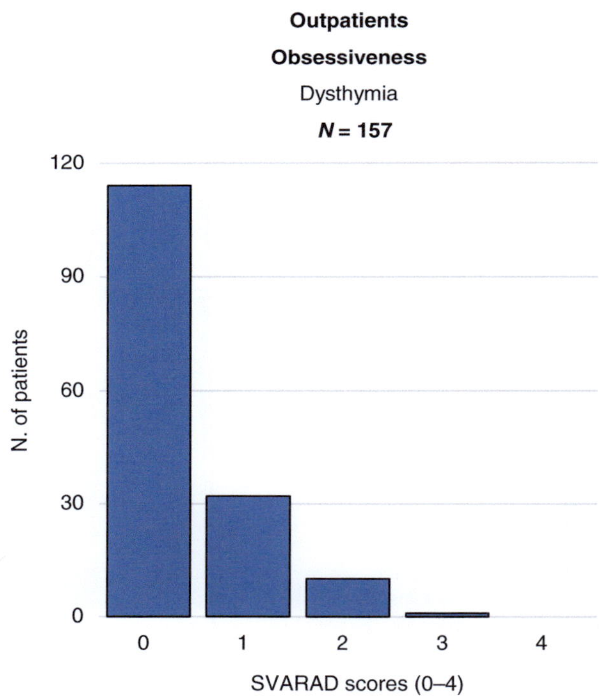

Fig. 8.13 Obsessiveness scores in the dysthymia outpatient group. Although mean scores are modest, findings suggest that the suffering of 27% of the patients includes an obsessive-compulsive component

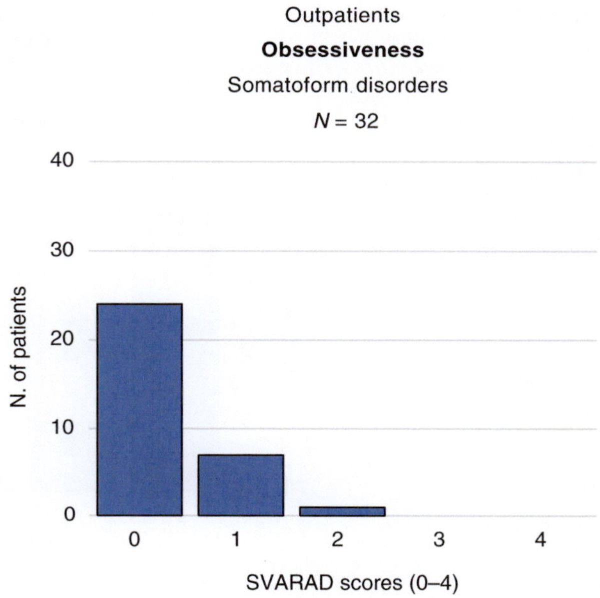

Fig. 8.14 Obsessiveness scores in the somatoform disorders outpatient group. Although mean scores are modest, findings suggest that the suffering of 26% of the patients includes an obsessive-compulsive component

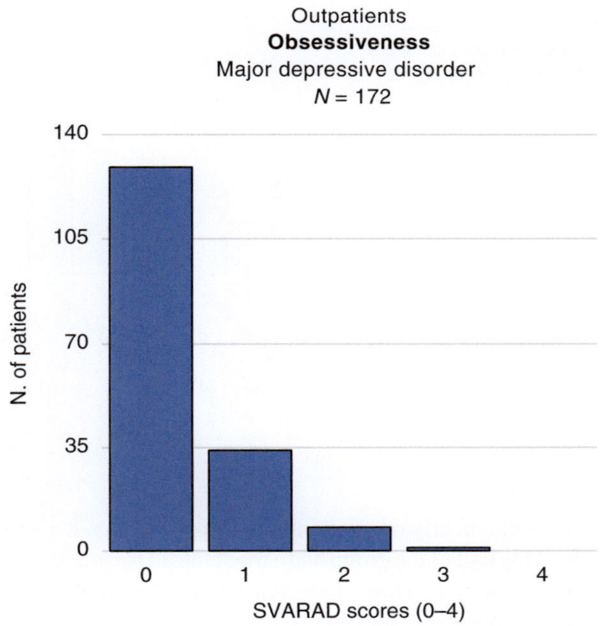

Fig. 8.15 Obsessiveness scores in the major depressive disorder outpatient group. Although mean scores are modest, findings suggest that the suffering of 25% of the patients includes an obsessive-compulsive component

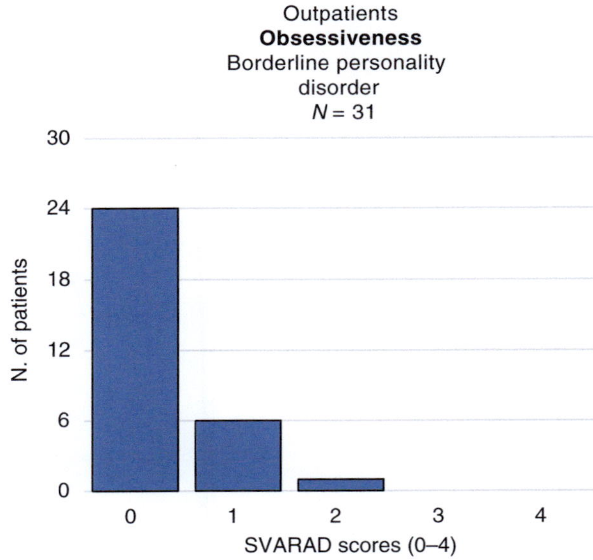

Fig. 8.16 Obsessiveness scores in the borderline personality disorder outpatient group. Although mean scores are modest, findings suggest that the suffering of 23% of the patients includes an obsessive-compulsive component

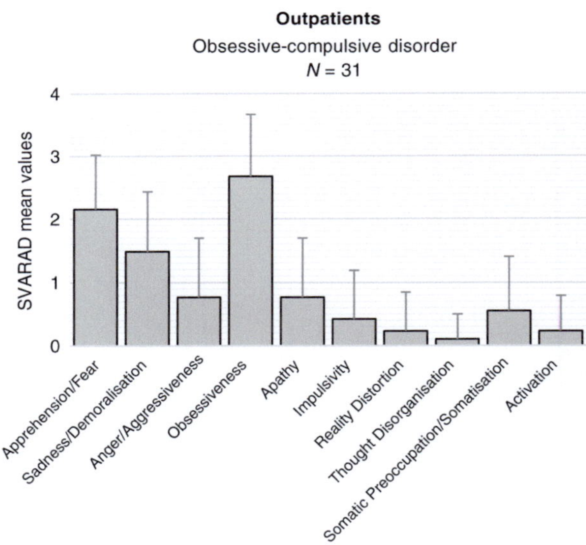

Fig. 8.17 SVARAD profile of outpatients with obsessive-compulsive disorder: Mean scores and standard deviations. Code type: 4-1-2 ($N = 31$)

Fear, while nearly one patient out of four displayed a severe degree of Apprehension/Fear. One half of the group had a moderate to severe degree of Sadness/Demoralisation. One patient out of four showed a moderate degree of Aggressiveness: that is, a clinically significant proportion.

8.9 Discussion

As expected, our findings showed low mean Obsessiveness dimension scores for several categories, with the exception of the OCD diagnostic category, both in the inpatient and outpatient groups. Regarding the OCD group, the SVARAD mean profiles for the inpatient and outpatient samples showed different clinically significant dimensional components for OCD. The OCD group consisted of 8 out of 867 inpatients and 31 out of 1124 outpatients. Given that seldom is OCD so severe as to require acute hospitalisation in Italian general hospital psychiatric wards, our OCD inpatient group represents patients with very severe symptoms, so that our findings may only be partially representative of the whole diagnostic category.

The first finding concerning low values of the Obsessiveness dimension in the whole inpatient and outpatient group reflects the fact that obsessiveness and iterativity are uncommon in psychiatric diagnoses other than OCD. This differs from Apprehension/Fear and Sadness/Demoralisation, which are widespread and present in various psychopathological disorders. In inpatients, after the OCD diagnostic category, the Obsessiveness dimension scores were the second and third most relevant dimensions in schizophrenia and bipolar disorder-depressive episode, respectively.

In the outpatient sample, Obsessiveness was the second and third most relevant dimension in eating disorders and delusional disorder, respectively (see Figs. 8.3 and 8.8). As concerns schizophrenia, our relatively high Obsessiveness SVARAD scores in some cases are not surprising: possible obsessive thoughts and iterative behaviours have often been described in the psychopathological literature for this diagnostic category, before and after the DSM-III [66, 68, 99]. Their presence, however, was not formally included in the DSM and ICD criteria for schizophrenia or delusional disorder. In light of this absence, SVARAD representation seems to be meaningful for psychopathological completeness and subsequent treatment. Two clinical cases might better illustrate this concept.

Case Vignette 1: Paranoid Schizophrenia

Mary is a 46-year-old woman with a DSM-IV-TR diagnosis of paranoid schizophrenia. She lives in a therapeutic community and has a middle-school diploma. She was hospitalised for 17 days. Her SVARAD dimensional profile shows the highest peak in the Obsessiveness dimension (score 4), followed by Apathy, Reality Distortion, and Thought Disorganisation (score 3), and finally Anger/Aggressiveness, Apprehension/Fear, and Activation (score 2) (Fig. 8.18). Psychopharmacological treatment was utilised to address the psychopathological needs of this patient, including the following treatments targeting the SVARAD dimensions: Reality Distortion (quetiapine RP 400 mg/day), Obsessiveness (sertraline 50 mg/day), Activation, Anger/Aggressiveness, and Apprehension/Fear (valproic acid CH 1000 mg/day, diazepam 20 mg/day, and clonazepam 6 mg/day).

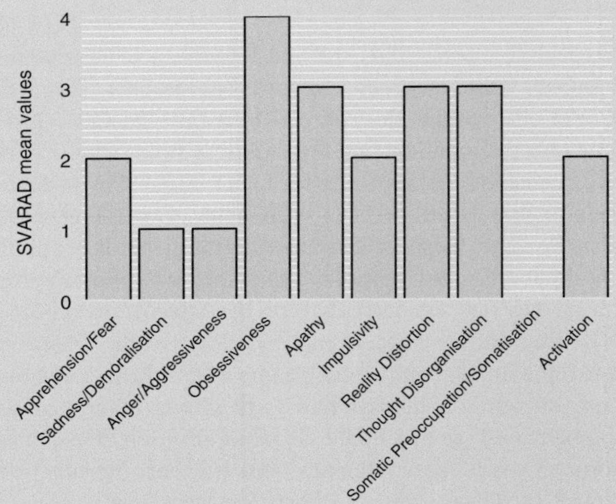

Fig. 8.18 A patient with paranoid schizophrenia. SVARAD Code type 4-7-8

Case Vignette 2: Psychotic Disorder Not Otherwise Specified

Michael is 29 years old, with a DSM-IV-TR diagnosis of Psychotic Disorder NOS. He lives with his parents, has a middle-school diploma, and is currently not working. He was hospitalised for 10 days. His SVARAD dimensional profile shows the highest peak in the Obsessiveness dimension (score 4) followed by Apprehension/Fear, Sadness/Demoralisation, Apathy, and Somatic Preoccupation/Somatisation (score 3). The Reality Distortion and Thought Disorganisation dimensions reached scores of 2. The patient did not fulfil DSM-IV-TR diagnostic criteria for obsessive-compulsive disorder or major depressive disorder. The patient was treated with olanzapine, a second-generation antipsychotic that may cause OCD symptoms OCDS (Fig. 8.19). Psychopharmacological treatment was utilised to address the psychopathological needs of this patient, including the following treatments targeting the SVARAD dimensions: Obsessiveness (fluoxetine 20 mg/day and aripiprazole 15 mg/day), Sadness/Demoralisation and Apathy (venlafaxine 150 mg/day), Reality Distortion (olanzapine 10 mg/day and aripiprazole 15 mg/day), and Apprehension/Fear (zolpidem 10 mg/day, gabapentin 900 mg/day, and clonazepam 5 mg/day).

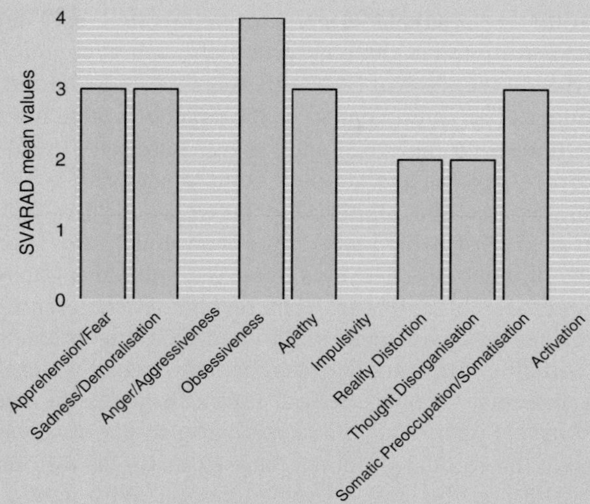

Fig. 8.19 A patient with psychotic disorder not otherwise specified. SVARAD Code type 4-1-10

It is also interesting to point out that the DSM-5 introduces three levels of insight concerning beliefs associated with OCD: good insight, poor insight, or no insight/delusional belief. A patient with an OCD diagnosis at this last level of insight should not be diagnosed—according to the DSM-5—as a psychotic disorder. The differential diagnosis suggested by the DSM-5 is to classify obsessions and compulsions without insight as schizophrenia or schizoaffective disorder only if the patients include in their clinical manifestation hallucinations or formal thought disorders (DSM-5, 2013, p. 241).

The issue of differential diagnosis between OCD and schizophrenia is, however, difficult in some clinical cases and has been under debate for many years and from several perspectives. Obsessions and compulsions in the past were seldom recognised by clinical psychiatry as a possible component in some clinical pictures of these disorders. As Fineberg [100] emphasises:

> In the past OCD was thought to have more in common with psychotic disorders than we recognize today…Like OCD, schizophrenia develops in early adulthood, runs a chronic course, and shows roughly equal gender ratios in clinical cohorts. Co-occurrence of OCD, bizarre grooming, and hoarding in schizophrenia is well recognized… it remains unclear whether the observed overrepresentation of obsessive-compulsive symptoms in schizophrenia reflects true comorbidity, more severe illness, or distinct neuropsychological substrates unique to this group.

Obsessive-compulsive symptoms of a neurotic disorder were viewed, according to the psychodynamic perspective, as the "last" defence from a psychotic decompensation and the emergence of a schizophrenic disorder. Their appearance during the clinical course of schizophrenia, conversely, might suggest the beginning of an improvement of the psychotic state towards a "less severe" and better integrated neurotic functioning. Bahnson (1966), in his psychodynamic model of "psychophysiological complementarity", for instance, proposes the hypothesis that the relationship between OCD and schizophrenia reflects a larger degree of behavioural regression under stress for the latter, as a consequence of external life events or severe internal intrapsychic conflicts [101]. It is interesting to point out that the SVARAD seems to be able to capture Obsessiveness components within non-OCD disorders, and this represents a relevant finding for complete psychopathological assessment and treatment of many patients. Of note, Bellodi [102] discusses the need for a fine cross-sectional psychopathological analysis but also the observation of the longitudinal course of the clinical picture of these cases for the differential diagnosis between OCD with poor insight and schizophrenia. With regard to our findings of mild mean Obsessiveness values in eating disorder outpatients, the issue of comorbidity between eating disorders and OCD has been widely discussed, as has been the presence of obsessive-compulsive symptoms in eating disorders. The DSM-5 considers obsessive-compulsive symptoms in the differential diagnosis of anorexia nervosa but not bulimia nervosa, suggesting that OCD comorbidity should only be considered if the individual shows obsessions and compulsions which are not related to food. Eating disorders have also been extensively discussed as belonging to the OCD spectrum [85, 86]. The following case vignette can better illustrate this concept.

Case Vignette 3: Anorexia Nervosa

Sara is a 30-year-old woman with a DSM-IV-TR diagnosis of anorexia nervosa. She lives with her parents and has a high school degree. She is currently studying at a university. She was hospitalised for a duration of 12 days. Her SVARAD dimensional profile showed the highest peak in the Obsessiveness dimension (score 4), followed by Sadness/Demoralisation, Apathy and Impulsivity (score 3), and Apprehension/Fear (score 2) (Fig. 8.20). The patient did not fulfil DSM-IV-TR diagnostic criteria for OCD or major depressive disorder. She was treated with clomipramine 150 mg/day (targeting Obsessiveness and Sadness/Demoralisation), topiramate 50 mg/day, and olanzapine 10 mg/day (targeting Impulsivity and Apprehension/Fear), together with medical treatment to support metabolic functions.

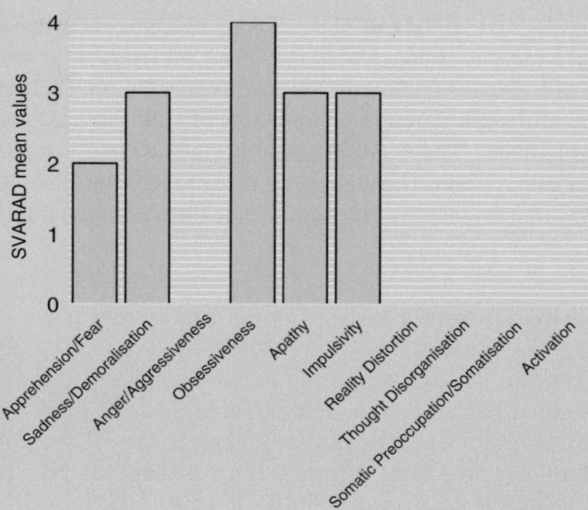

Fig. 8.20 A patient with anorexia nervosa. SVARAD Code type 4-2-6

Considering the above discussion about psychotic disorders, eating disorders, and OCD symptoms and the formulation of clear-cut criteria and categorical distinction by the DSM-5, it is interesting to point out how the routine use of the SVARAD might be useful in the clinical setting. The SVARAD allows the recognition of Obsessiveness components in both disorders, even if they do not fully satisfy the criteria for classification of OCD as a comorbidity. Of some interest is the fact that the SVARAD Obsessiveness dimension displays a significant role in major depression. Although not relevant for this diagnostic group as a whole, the recognition of a major depression subgroup with significant Obsessiveness components is of peculiar interest because of its potential role for assessment of the intrapsychic functioning and the choice of psychopharmacological treatment, perhaps suggesting 5HT-ergic (such as clomipramine and SSRI) antidepressants as better than noradrenergic antidepressants [103]. In the OCD inpatients, we found that the Obsessiveness dimension displayed the highest score, followed by Apprehension/Fear, Sadness/Demoralisation, and Apathy (all with mean scores above 2). Furthermore, Obsessiveness was negatively correlated with Impulsivity ($r = -0.71$, $p = 0.038$) and Activation ($r = -0.629$, $p = 0.038$), and positively correlated with Apprehension/Fear (Rho = 0.147, $p < 0.001$), Sadness/Demoralisation (Rho = 0.148, $p < 0.001$), and Apathy (Rho = 0.51, $p < 0.001$). In the OCD outpatients, the SVARAD dimensional profile also showed the highest value for the Obsessiveness dimension, followed by Apprehension/Fear, Sadness/Demoralisation, Apathy, and Anger/Aggressiveness. Unlike in the inpatient OCD group, Apathy mean scores were below 1, suggesting lesser impairment of affective reactivity and more active involvement with life. One can argue that these results are a reflection of a less severe psychopathology than that of the inpatient group. In addition, the outpatient group showed, as the fourth peak, Anger/Aggressiveness: this finding could confirm the association of anger with symptom subtypes in severe OCD outpatients [104]. The two SVARAD dimensional mean profiles for OCD inpatients and outpatients reflect the main different components of suffering that the clinician can observe in these patients. These components are not formally included in DSM-5 and ICD diagnostic criteria. The clinical picture of OCD patients—even when cases fulfil the DSM criteria—can significantly differ from individual to individual.

Case Vignette 4: Obsessive-Compulsive Disorder

Marc is 69 years old, with a DSM-IV-TR diagnosis of obsessive-compulsive disorder. He lives alone and is an unemployed university graduate. He was hospitalised for a duration of 16 days. His SVARAD profile shows two main peaks (score 4) in Obsessiveness and Sadness/Demoralisation, followed by a third peak in Apathy (score 3) and a fourth peak in Apprehension/Fear (score 2). The patient does not fulfil the DSM criteria for a depressive episode or major depressive disorder (Fig. 8.21).

The psychopharmacological treatment addressed the psychopathological needs of this patient, including the following treatments targeting the profile's dimensions: Obsessiveness (sertraline 50 mg/day, aripiprazole 10 mg/day), Sadness/Demoralisation and Apathy (venlafaxine 150 mg/day), and Apprehension/Fear (diazepam 20 mg/day and quetiapine SR 100 mg/day).

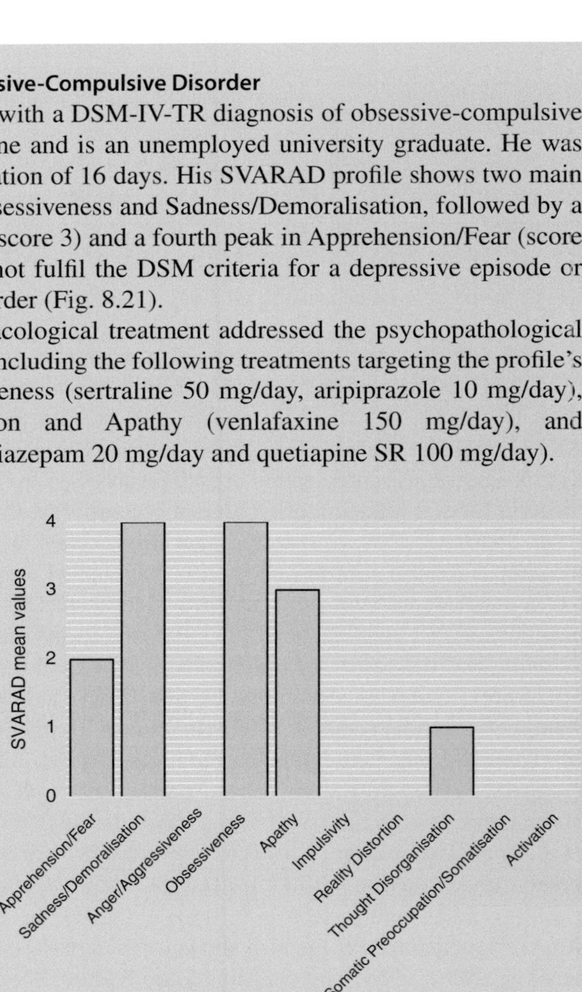

Fig. 8.21 A patient with obsessive-compulsive disorder. SVARAD Code type 4-2-5

Case Vignette 5: Obsessive-Compulsive Disorder with Comorbid Diagnosis of Psychotic Disorder NOS

Colette is 37 years old, with a DSM-IV-TR diagnosis of OCD with comorbid psychotic disorder NOS. She currently lives with her parents and is unemployed. She was hospitalised for a duration of 18 days. The patient also takes a full dose of clozapine, and this may induce the onset of obsessive symptoms. Her SVARAD profile shows two main peaks (score 4) in Obsessiveness and Apathy, followed by Apprehension/Fear, Sadness/Demoralisation, Reality Distortion, and Thought Disorganisation (score 2) (Fig. 8.22). The psychopathological needs of this patient were addressed through psychopharmacological treatment, including the following drug treatments targeting the profile's dimensions: Reality Distortion (clozapine 500 mg/day), Sadness/Demoralisation and Apathy (lamotrigine 150 mg/day, as a mood stabiliser with mild antidepressant properties), Apprehension/Fear (zolpidem 7.5 mg/day and diazepam 4 mg/day), and Obsessiveness (clomipramine 150 mg/day).

Fig. 8.22 A patient with obsessive-compulsive disorder with comorbid diagnosis of psychotic disorder not otherwise specified. SVARAD Code type 4-5-7

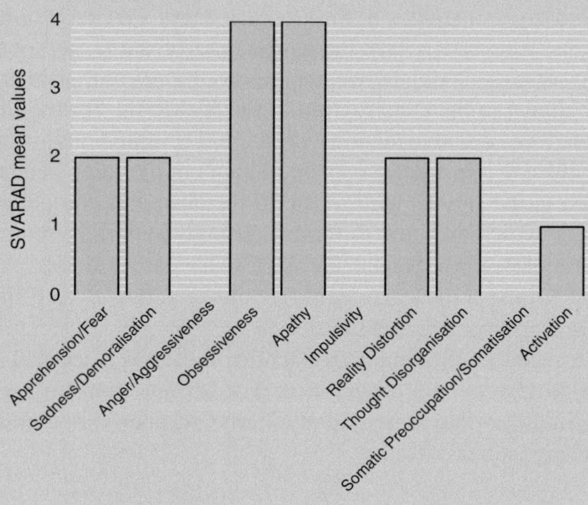

Another issue resulting from our findings is the presence of the Obsessiveness dimension at mild (score 1) to moderate (score 2) levels in several DSM-IV diagnostic groups, as well as mild to severe (score 3) levels in other groups. As previously reported, in outpatients, one patient out of three shows an Obsessiveness score between mild and severe for the following disorders: eating disorders (40%), bipolar disorder and depressive episode (36%), delusional disorder (33.8%), and schizophrenia (30%).

One outpatient out of four shows mild to severe Obsessiveness for the following disorders: psychotic NOS (27%), dysthymia (27%), somatoform disorders (26%), major depressive disorder (25%), anxiety disorder NOS (24%), and finally, borderline personality disorder (23%). In a similar way, although mean Obsessiveness scores were low in the whole inpatient group, 25.1% of major depression patients, 20% of bipolar patients-depressive episode, and 23.3% of schizophrenic patients have an Obsessiveness SVARAD score between 1 (mild) and 4 (very severe).

In both inpatient and outpatient groups, some degree of Obsessiveness is present in a small but significant proportion of cases, providing suggestions for clinical attention and more precise psychopharmacological targeting and related interventions. The construction of the therapeutic alliance should take into account the possible presence of patient characteristics like doubtfulness, inflexibility, preciseness, checking or preventing behaviours, presence of obsessions or compulsions, rumination, and some tendency towards slowness, ranging from mild to severe. Although mean Obsessiveness scores seemed low, in one third to one quarter of patients from several diagnostic groups, Obsessiveness played a role, from mild to severe, that could be considered as helpful for better comprehension and treatment.

Finally, our studies of the Obsessiveness SVARAD dimension have several limitations. The first is that the assessment of the Obsessiveness dimension does not allow distinguishing between obsessions and compulsions. This was a methodological choice in the construction of the SVARAD, for the sake of simplicity, because obsessions, iterative manifestations, and compulsions are clinically often—although not always—presented as concomitants in the same patient. This represents a limitation of the studies with regard to the obsessive compulsive spectrum of these disorders, which may manifest more strongly in iterative/compulsive behaviour than in the obsessive component. In all of these cases, further psychopathological assessment with specific instruments should be added, even if they require more complex designs, personnel, and time.

A second limitation is the fact that the cases discussed in this chapter are limited to a cross-sectional observation. Further analyses are needed, taking into account longitudinal course and, particularly, response to treatment, via pre- and post-discharge assessments.

References

1. Kraepelin E. Psychiatrie. Ein Lehrbuch für Studierende und Ärzte Klinische Psychiatrie. Leipzig: Barth; 1915.
2. Jaspers K. Allgemeine Psychopathologie. 5th ed. Berlin: Springer; 1948.
3. Krafft-Ebing R. Űber Geisterstörungen durch Zwangsvorstellung. Allg Z Psychiatr. 1879;35:303–28.
4. Westphal C. Űber Zwangsvorstellung. Arch Psychiatr Nervenkr. 1872;3:138–61.
5. Janet P. Les Obsessions et la Psychasthenié. Paris: Alcan; 1919.
6. Falret JP. Obsession intellectuelles et emotives. Arch Neurolog. 1889;2:274–93.
7. Berrios GE. The history of mental symptoms. Cambridge: Cambridge Univeristy Press; 1996.

8. Hollander E, Schiffman E, Cohen B, Rivera-Stein MA, Rosen W, Gorman JM, Fyer AJ, Papp L, Liebowitz MR. Signs of central nervous system dysfunction in obsessive-compulsive disorder. Arch Gen Psychiatry. 1990;47(1):27–32.
9. Hollander E. Obsessive-compulsive spectrum disorders: an overview. Psychiatr Ann. 1993;23(7):355–8.
10. American Psychiatric Association. Diagnostic and statistical manual of mental disorders, 3rd edition, revised. Washington, DC: American Psychiatric Association; 1987.
11. American Psychiatric Association. Diagnostic and statistical manual of mental disorders. 5th ed. Washington, DC: American Psychiatric Association; 2013.
12. Torres AR, Prince MJ, Bebbington PE, Bhugra D, Brugha TS, Farrell M, Jenkins R, Lewis G, Meltzer H, Singleton N. Obsessive-compulsive disorder: prevalence, comorbidity, impact, and help-seeking in the British National Psychiatric Morbidity Survey of 2000. Am J Psychiatry. 2006;163(11):1978–85.
13. Crino R, Slade T, Andrews G. The changing prevalence and severity of obsessive-compulsive disorder criteria from DSM-III to DSM-IV. Am J Psychiatry. 2005;162(5):876–82.
14. Ruscio AM, Stein DJ, Chiu WT, Kessler RC. The epidemiology of obsessive-compulsive disorder in the National Comorbidity Survey Replication. Mol Psychiatry. 2010;15(1):53–63.
15. Castle DJ, Phillips KA. Obsessive–compulsive spectrum of disorders: a defensible. Aust N Z J Psychiatry. 2006;40(2):114–20.
16. Fullana MA, Mataix-Cols D, Caspi A, Harrington H, Grisham JR, Moffitt TE, Poulton R. Obsessions and compulsions in the community: prevalence, interference, help-seeking, developmental stability, and co-occurring psychiatric conditions. Am J Psychiatry. 2009;166(3):329–36.
17. Clark DA. Innovation in obsessive compulsive disorder: A commentary. J Behav Ther Exp Psychiatry. 2015;49(Pt B):129–32.
18. Berry LM, Laskey B. A review of obsessive-compulsive thoughts in the general population. J Obsessive Compulsive Relat Disord. 2012;1:125–32.
19. Berthier ML, Kulisevsky J, Gironell A, Heras JA. Obsessive-compulsive disorder associated with brain lesions: clinical phenomenology, cognitive function, and anatomic correlates. Neurology. 1996;47:353–61.
20. Carmin CN, Wiegartz PS, Yunus U, Gillock KL. Treatment of late-onset OCD following basal ganglia infarct. Depress Anxiety. 2002;15:87–90.
21. Murphy DL, Timpano KR, Wheaton MG, Greenberg BD, Miguel EC. Obsessive-compulsive disorder and its related disorders: a reappraisal of obsessive-compulsive spectrum concepts. Dialogues Clin Neurosci. 2010;12:131–48.
22. Bolton D, Gibb W, Lees A, Raven P, Gray JA, Chen E, et al. Neurological soft signs in obsessive compulsive disorder: standardised assessment and comparison with schizophrenia. Behav Neurol. 1998;11:197–204.
23. Milad MR, Rauch SL. Obsessive compulsive disorder: beyond segregated cortico-striatal pathways. Trends Cogn Sci. 2012;16(1):43–51.
24. Breiter HC, Rauch SL. Functional MRI and the study of OCD: From symptom provocation to cognitive-behavioral probes of cortico-striatal systems and the amygdala. NeuroImage. 1996;4(3 Pt 3):127–38.
25. Harrison BJ, Soriano-Mas C, Pujol J, Ortiz H, López-Solà M, Hernández-Ribas R, Deus J, Alonso P, Yücel M, Pantelis C, Menchon JM, Cardoner N. Altered corticostriatal functional connectivity in obsessive-compulsive disorder. Arch Gen Psychiatry. 2009;66(11):1189–200.
26. Szeszko PR, Robinson D, Alvir JM, Bilder RM, Lencz T, Ashtari M, et al. Orbital frontal and amygdala volume reductions in obsessive-compulsive disorder. Arch Gen Psychiatry. 1999;56:913–9.
27. Rotge J-Y, Guehl D, Dilharreguy B, Tignol J, Bioulac B, Allard M, et al. Meta-analysis of brain volume changes in obsessive-compulsive disorder. Biol Psychiatry. 2009;65(1):75–83.

28. Saxena S, Rauch SL. Functional neuroimaging and the neuroanatomy of obsessive-compulsive disorder. Psychiatr Clin North Am. 2000;23:563–86.
29. Saxena S, Bota RG, Brody AL. Brain-behavior relationships in obsessive-compulsive disorder. Semin Clin Neuropsychiatry. 2001;6(2):82–101.
30. Whiteside SP, Port JD, Abramowitz JS. A meta-analysis of functional neuroimaging in obsessive-compulsive disorder. Psychiatry Res. 2004;132(1):69–79.
31. Nordahl TE, Benkelfat C, Semple WE, Gross M, King AC, Cohen RM. Cerebral glucose metabolic rates in obsessive compulsive disorder. Neuropsychopharmacology. 1989;2:23–8.
32. Benkelfat C, Nordahl TE, Semple WE, King AC, Murphy DL, Cohen RM. Local cerebral glucose metabolic rates in obsessive-compulsive disorder. Patients treated with clomipramine. Arch Gen Psychiatry. 1990;47:840–8.
33. Perani D, Colombo C, Bressi S, Bonfanti A, Grassi F, Scarone S, et al. [18F] FDG PET study in obsessive-compulsive disorder A clinical/metabolic correlation study after treatment. Br J Psychiatry. 1995;166:244–50.
34. Hoehn-Saric R, Schlaepfer TE, Greenberg BD, McLeod DR, Pearlson GD, Wong SH. Cerebral blood flow in obsessive-compulsive patients with major depression: effect of treatment with sertraline or desipramine on treatment responders and non-responders. Psychiatry Res. 2001;108:89–100.
35. Hansen ES, Hasselbalch S, Law I, Bolwig TG. The caudate nucleus in obsessive-compulsive disorder. Reduced metabolism following treatment with paroxetine: a PET study. Int J Neuropsychopharmacol. 2002;5:1–10.
36. Baxter LR Jr, Schwartz JM, Bergman KS, Szuba MP, Guze BH, Mazziotta JC, et al. Caudate glucose metabolic rate changes with both drug and behavior therapy for obsessive-compulsive disorder. Arch Gen Psychiatry. 1992;49:681–9.
37. Brody AL, Saxena S, Schwartz JM. FDG-PET predictors of response to behaviour therapy and pharmacotherapy in obsessive compulsive disorder. Psychiatry Res. 1998;84:1–6.
38. Kang DH, Kwon JS, Kim JJ, Youn T, Park HJ, Kim MS, et al. Brain glucose metabolic changes associated with neuropsychological improvements after 4 months of treatment in patients with obsessive-compulsive disorder. Acta Psychiatr Scand. 2003;107:291–7.
39. Arnsten Amy FT. Catecholamine influences on dorsolateral prefrontal cortical networks. Biol Psychiatry. 2011;69(12):89–99.
40. Apostolova I, Block S, Buchert R, Osen B, Conradi M, Tabrizian S, et al. Effects of behavioral therapy or pharmacotherapy on brain glucose metabolism in subjects with obsessive-compulsive disorder as assessed by brain FDG PET. Psychiatry Res. 2010;184:105–16.
41. Ho Pian KL, van Megen HJ, Ramsey NF, Mandl R, van Rijk PP, Wynne HJ, et al. Decreased thalamic blood flow in obsessive-compulsive disorder patients responding to fluvoxamine. Psychiatry Res. 2005;138:89–97.
42. Yücel M, Wood SJ, Wellard RM, Harrison BJ, Fornito A, Pujol J, et al. Anterior cingulate glutamate-glutamine levels predict symptom severity in women with obsessive-compulsive disorder. Aust N Z J Psychiatry. 2008;42:467–77.
43. Brennan BP, Rauch SL, Jensen JE, Pope HG Jr. A critical review of magnetic resonance spectroscopy studies of obsessive-compulsive disorder. Biol Psychiatry. 2013;73(1):24–31.
44. Adler CM, McDonough-Ryan P, Sax KW, Holland SK, Arndt S, Strakowski SM. fMRI of neuronal activation with symptom provocation in unmedicated patients with obsessive compulsive disorder. J Psychiatr Res. 2000;34:317–24.
45. Nakao T, Nakagawa A, Yoshiura T, Nakatani E, Nabeyama M, Yoshizato C, et al. Brain activation of patients with obsessive-compulsive disorder during neuropsychological and symptom provocation tasks before and after symptom improvement: a functional magnetic resonance imaging study. Biol Psychiatry. 2005;57:901–10.
46. Kang DH, Jang JH, Han JY, Kim JH, Jung WH, Choi JS, et al. Neural correlates of altered response inhibition and dysfunctional connectivity at rest in obsessive-compulsive disorder. Prog Neuro-Psychopharmacol Biol Psychiatry. 2013;40:340–6.
47. Tian L, Meng C, Jiang Y, Tang Q, Wang S, Xie X, et al. Abnormal functional connectivity of brain network hubs associated with symptom severity in treatment-naive patients with obses-

sive compulsive disorder: a resting-state functional MRI study. Prog Neuropsychopharmacol Biol Psychiatry. 2016;66:104–11.
48. Bernstein GA, Mueller BA, Schreiner MW, Campbell SM, Regan EK, Nelson PM, et al. Abnormal striatal resting-state functional connectivity in adolescents with obsessive-compulsive disorder. Psychiatry Res. 2016;247:49–56.
49. Nakatani E, Nakgawa A, Ohara Y, Goto S, Uozumi N, Iwakiri M, et al. Effects of behavior therapy on regional cerebral blood flow in obsessive-compulsive disorder. Psychiatry Res. 2003;124:113–20.
50. Freyer T, Klöppel S, Tüscher O, Kordon A, Zurowski B, Kuelz AK, et al. Frontostriatal activation in patients with obsessive-compulsive disorder before and after cognitive behavioral therapy. Psychol Med. 2011;41:207–16.
51. Rauch SL, Savage CR. Neuroimaging and neuropsychology of the striatum. Bridging basic science and clinical practice. Psychiatr Clin North Am. 1997;20:741–68.
52. Fineberg NA, Sivakumaran T, Roberts A, Gale T. Adding quetiapine to SRI in treatment-resistant obsessive-compulsive disorder: a randomized controlled treatment study. Int Clin Psychopharmacol. 2005;20(4):223–6.
53. Fineberg NA, Gale TM, Sivakumaran T. A review of antipsychotics in the treatment of obsessive compulsive disorder. J Psychopharmacol. 2006;20(1):97–103.
54. Gillan CM, Fineberg NA, Robbins TW. A trans-diagnostic perspective on obsessive-compulsive disorder. Psychol Med. 2017;47(9):1528–48.
55. Gillan CM, Papmeyer M, Morein-Zamir S, Sahakian BJ, Fineberg NA, Robbins TW, et al. Disruption in the balance between goal-directed behavior and habit learning in obsessive-compulsive disorder. Am J Psychiatry. 2011;168:718–26.
56. Voon V, Reiter A, Sebold M, Groman S. Model-based control in dimensional psychiatry. Biol Psychiatry. 2017;82(6):391–400.
57. Fontenelle LF, Oostermeijer S, Harrison BJ, Pantelis C, Yucel M. Obsessive-compulsive disorder, impulse control disorders and drug addiction: common features and potential treatments. Drugs. 2011;71:827–40.
58. Sjoerds Z, de Wit S, van den Brink W, Robbins TW, Beekman AT, Penninx BW, Veltman DJ. Behavioral and neuroimaging evidence for overreliance on habit learning in alcohol-dependent patients. Transl Psychiatry. 2013;3(12):e33.
59. Voon V, Derbyshire K, Rück C, Irvine MA, Worbe Y, Enander J, Schreiber LR, Gillan C, Fineberg NA, Sahakian BJ, Robbins TW, Harrison NA, Wood J, Daw ND, Dayan P, Grant JE, Bullmore ET. Disorders of compulsivity: a common bias towards learning habits. Mol Psychiatry. 2015;20(3):345–52.
60. Pancheri P, et al. Validazione della scala per la valutazione rapida dimensionale "SVARAD". Riv Psichiatr. 1999;34:84–93.
61. Bellodi L. Diagnosi differenziale e casi clinici. In: Smeraldi E, editor. Il disturbo Ossessivo-Compulsivo e il suo spettro, vol. 2003. Milano: Masson; 2003. p. 303–26.
62. Berman I, Merson A, Viegner B, Losonczy MF, Pappas D, Green AI. Obsessions and compulsions as a distinct cluster of symptoms in schizophrenia: a neuropsychological study. J Nerv Ment Dis. 1998;186:150–6.
63. Bland RC, Newman SC, Orn H. Schizophrenia: lifetime comorbidity in a community sample. Acta Psychiatr Scand. 1987;75:383–91.
64. de Haan L, Dudek-Hodge C, Verhoeven Y, Denys D. Prevalence of psychotic disorders in patients with obsessive-compulsive disorder. CNS Spectr. 2009;14(8):415–7.
65. Porto L, Bermanzohn PC, Pollack S, Morrissey R, Siris SG. A profile of obsessive-compulsive symptoms in schizophrenia. CNS Spectr. 1997;2:21–5.
66. Bottas A, Cooke RG, Richter MA. Comorbidity and pathophysiology of obsessive-compulsive disorder in schizophrenia: is there evidence for a schizo-obsessive subtype of schizophrenia? J Psychiatry Neurosci. 2005;30(3):187–93.
67. Poyurovsky M, Koran LM. Obsessive-compulsive disorder (OCD) with schizotypy vs. schizophrenia with OCD: diagnostic dilemmas and therapeutic implications. J Psychiatr Res. 2005;39:399–408.

68. Rosen I. The clinical significance of obsessions in schizophrenia. J Ment Sci. 1957;103:773–85.
69. Hwang MY, Morgan JE, Losconzcy MF. Clinical and neuropsychological profiles of obsessive–compulsive schizophrenia: a pilot study. J Neuropsychiatry Clin Neurosci. 2000;12:91–4.
70. Niendam TA, Berzaka J, Cannon TD, Bearden CE. Obsessive compulsive symptoms in the psychosis prodrome: correlates of clinical and functional outcome. Schizophr Res. 2009;108(1–3):170–5.
71. Öngür D, Goff DC. Obsessive-compulsive symptoms in schizophrenia: associated clinical features, cognitive function and medication status. Schizophr Res. 2005;75:349–62.
72. Lim M, Park DY, Kwon JS, Joo YH, Hong KS. Prevalence and clinical characteristics of obsessive-compulsive symptoms associated with atypical antipsychotics. J Clin Psychopharmacol. 2007;27(6):712–3.
73. Tiihonen J, Lonnqvist J, Wahlbeck K, Klaukka T, Niskanen L, Tanskanen A, Haukka J. 11-year follow-up of mortality in patients with schizophrenia: a population-based cohort study (FIN11 study). Lancet. 2009;374(9690):620–7.
74. Englisch S, Esslinger C, Inta D, et al. Clozapine-induced obsessive-compulsive syndromes improve in combination with aripiprazole. Clin Neuropharmacol. 2009;32:4,227–9.
75. Kim SW, Shin IS, Kim JM, Yang SJ, Hwang MY, Yoon JS. Amisulpride improves obsessive-compulsive symptoms in schizophrenia patients taking atypical antipsychotics: an open-label switch study. J Clin Psychopharmacol. 2008;28(3):349–52.
76. Tundo A, Necci R. Cognitive-behavioural therapy for obsessive-compulsive disorder co-occurring with psychosis *systematic review of evidence*. World J Psychiatry. 2016;6(4):449–55.
77. Adam Y, Meinlschmidt G, Gloster AT, Lieb R. Obsessive-compulsive disorder in the community: 12-month prevalence, comorbidity and impairment. Soc Psychiatry Psychiatr Epidemiol. 2012;47(3):339–49.
78. Leone JE, Sedory EJ, Gray KA. Recognition and treatment of muscle dysmorphia and related body image disorders. J Athl Train. 2005;40(4):352–9.
79. Davidson J. Diabulimia: how eating disorders can affect adolescents with diabetes. Nurs Stand. 2014;29(2):44–9.
80. Roosen KM, Mills JS. Exploring the motives and mental health correlates of intentional food restriction prior to alcohol use in university students. J Health Psychol. 2015;20(6):875–86.
81. Mathieu J. What is pregorexia? J Am Diet Assoc. 2009;109(6):976–9.
82. Mathieu J. What is orthorexia? J Am Diet Assoc. 2005;105(10):1510–2.
83. Donini LM, Marsili D, Graziani MP, Imbriale M, Cannella C. Orthorexia nervosa: a preliminary study with a proposal for diagnosis and an attempt to measure the dimension of the phenomenon. Eat Weight Disord. 2004;9(2):151–7.
84. Dunn TM, Bratman S. On orthorexia nervosa: a review of the literature and proposed diagnostic criteria. Eat Behav. 2015;21:11–7.
85. Halmi KA, Tozzi F, Thornton LM, et al. The relation among perfectionism, obsessive-compulsive personality disorder and obsessive-compulsive disorder in individuals with eating disorders. Int J Eat Disord. 2005;38(4):371–4.
86. Dell'Osso L, Abelli M, Carpita B, Pini S, Castellini G, Carmassi C, Ricca V. Historical evolution of the concept of anorexia nervosa and relationships with orthorexia nervosa, autism, and obsessive-compulsive spectrum. Neuropsychiatr Dis Treat. 2016;12:1651–60.
87. McKay D, Sookman D, Neziroglu F, Wilhelm S, Stein DJ, Kyrios M, Matthews K, Veale D. Accreditation task force of the Canadian Institute for Obsessive Compulsive Disorders. Efficacy of cognitive-behavioral therapy for obsessive-compulsive disorder. Psychiatry Res. 2015;227(1):104–13.
88. Fullana MA, Zhu X, Alonso P, Cardoner N, Real E, López-Solà C, Segalàs C, Subirà M, Galfalvy H, Menchón JM, Simpson HB, Marsh R, Soriano-Mas C. Basolateral amygdala-ventromedial prefrontal cortex connectivity predicts cognitive behavioural therapy outcome in adults with obsessive-compulsive disorder. J Psychiatry Neurosci. 2017;42(5):160215.
89. Külz A, Barton B, Voderholzer U. Third wave therapies of cognitive behavioral therapy for obsessive compulsive disorder: a reasonable add-on therapy for CBT? State of the art. Psychother Psychosom Med Psychol. 2016;66(3–4):106–11.

90. Downar J, Geraci J, Salomons TV, Dunlop K, Wheeler S, McAndrews MP, et al. Anhedonia and reward-circuit connectivity distinguish nonresponders from responders to dorsomedial prefrontal repetitive transcranial magnetic stimulation in major depression. Biol Psychiatry. 2014;76:176–85.
91. Dunlop K, Woodside B, Olmsted M, Colton P, Giacobbe P, Downar J. Reductions in cortico-striatal hyperconnectivity accompany successful treatment of obsessive-compulsive disorder with dorsomedial prefrontal rTMS. Neuropsychopharmacology. 2016;41:1395–403.
92. Denys D, Mantione M, Figee M, van den Munckhof P, Koerselman F, Westenberg H, et al. Deep brain stimulation of the nucleus accumbens for treatment-refractory obsessive-compulsive disorder. Arch Gen Psychiatry. 2010;67:1061–8.
93. Aouizerate B, Cuny E, Martin-Guehl C, Guehl D, Amieva H, Benazzouz A, et al. Deep brain stimulation of the ventral caudate nucleus in the treatment of obsessive—compulsive disorder and major depression. J Neurosurg. 2004;101:682–6.
94. Ruffini C, Locatelli M, Lucca A, Benedetti F, Insacco C, Smeraldi E. Augmentation effect of repetitive transcranial magnetic stimulation over the orbitofrontal cortex in drug-resistant obsessive-compulsive disorder patients: a controlled investigation. Prim Care Companion J Clin Psychiatry. 2009;11:226–30.
95. Nauczyciel C, Le Jeune F, Naudet F, Douabin S, Esquevin A, Vérin M, et al. Repetitive transcranial magnetic stimulation over the orbitofrontal cortex for obsessive-compulsive disorder: a double-blind, crossover study. Transl Psychiatry. 2014;4:e436.
96. Mondino M, Haesebaert F, Poulet E, Saoud M, Brunelin J. Efficacy of cathodal transcranial direct current stimulation over the left orbitofrontal cortex in a patient with treatment-resistant obsessive-compulsive disorder. J ECT. 2015;31:271–2.
97. Bation R, Poulet E, Haesebaert F, Saoud M, Brunelin J. Transcranial direct current stimulation in treatment-resistant obsessive compulsive disorder: an open-label pilot study. Prog Neuro-Psychopharmacol Biol Psychiatry. 2016;65:153–7.
98. Fettes P, Schulze L, Downar J. Cortico-striatal-thalamic loop circuits of the orbitofrontal cortex: promising therapeutic targets in psychiatric illness. Front Syst Neurosci. 2017;11:25.
99. Cunill R, Castells X, Simeon D. Relationships between obsessive-compulsive symptomatology and severity of psychosis in schizophrenia: a systematic review and meta-analysis. J Clin Psychiatry. 2009;70(1):70–82.
100. Fineberg NA, Saxena S, Zohar J, Craig KJ. Obsessive-compulsive disorder. In: Hollander E, Zohar J, Sirovatka PJ, Regier DA, editors. Obsessive-compulsive spectrum disorders. Arlington: American Psychiatric Publishing; 2011. p. 9–10.
101. Bahnson CB, Kissen DM. Psychophysiological aspects of cancer. Ann N Y Acad Sci. 1966;125:775–1055.
102. Bellodi L, Erzegovesi S. Disturbi dell'alimentazione. In: Smeraldi E, editor. Il disturbo Ossessivo-Compulsivo e il suo spettro. Milano: Masson; 2003. p. 489–544.
103. Hirschtritt ME, Bloch MH, Mathews CA. Obsessive-compulsive disorder: advances in diagnosis and treatment. JAMA. 2017;317(13):1358–67.
104. Piacentino D, Pasquini M, Tarsitani L, Berardelli I, Roselli V, Maraone A, Biondi M. The Association of Anger with symptom subtypes in severe obsessive-compulsive disorder outpatients. Psychopathology. 2016;49(1):40–6.

9 Psychopathological Dimensions and the Clinician's Subjective Experience

Mauro Pallagrosi, Angelo Picardi, Laura Fonzi, and Massimo Biondi

9.1 Introduction

The dimensional approach to diagnosis, as previously discussed in this book, constitutes a substantial effort at refining psychopathological assessment and tailoring therapeutic intervention. This perspective on mental illness seems to account for an issue largely neglected by categorical nosology, namely, the uniqueness and individual specificity of psychic suffering. Indeed, by giving value to the unique expression of mental illness, this approach highlights the importance of the interconnection between a certain clinical fixed picture (category) and the dynamic *colour* of the individual pathological experience (dimensions). Such exploration is still carried out in the framework of a strictly objective and empirically supported approach towards classification [1]. In fact, dimensional assessment, like categorical assessment, relies on standardised profiles obtained through reliable, validated instruments (i.e. the SVARAD or the dimensions derived from general psychopathology assessment tools such as the *Brief Psychiatric Rating Scale*). This is a remarkable advantage, and it has led to the inclusion of some dimensional criteria in the DSM-5 Appendix [2].

M. Pallagrosi (✉)
Department of Human Neurosciences, Sapienza University of Rome, Rome, Italy
e-mail: mauro.pallagrosi@uniroma1.it

A. Picardi
Centre for Behavioural Sciences and Mental Health, Italian National Institute of Health, Rome, Italy
e-mail: angelo.picardi@iss.it

L. Fonzi
Training Institute, Italian Psychoanalytical Society, Rome, Italy

M. Biondi
Department of Human Neurosciences, Policlinico Umberto I Hospital, Sapienza University of Rome, Rome, Italy
e-mail: massimo.biondi@uniroma1.it

M. Biondi et al. (eds.), *Dimensional Psychopathology*,
https://doi.org/10.1007/978-3-319-78202-7_9

In light of the limitations of the categorical approach to diagnosis, the dimensional approach has highlighted the advantages of a more open framework that is able to encompass different perspectives. This has encouraged us to move a step further in the direction of grasping the multifaceted nature of psychiatric assessment. In fact, as stated above, the dimensional evaluation still operates within the objectifying framework of descriptive and syndromal psychopathology, which doesn't explicitly take into account the subjective features of the clinical encounter. In contrast, phenomenological psychopathology specifically addresses this issue, as it deals with the necessity of understanding both the characteristics of the patient's lived experience and the relational determinants of clinical interaction [3, 4]. The latter, in particular, constitutes a peculiar field of in-depth analysis, since many psychopathologists consider what happens between clinician and patient as a central aspect of psychiatric assessment, which provides unique and valuable data on the patient's way of *being-in-the-world.*

The phenomenological perspective, in fact, posits that when a clinician meets a patient, he or she is quickly dragged into the patient's personal way of living and experiencing reality. Thus, the clinician's feeling itself can be an actual clinical fact. As finely stated by Fuchs [4], "the psychiatrist's own subjective experience functions [...] as a complement to the patient's inner world and his way of relating to others". In this field, Jaspers [5], who first conceptualised empathy (*Einfühlung*) as the epistemic means by which a deep knowledge of the patient can be gathered, started a long and fruitful tradition in the study of clinicians' *intuition* [6–8].

Clinical intuition can be defined as a pre-reflective, immediate, holistic grasp of the patient's way of being in relation to others and is embedded in the psychiatrist's inner experience. Historically, many scholars have attempted to describe it, developing different but comparable articulations of the concept of such an ineffable *diagnostic feeling*. They all share the thesis that the clinician can identify psychopathological phenomena, even subtle ones, by analysing his or her own feelings or perceptions. Maybe the most popular description in this field is Rümke's *Praecox Gefühl* [9], a sort of sense of *schizophrenicity* which, according to the author, arises from the impact on the clinician of some typical "dimensions" of schizophrenic condition: lack of exchange of affect, poor rapprochement instinct, and changes in motor behaviour and speech. While it represents a popular theoretical construct, somehow implicitly kept in psychiatrists' minds [10, 11], many other seminal contributions, equally focused on the interaction with patients affected by schizophrenia, are worthy of examination. The fine descriptions of the feeling of being rebounded back into oneself [12], of the disturbing lack of the possibility to share the same reality [13], and of the perception of an absence of vitality [14] are still key illustrations of the feelings emerging from the encounter with the psychotic world. They also gave rise to a more extensive phenomenology of the intersubjective world and to accurate conceptualisations of the diagnostic value of the clinician's feeling and intuition, such as *Diagnosis through Intuition* [13], *Diagnostic par Pénétration* [14, 15], *Gefühl Diagnose* [12, 16], and *Atmospheric Diagnosis* [17].

Psychoanalysts have given specific attention to relational phenomena and actually have assigned a technical value to the examination of their own feelings with the patients. In particular, starting from the seminal contribution of Paula Heimann [18], many psychoanalytic authors have pointed out the relevance of the *insights* arising from the unconscious interaction between the analyst and the analysed patient. The main examples of this tradition are the classical notion of "diagnostic" use of *countertransference* [18–20], studies on the phenomenon of *projective identification* [21, 22], and research on *empathic knowledge* [23, 24]. However, despite the early warning by Winnicott [25], research and theorising in this field remained confined to psychoanalysis and did not substantially influence psychiatric clinical settings.

In this chapter, we first describe an empirically supported way of following the phenomenological call towards the use of the clinician's feelings in the diagnostic process. Then, we describe a study that has specifically investigated the relation between psychopathological dimensions and the pattern of the clinician's subjective experience in everyday psychiatric settings. Finally, we discuss the questions raised by our empirical research and illustrate the theoretical and methodological implications of an integrated view of psychiatric assessment.

9.2 Empirical Investigation of the Clinician's Subjective Experience

9.2.1 The Assessment of Clinician's Subjective Experience (ACSE): Background and Development

The clinician's subjective experience, as we stated above, represents one of the main epistemic means by which a deep understanding of the patient's inner world can be achieved. Indeed, it can and should be viewed as a useful object of investigation, even through quantitative empirical research. Nevertheless, the psychiatric field has shown some difficulties accepting the power and subtleties of subjective experiential data, and only a handful of studies have attempted to investigate the intersubjective dimension and its clinical correlates through standardised methods. On the one hand, a few studies on the value of the *Praecox Gefühl*, performed in clinical psychiatric settings, have yielded inconclusive results [26, 27]. On the other hand, in psychotherapy research, a number of studies have explored the relation between the therapist's feelings and the patient's diagnosis or symptoms, but they did not specifically investigate the potential diagnostic value of these feelings in psychiatric clinical settings [28–33]. Psychoanalytic studies, which have focused on the relation between *countertransference* and diagnosis [34–38], have also dealt with psychotherapeutic settings and long-lasting relationships, but have not directly considered the issue of psychiatric assessment, with its intersubjective and intuitive correlates.

In an effort to fill this gap, we have recently developed and validated a standardised instrument specifically designed to describe the clinicians' subjective experience during their interaction with a patient [39]. This instrument—the *Assessment of Clinician's Subjective Experience (ACSE)*—is rooted both in the phenomenological perspective and in everyday clinical practice. It is a self-completed instrument, consisting of 46 items, each rated on a 5-point scale ranging from 0 to 4. It yields scores on five scales, named Tension, Difficulty in Attunement, Engagement, Disconfirmation, and Impotence. These scales have been factorially derived, have shown convergent validity with changes in *Profile of Mood States (POMS)* scores during the visit, and have displayed high internal consistency and test-retest reliability [39].

The Tension scale includes items indicating physical tension and clumsiness, reduced spontaneity, and feelings of worry, nervousness, and alarm (e.g. "I felt tense in moments of silence", "I maintained a rigid posture", "I was afraid that the patient could act unpredictably"). The Difficulty in Attunement scale consists of items describing difficulty in establishing emotional contact, being empathic, understanding the patient's experience, and communicating with the patient (e.g. "At the beginning of the interview I struggled to establish an emotional connection with the patient", "I found it difficult to follow the train of thoughts expressed by the patient", "I perceived a discordance between the way in which the patient experienced some of his/her life events and the way in which I would have experienced them"). The Engagement scale contains items describing the degree of the psychiatrist's involvement with the patient, such as feelings of boredom, indifference, detachment, lack of attention and, conversely, desire to take care of the patient, feelings of involvement in the patient-physician relationship, emotional closeness, and tenderness (e.g. "I experienced a feeling of tenderness towards the patient", "I felt emotionally close to the patient"). The Disconfirmation scale includes items describing a failure to establish an authentic relationship with the patient and feelings of being manipulated, rejected, criticised, or devalued by the patient (e.g. "I felt depreciated by the patient", "I felt judged by the patient", "I felt rejected by the patient", "I felt that I did not exist for the patient"). The Impotence scale consists of items indicating feelings of helplessness, frustration, desolation, emptiness, loneliness, and being drained (e.g. "I felt a sense of loneliness", "I felt a sense of emptiness", "At the end of the interview I felt a sense of impotence").

9.2.2 ACSE and Psychiatric Diagnosis

In our first study, we attempted to explore the relationship between profiles of the clinician's subjective experience and psychiatric diagnosis, expressed by a number of major DSM- and ICD-like categories [40].

The study was performed in several psychiatric inpatient and outpatient units in Rome, Italy. The clinicians completed the ACSE questionnaire and other standardised assessment instruments when they evaluated a previously unknown patient. All adult patients diagnosed with schizophrenia, Cluster B personality disorder, manic or mixed bipolar I episode, and unipolar depression or anxiety disorder were included in the study, for a total of 422 patients evaluated by 35 clinicians.

We found a significant and theoretically consistent relationship between the clinicians' patterns of subjective experience during the first visit and patients' clinical diagnoses (Fig. 9.1), both in univariate and multivariate analyses. In particular, in multivariate models controlling for patient's age and education, symptom severity,

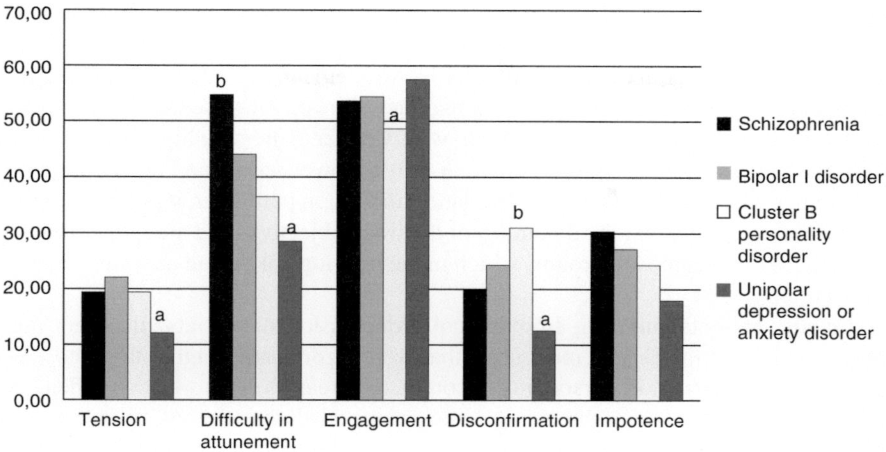

Fig. 9.1 Clinician's subjective experience scores* by diagnosis. * ACSE scores were converted to a common metric (i.e., the percentage of the maximum possible score on each scale). **a** significantly lower than the other groups in multivariate analysis. **b** significantly higher than the other groups in multivariate analysis

clinician's sex, duration of the visit, and setting, the categorical diagnosis remained a significant predictor of scores on all ACSE scales, except for Impotence.

The clinicians reported significantly higher levels of Difficulty in Attunement with schizophrenic patients than with all other patients, including those with a bipolar manic or mixed episode. We have hypothesised that experiencing a failure in emotional contact and empathic understanding represents a specific characteristic of the quality of the interaction with these patients, which does not depend on the severity of the clinical picture, even when psychotic symptoms are present. This seems intriguing, as the Difficulty in Attunement dimension strongly resembles the experience described by many psychopathologists as a feeling of *extraneity* and lack of *shared foundations* with the other. As previously described (Sect. 9.1), it is in this feeling that the conceptualisations of an intuitive diagnosis are grounded.

Patients with Cluster B personality disorders displayed significantly higher scores on the Disconfirmation scale than all the other patient groups. They also showed significantly lower scores on the Engagement scale, which was also rather expected, consistently with a reaction characterised by negative feelings. Such a profile of subjective experience, characterised mainly by feelings of detachment, boredom, and being rejected, devalued, and manipulated, is consistent with what is informally reported by many psychiatrists regarding their everyday clinical practice with these patients. Indeed, this sort of "typical" reaction also resembles the one identified by means of other instruments [29, 34–37] and is reminiscent of some psychoanalytical *countertransference* descriptions [20, 41].

Patients suffering from a manic or mixed episode showed no "marker" ACSE dimension. However, the findings suggested a distinct profile of clinician's subjective experience, as these patients significantly differed from each of the other groups on at least one dimension. In comparison with patients with unipolar depression and anxiety, those with a bipolar manic or mixed episode were characterised by higher scores for Tension, Impotence, Difficulty in Attunement, and Disconfirmation. As compared with patients with schizophrenia, they showed lower scores on Difficulty in Attunement. With respect to patients with Cluster B personality disorders, the patients with a manic or mixed episode displayed lower scores on Disconfirmation and higher scores on Engagement. The latter finding, in particular, was challenging, because it suggested that the dynamics of the intersubjective field are quite dissimilar in these two diagnostic groups, which in the nosological debate are often seen as overlapping.

Finally, patients suffering from unipolar depression or anxiety disorders displayed scores on the Engagement scale that were significantly higher than those of patients with Cluster B personality disorder, whereas they showed significantly lower scores than the other three diagnostic groups on all the other ACSE scales. We interpreted this less pronounced reaction elicited in clinicians as in line with the hypothesis that the interaction with these patients is less distressing due to the "reassuring" help-seeking attitude that they commonly show, which is typical of the doctor-patient relationship in medicine.

In conclusion, for each patient group, we observed a fairly distinct profile of clinician's reaction, a sort of "average expectable response" towards patients belonging to different broad diagnostic categories. These findings suggested that, at least for those categories, the clinician's subjective experience as measured by ACSE could contribute to the process of differentiating patients into categorically defined syndromes. In other words, despite the general idea that subjective elements are confounding factors for an accurate and reliable diagnosis, our first study supported the tenet of classical psychopathology that the clinician's subjective experience plays a significant role in the diagnostic process.

9.3 ACSE and Psychopathological Dimensions

Given the valuable diagnostic potential of the dimensional approach when compared with the categorical one, as discussed in previous chapters, we reasoned that the analysis of the relation between the clinician's subjective experience and psychopathological dimensions would represent a further critical step in our investigation.

First, we carried out a large study aimed at identifying which psychopathological dimensions, derived from a widely used psychiatric rating scale, are associated with certain reactions in clinicians. Some of the findings of this study have been presented in a recent paper [42]. While that paper presented analyses aimed at identifying which psychopathological dimensions are independently associated with each facet of clinician's subjective experience, the present chapter presents analyses aimed at delineating the pattern of clinician's subjective experience independently associated with each psychopathological dimension. We also sought to determine whether the clinician's subjective experience adds significant information about a given psychopathological dimension over and above the information provided by all the other dimensions. Then, we compared the findings of the study on psychopathological dimensions with those of the previous study on categorical diagnoses. Finally, we made some hypotheses on the potential associations between the clinician's subjective experience and psychopathological dimensions as assessed with a dimensionally oriented instrument such as the SVARAD, in order to propose a more specific integration between dimensional assessment and the intersubjective perspective.

9.3.1 Description of the Study

9.3.1.1 Setting and Participants

The study was carried out in several psychiatric inpatient and outpatient units of the National Health Service in Rome, Italy. The clinicians working in these units were asked to complete a number of assessment instruments when they saw a new patient for clinical and diagnostic evaluation. To be included in this study, patients had to meet the following criteria: (1) at least 18 years of age, (2) Italian nationality (to prevent problems in mutual understanding due to language difficulties in foreign patients), (3) absence of intellectual disability or substantial cognitive impairment, (4) absence of substance use disorder, and (5) absence of major medical illness.

9.3.1.2 Assessment

A standardised form was used to gather information about demographic variables, setting and duration of the visit, and patient clinical diagnosis according to DSM-IV-TR [43] or ICD-10 [44] criteria. After the visit, the clinician completed the Assessment of Clinician's Subjective Experience (ACSE) instrument and the 24-item version of the Brief Psychiatric Rating Scale (BPRS).

The 24-item BPRS [45, 46] is an expanded version of the original 16- and 18-item versions of the instrument [47, 48]. We used an Italian version of the BPRS [49] that has high reliability [50] as it is based on the BPRS manual of administration [46], with defined anchor points, detailed probe questions, and rules for scoring. The items are scored on a 7-point severity scale. Higher scores indicate greater severity of psychiatric symptoms.

The BPRS does not provide subscales, as neither the 18- nor the 24-item versions were designed with a specific scale structure in mind. The instrument was developed to assess a wide variety of psychiatric symptoms, and the items were selected for breadth of coverage rather than as indicators of specific psychopathological dimensions. However, many factor analytic studies have been performed, and a recent meta-analysis of this literature [51] found evidence for four core dimensions underlying the 24-item version. These four dimensions are Affect (anxiety, depression, suicidality, guilt), Positive Symptoms (suspiciousness, hallucinations, unusual thought content, and grandiosity), Negative Symptoms (blunted affect, emotional withdrawal, and motor retardation), and Activation (elevated mood, excitement, distractibility, and motor hyperactivity). In addition, a fifth dimension named

Disorganisation (self-neglect, disorientation, conceptual disorganisation, mannerisms, and posturing) deserves to be considered, as it was present in several factor analyses. In this study, all these BPRS dimensions showed adequate reliability in terms of internal consistency.

9.3.1.3 Statistical Analyses

All statistical analyses were performed using SPSS for Windows, version 22.0 (SPSS Inc., Chicago, IL). All tests were two-tailed, with alpha set at 5%.

First, to examine the association between ACSE scores and study variables, analysis of variance was used for categorical variables, and Pearson's correlation coefficient was used for continuous variables. Subsequently, multiple linear regression analysis was used to identify the aspects of clinician's subjective experience that were independently associated with psychopathological dimensions. In each model, the relevant BPRS dimension served as the dependent variable, while scores on the ACSE scales and a number of patient (age, sex, and education), clinician (age and sex), and context (setting and duration of the visit) variables were entered as independent variables.

Finally, five hierarchical regression models were constructed with the aim of determining if adding information about the clinician's subjective experience improved prediction of patient scores on each psychopathological dimension beyond that afforded by demographic and context variables and by the scores on all the other psychopathological dimensions. Each model included one BPRS dimension as dependent variable and three blocks of independent variables. The first block included demographic and context variables (patient and clinician age and sex, patient education, setting, and duration of the visit); the second included the scores on all the other BPRS dimensions except the one under examination; the third included the clinician's subjective experience during the visit (scores on all ACSE scales). In all, 16 variables grouped into three blocks were considered in the analysis.

In all regression analyses, outliers in each solution were identified by examination of standardised residuals. A criterion of $p < .001$ was used to define outliers [52].

9.3.2 Results

Overall, 30 psychiatrists and 15 senior psychiatry residents with different theoretical backgrounds and attitudes were involved in the study. The mean number of patients rated per clinician was 17.4 (range 4–40). Altogether, the sample of clinicians comprised 18 males (40.0%) and 27 females (60.0%). The mean age of psychiatrists and residents was 40.7 ± 10.0 and 30.9 ± 2.8, respectively. The mean post-residency experience of the 30 psychiatrists was 10.0 ± 9.0 years.

They recruited a total of 783 patients, of whom 44.6% were evaluated in outpatient clinics and 55.4% in hospital settings (acute inpatient wards or emergency rooms); the mean duration of the visit was 42.1 ± 15.8 minutes. The patients were composed of 348 men (44.5%) and 434 women (55.5%), with a mean age of

42.9 ± 15.2. Given the large inclusion criteria and the number of different settings involved in the study, almost all of the common diagnoses were represented in the sample, with 228 (29.1%) patients diagnosed with a psychotic disorder, 293 (37.4%) patients diagnosed with a mood (unipolar or bipolar) disorder, 77 (9.8%) patients diagnosed with an anxiety disorder, and 185 (23.6%) patients with other or no Axis I disorders. Regarding Axis II, 165 (21.1%), 21 (2.7%), and 28 (3.6%) of the patients received a diagnosis of Cluster B, Cluster A, and Cluster C personality disorders, respectively. Most patients ($N = 537$, 68.6%) did not receive an Axis II diagnosis. The BPRS mean total score was 50.0 ± 15.9.

Univariate analysis revealed several significant associations between psychopathological dimensions, clinician's subjective experience, and various demographic and context variables. Therefore, multiple regression analysis was performed to control for the main potential confounders identified in univariate analysis. This analysis was performed on 754 patients with complete data for all the variables of interest. We constructed five regression models (one for each BPRS dimension) including the ACSE scores and all the other variables that were found to be associated with BPRS dimensions. For the five BPRS dimensions included as dependent variables (BPRS Affect, Positive symptoms, Negative symptoms, Activation, and Disorganisation), a total of 3, 3, 4, 18, and 15 outliers, respectively, were identified and removed from the models. The results of multiple regression analysis, including zero-order correlations, are reported in Table 9.1.

The strongest independent predictor of Positive Symptoms was Difficulty in Attunement, which explained a percentage as high as 13% of unique variance. Among ACSE scales, higher Tension and Engagement and lower Disconfirmation also emerged as significant, though modest, independent predictors. Other significant predictors were context and demographic variables, such as hospital setting, older clinician age, and higher patient education.

Impotence was the strongest predictor of Negative Symptoms and explained 9% of unique variance. Other strong predictors were higher Difficulty in Attunement and lower Disconfirmation, which explained 5% and 4% of unique variance, respectively. Lower Engagement was also a significant, though modest, predictor, as well as clinician older age and male sex.

Difficulty in Attunement was by far the strongest predictor of Disorganisation, explaining 11% of unique variance. Lesser predictors were Disconfirmation, Impotence, patient older age and less education, clinician older age and male sex, and hospital setting.

Concerning Activation, all ACSE scales were found to be significant, though modest or moderate, independent predictors, with Impotence displaying a negative association and the other scales a positive association. Other significant predictors were hospital setting and older clinician age.

Although the regression model with Affect as a dependent variable was significant ($p < 0.001$), it explained a relatively low proportion of variance, and the predictors were less strong, as compared with the other four models. Affect was found to be positively associated with clinician's Engagement, Disconfirmation, and Impotence and negatively associated with Tension and Difficulty in Attunement.

Table 9.1 Multiple regression of BPRS psychopathological dimensions on ACSE scores, patient age, sex and education, clinician age and sex, setting, and duration of the visit

	Affect			Positive symptoms			Negative symptoms			Activation			Disorganisation		
F	6.9[c] (df 12, 738)			39.4[c] (df 12, 738)			26.9[c] (df 12, 737)			17.9[c] (df 12, 723)			30.7[c] (df 12, 726)		
R	0.32			0.62			0.55			0.48			0.58		
R^2	0.10			0.39			0.30			0.23			0.34		
Adjusted R^2	0.09			0.38			0.29			0.22			0.33		
	R	β	Sr²	R	β	Sr²	r	β	Sr²	r	β	Sr²	r	β	Sr²
Tension	−0.06	−0.10	0.00	0.34	0.12[b]	0.01	0.16	−0.09	0.00	0.29	0.14[b]	0.01	0.22	−0.02	0.00
Difficulty in attunement	−0.06	−0.12[a]	0.01	0.54	0.48[c]	0.13	0.40	0.30[c]	0.05	0.33	0.22[c]	0.03	0.50	0.44[c]	0.11
Engagement	0.10	0.12[b]	0.01	0.03	0.12[c]	0.01	−0.07	−0.13[b]	0.01	−0.02	0.14[c]	0.01	−0.04	0.02	0.00
Disconfirmation	0.08	0.12[a]	0.01	0.19	−0.10	0.00	0.07	−0.29[c]	0.04	0.27	0.16[b]	0.01	0.12	0.13[b]	0.01
Impotence	0.08	0.12[b]	0.01	0.30	0.00	0.00	0.40	0.40[c]	0.09	0.16	−0.11[a]	0.01	0.30	0.11[b]	0.01
Patient age	−0.01	−0.02	0.00	−0.03	0.02	0.00	0.01	−0.02	0.00	0.00	0.06	0.00	0.12	0.14[c]	0.02
Patient sex	−0.03	−0.03	0.00	0.03	0.00	0.00	0.08	0.03	0.00	0.08	0.05	0.00	0.01	−0.02	0.00
Patient education	0.10	0.02	0.00	−0.01	0.08[b]	0.01	−0.01	0.04	0.00	−0.02	−0.03	0.00	−0.17	−0.10[b]	0.01
Clinician sex	0.45	0.06	0.00	0.15	0.05	0.00	0.11	0.12[c]	0.01	0.15	0.02	0.00	0.12	0.08[a]	0.01
Clinician age	0.07	0.16[c]	0.02	0.10	0.08[a]	0.01	0.18	0.16[c]	0.02	0.28	0.24[c]	0.05	0.11	0.11[b]	0.01
Setting	0.78	0.10[a]	0.01	0.36	0.24[c]	0.04	0.11	0.02	0.00	0.15	0.11[b]	0.01	0.28	0.18[c]	0.02
Duration of the visit	0.04	0.16[c]	0.02	−0.14	−0.02	0.00	−0.06	−0.02	0.00	−0.01	0.03	0.00	−0.12	0.02	0.00

r = zero-order correlation; β = standardised regression coefficient; Sr² = squared semipartial correlation

[a]$p < 0.05$

[b]$p < 0.01$

[c]$p < 0.001$

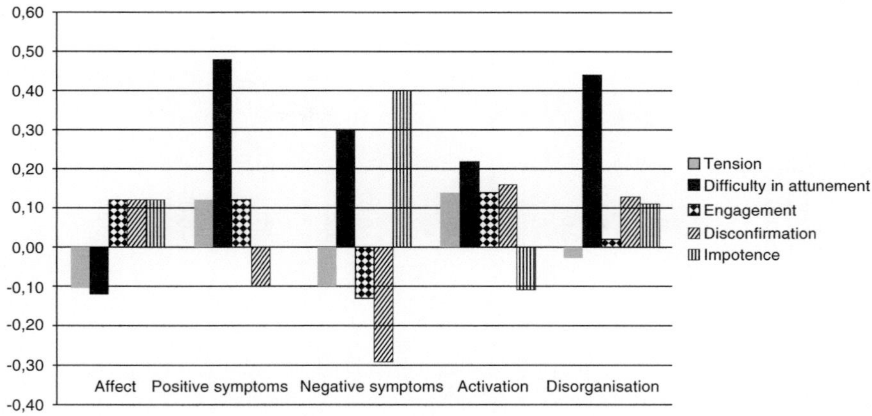

Fig. 9.2 Patterns of independent associations (beta coefficients) between each clinician's subjective experience facet and psychopathological dimensions. Each bar represents the independent association, expressed by the standardised regression coefficient, between each BPRS dimension and each ACSE scale

Other significant predictors were clinician age, hospital setting, and duration of the visit.

Overall, the analysis revealed that each psychopathological dimension is characterised by a distinct pattern of independent associations with the clinician's subjective experience, as graphically illustrated in Fig. 9.2.

Regarding hierarchical regression analysis, for the five BPRS dimensions included as dependent variables (Affect, Positive Symptoms, Negative Symptoms, Activation, and Disorganisation), a total of 4, 7, 9, 23, and 20 outliers, respectively, were identified and removed from the models. This analysis showed that for all psychopathological dimensions except Disorganisation, the ACSE scores significantly improved prediction beyond that afforded by demographic and context variables and by all the other psychopathological dimensions. Positive and negative symptoms, in particular, showed a marked improvement in prediction when ACSE scores were added to the regression model (Table 9.2). These results suggest that not only during a categorical assessment but also during a dimensional assessment, subjective experience provides the clinician with perceptual and intuitive information that otherwise might get lost.

9.3.3 Discussion

The main limitation of our studies is the nonindependence of the assessments, as the same clinician completed the ACSE and evaluated the patient for the purpose of categorical diagnosis or dimensional assessment. However, the psychiatrists involved in the studies were not aware of the studies' objectives. Therefore, it is unlikely that they deliberately used their own feelings for rating psychopathology or

Table 9.2 Hierarchical regression analyses predicting each psychopathological dimension from demographic and context variables, all other psychopathological dimensions, and ACSE scales

	Affect			Positive symptoms			Negative symptoms			Activation			Disorganisation		
	Final β	R^2	Δ*F* (df)	Final β	R^2	Δ*F* (df)	Final β	R^2	Δ*F* (df)	Final β	R^2	Δ*F* (df)	Final β	R^2	ΔF (df)
Step 1: Demographic and context variables		0.04	7.7[c] (7, 742)		0.15	18.0[c] (7, 739)		0.05	5.8[c] (7, 737)		0.12	13.6[c] (7, 723)		0.14	17.3[c] (7, 726)
Patient sex	−0.02			0.00			0.06[a]			0.07[a]			−0.04		
Patient age	−0.01			−0.01			-0.06[a]			0.02			0.11[c]		
Patient education	0.03			0.12[c]			0.05			0.00			−0.12[c]		
Clinician sex	0.06			0.04			0.09[b]			0.05			0.02		
Clinician age	0.19[b]			−0.01			0.17[c]			0.24[c]			−0.02		
Setting	−0.17[c]			0.16[c]			−0.02			0.01			0.09[b]		
Duration of the visit	0.16[c]			−0.02			−0.04			0.02			0.04		
Step 2: Other psychopathological dimensions		0.14	15.9[c] (4, 738)		0.49	125.1[c] (4, 735)		0.40	108.8[c] (4, 733)		0.38	78.2[c] (4, 719)		0.52	144.4[c] (4, 722)
	−0.28[c]			−0.16[c]			0.05			−0.05			−0.02		
	0.09[a]			0.06			0.08[a]			0.26[c]			0.29[c]		
	−.08			0.24[c]			−0.37[c]			−0.39[c]			0.37[c]		
	−0.02			0.28[c]			0.43[c]			0.33[c]			0.25[c]		
Step 3: Clinician's subjective experience		0.17	5.5[c] (5, 733)		0.55	19.3[c] (5, 730)		0.50	29.1[c] (5, 728)		0.42	8.5[c] (5, 714)		0.53	2.0 (5, 717)
Tension	−0.06			0.05			−0.04			0.08[a]			−0.05		

Difficulty in attunement	0.01			0.29[c]			0.20[c]			0.09[a]			0.10[b]		
Engagement	0.18[c]			0.12[c]			−0.08[b]			0.07[a]			−0.03		
Disconfirmation	0.14[b]			−0.05			−0.19[c]			0.13[b]			−0.05		
Impotence	0.08			−0.03			0.30[c]			−0.03			0.02		

For each model, the table displays R^2, and variation in F (ΔF) for each step, and the final standardised regression coefficients (beta) for each variable. The final beta weights for BPRS dimensions are listed in the order they are listed in the first row, with exclusion of the predicted dimension (e.g. the first beta in the Affect column relates to Positive Symptoms, the first beta in the Positive Symptoms column pertains to Affect and the second to Negative Symptoms, etc.)

[a]$p < 0.05$

[b]$p < 0.01$

[c]$p < 0.001$

making diagnoses to a greater than usual extent as a consequence of participating in the study. The nonindependence of assessments should not therefore detract from the finding of a significant relation between the dimensional or categorical assessment made by a psychiatrist and his or her pattern of subjective experience during the interaction with the patient.

These studies suggest that there is a meaningful connection between the clinician's feelings and the patient's mental suffering. This holds true from both a categorical and a dimensional point of view, which indirectly confirms the interconnections between these two perspectives on mental illness. If we look at the results of the two studies as a whole, in fact, we can identify some coherent and quite expectable patterns of association between the characteristics of patients' pathological experience and the clinicians' perception of the first interaction with them.

For the clinicians who took part in the studies, the encounter with chronic psychotic patients seemed to be characterised by a struggling attempt to establish an empathic relationship. This difficulty was not fully explained by the overall psychopathological severity and was not influenced by the degree of affective involvement. Difficulty in Attunement, which showed predictive value with regard to the schizophrenia category (a sort of "operational" *Praecox Gefühl*), was also by far the strongest independent predictor of the BPRS dimensions of Positive Symptoms and Disorganisation. Taken together, these results suggest that the clinician's reaction to psychopathological dimensions like reality distortion, and to clinical phenomena such as mannerisms and psychic disarticulation, may significantly account for the unique experience of being with a schizophrenic. Moreover, as suggested by the hierarchical model, this particular sense of a patient's *psychoticity* seems not to depend on other sources of clinical information (demographic or psychopathological) and is arguably grounded in a clinician's gestaltic perception. This finding corroborates the notion advanced by classical psychopathologists, who considered such perception as an unfiltered intuition of the schizophrenic patient's peculiar way of *being-in-the-world*. The finding is also consistent with the pre-reflective and insight-based nature of this intuition.

For patients with Cluster B personality disorder, our study found the clinicians' reactions to be characterised by the highest level of Disconfirmation and the lowest degree of Engagement among all the clinical interactions that we examined. Concerning psychopathological dimensions, Disconfirmation was a substantial independent predictor only for lower Negative Symptoms and a modest one for all the other dimensions. Similarly, Engagement was a modest or negligible predictor of all BPRS dimensions. The finding that these two ACSE scales displayed a strong categorical association and modest dimensional associations, in multivariate as well as in univariate analysis, is intriguing. One may hypothesise that the BPRS, even in its dimensional structure, does not grasp the psychopathological dimensions that could best explain the clinician's experience of detachment and disconfirmation when encountering a patient with Cluster B personality disorder. Possibly, patient characteristics such as impulsivity, explicit or implicit aggressiveness, and negative attitude towards others might better account for the interpersonal dynamics of the encounter with these patients and for the related clinician's lived experience.

Although no ACSE dimension specifically characterised the patients' suffering from a manic or mixed episode, they displayed a distinct profile of clinician's subjective experience, because they significantly differed from each of the other diagnostic groups in at least one dimension. Higher Difficulty in Attunement distinguished these patients from those with schizophrenia. Lower Engagement and higher Disconfirmation discriminated them from those with Cluster B personality disorders. In comparison with patients with unipolar depression and anxiety, higher scores on Tension, Impotence, Difficulty in Attunement, and Disconfirmation characterised those with a manic or mixed episode. From a dimensional perspective, in univariate analysis, clinicians' Tension was moderately correlated with Positive Symptoms, Activation, and Disorganisation, whereas in multivariate analysis, it displayed only modest independent associations with Activation and Positive Symptoms. This finding suggests that Tension, when considered together with the other aspects of clinician's subjective experience, has poor discriminating ability between subtle psychopathological differences. Rather, it seems to account for a general sense of impending threat and danger, which, regardless of diagnosis, may guide the intervention and indicate the need to establish a quiet and containing relationship. On the other hand, Impotence showed a strong independent association with Negative Symptoms and modest associations with Disorganisation, Affect, and lower Activation. Thus, regardless of the peculiar nature of the diagnosable mental disorder, clinicians' experiences of impotence seemed to be elicited mainly by patients' traits reflecting isolation and blunting of affect and motor behaviour. These characteristics basically pertain to chronicity of severe mental illness, and their correspondence with clinicians' feelings of frustration and helplessness may account for a remarkable aspect of the therapeutic relationship in terms of a clinician's confidence in treatment resources and success. It is not surprising, in light of the above, that the clinicians reported low levels of Impotence and Tension, as well as Difficulty in Attunement and Disconfirmation, when they saw patients with anxiety disorders or unipolar depression, as these patients represent a less challenging patient population in many respects.

As already mentioned, the use of an instrument explicitly designed to dimensionally describe patient psychopathology would allow a more in-depth analysis of the relation between psychopathological dimensions and the clinician's subjective experience. We are attempting to address this issue in an ongoing study involving the use of the ACSE and the SVARAD. While results are not yet available for this study, some plausible hypotheses can be made on the basis of research data on the correlation between BPRS dimensions and SVARAD items. These unpublished data come from a study on individual differences and psychopathology [53–55] and from our previously mentioned ongoing study. In the first of these studies, 151 psychiatric inpatients were administered the SVARAD together with several other assessment instruments, among which was the 24-item BPRS. In the second study, 105 psychiatric inpatients and outpatients were administered the SVARAD and a number of other assessment instruments, among which was the 24-item BPRS. In both these patient samples, we observed several strong and highly significant correlations ($p < 0.001$) between the BPRS and SVARAD dimensions. The BPRS Positive Symptoms dimension was highly correlated with the SVARAD Reality Distortion and Thought Disorganisation dimensions ($r = 0.72$ and 0.47, respectively, in the first sample; $r = 0.88$ and 0.73, respectively, in the second sample). Similarly, BPRS Disorganisation was highly correlated with SVARAD Reality Distortion and Thought Disorganisation ($r = 0.34$ and 0.50, respectively, in the first sample; $r = 0.62$ and 0.78, respectively, in the second sample). BPRS Affect showed substantial correlations with SVARAD Apprehension/Fear, Sadness/Demoralisation, and Apathy ($r = 0.65$, 0.73, and 0.58, respectively, in the first sample; $r = 0.41$, 0.72, and 0.35, respectively, in the second sample). While also being somewhat correlated with BPRS Affect, SVARAD Apathy showed a stronger association with BPRS Negative Symptoms ($r = 0.69$ and 0.59 in the first and second sample, respectively). BPRS Activation was highly correlated with SVARAD Activation ($r = 0.65$ and 0.81 in the first and second sample, respectively) and, to a lesser extent, with Thought Disorganisation ($r = 0.32$ and 0.42 in the first and second sample, respectively). The remaining SVARAD dimensions, i.e. Anger/Aggressiveness, Obsessiveness, Impulsivity, and Somatic Preoccupation/Somatisation, did not consistently show substantial correlations with BPRS dimensions, in both studies. This suggests that the BPRS, even when used in the form of its factorially derived dimensions, does not capture the same wide spectrum of psychopathology that is covered by the SVARAD.

From these findings, it might be hypothesised that the pattern of associations between the ACSE scales and the SVARAD dimensions that are fairly comparable with the BPRS ones should be predictable to a certain degree. For example, one would expect Difficulty in Attunement to be correlated with SVARAD Reality Distortion and Thought Disorganisation and Impotence and Disconfirmation to be correlated with SVARAD Apathy. It may also be hypothesised that the SVARAD dimensions that do not have a clear counterpart in the BPRS might better account for the clinician's feelings of Engagement, Tension, and Disconfirmation. Therefore, the study of the relation between the clinician's subjective experience and psychopathological dimensions as assessed by the SVARAD may contribute to improved understanding of the association between clinicians' feelings and categorical diagnosis and to a further clarification of the relation between patients' dimensional profiles and clinicians' reactions to them.

9.4 Categorical, Dimensional, and Intersubjective Perspectives: A Possible Path to the Patient's *Three-Dimensionality*

In this chapter, we have attempted to introduce, starting from theoretical considerations and continuing through empirical data, the potential of a clinical vision that encompasses categorical, dimensional, and intersubjective perspectives. In particular, we have made an effort to complement the personalising attitude underlying the dimensional approach with a change in the observation angle of the clinical encounter. Essentially, we consider that the clinician's subjective experience tells us something about the patient. This shift in perspective is based on the phenomenological concept of a co-constructed reality, which assumes that human beings live immersed in the field of a *being-with-the-other* and that their individual experience is indeed the result of an intersubjective intertwining.

Our studies supported this position, as they corroborated a multifaceted view of clinician-patient interactions, in which the clinician's subjective experience revealed its overdetermined nature. With respect to the clinical situation, in fact, some of the clinician's feelings seemed to particularly refer to the core features of clinical entities, i.e. the feeling of an empathic disruption in relation to the schizophrenic intersubjective world disintegration. In other cases, clinicians' feelings seemed to better grasp relational or individual dimensions that cross-cut the diagnostic categories, i.e. the feeling of impotence in relation to the apparent freezing of any evolving possibility or the feeling of tension in relation to interpersonal dimensions suggesting the risk of unpredictable outbursts.[1]

In our research, we attempted—for the first time—to examine, from the clinician's perspective, the *process* activated by the clinician-patient encounter that culminates in a diagnostic synthesis. Our results have led us to observe, in everyday clinical settings, that psychiatric assessment seems indeed to result from different and only artificially divisible phases:

1. First, a *second-person* phase [3, 4], in which the clinician grasps, through a personal involvement in the interaction, the *Gestalt*, the wholeness of the clinical picture; this understanding starts from the initial pre-reflective feeling towards the *presence* of the patient.
2. Second, a *third-person* phase, in which the clinician organises the symptoms and signs in meaningful entities through a more reflective and objectifying process, both in a dimensional and a categorical perspective.

This view may align with psychiatrists' growing attention to the development of the ability to draw individual and unique profiles of patients' needs and to offer person-centred therapeutic interventions. It indeed adds value to the substantial body of knowledge presented in this book, as it complements the concept of the dimensionality of the patient's psychopathology with a more extensive dimensionality: the dimensionality of the clinician's experience and, above all, the dimensionality of the inter-human encounter. Our research may also contribute by raising

[1] It is in view of these empirical observations that we suggest looking at the ACSE as a sort of *phenomenological probe*, which, exploring the space between the clinician and the patient, could inform about the complex nature of the patient's way of *being-with-the-other*. One may hypothesise that Tension generally accounts for the spatial coordinates of the encounter, while Difficulty in Attunement accounts for decreased empathic possibilities, Engagement for the development of sympathy, Disconfirmation for the effects of a nullifying attitude towards the other, and Impotence for the temporality emerging from the synchronisation or desynchronisation of two different *times*. The reader is referred to Chap. 10 for a thorough examination of the concept of temporality.

psychiatrists' awareness of the relational aspects of the therapeutic relationship and thereby improving the ability of trained psychiatrists to foster a therapeutic alliance, even with those patients who are treated with medication only.

This new line of research might promote the scientific investigation of the deep relation between intersubjectivity and mental disorders. Research in social neuroscience supports the view that the human brain evolved to be a "social brain", which is consistent with the phenomenological view that the human brain or mind never works in isolation but in relation to others. To the extent that neuroscience, phenomenological psychopathology, and diagnostic classification work together in the future, many of our current diagnostic categories might have to be reconceptualised as disorders of an embodied, intersubjective human self, embedded in interactions as social agent [56].

References

1. Potuzak M, Ravichandran C, Lewandowski KE, Ongür D, Cohen BM. Categorical vs dimensional classifications of psychotic disorders. Compr Psychiatry. 2012;53:1118–29.
2. APA. Diagnostic and statistical manual of mental disorders. 5th ed. Arlington, VA: American Psychiatric Publishing; 2013.
3. Stanghellini G. The grammar of the psychiatric interview. A plea for the second-person mode of understanding. Psychopathology. 2007;40:69–74.
4. Fuchs T. Subjectivity and intersubjectivity in psychiatric diagnosis. Psychopathology. 2010;43:268–74.
5. Jaspers K. Allgemeine Psychopathologie. Berlin: Springer; 1913.
6. Kraus A. The significance of intuition for the diagnosis of schizophrenia. In: Maj M, Sartorius N, editors. Schizophrenia. Chichester: Wiley; 1999. p. 47–9.
7. Srivastava A, Grube M. Does intuition have a role in psychiatric diagnosis? Psychiatry Q. 2009;80:99–106.
8. Parnas J. A disappearing heritage. The clinical core of schizophrenia. Schizophr Bull. 2011;37(6):1121–30.
9. Rümke HC. Das Kernsymptom der Schizophrenie und das 'Praecox Gefuhl'. Z Gesamte. Neurol Psychiatr. 1941;102:168–75.
10. Sagi GA, Schwartz M. The "praecox feeling" in the diagnosis of schizophrenia: a survey of Manhattan psychiatrists. Schizophr Res. 1989;2:35.
11. Dimic S, Wildgrube C, McCabe R, Hassan I. Non-verbal behaviour of patients with schizophrenia in medical consultations – a comparison with depressed patients and association with symptom levels. Psychopathology. 2010;43:216–22.
12. Binswanger L. Welche Aufgaben ergeben sick für die Psychiatrie aus den Fortschritten der neueren Psychologie? Z Gesamte Neurol Psychiatr. 1924;91(1):402–36.
13. Wyrsch J. Über die Intuition bei der Erkennung der Schizophrenen. Schweiz Med Wochenschr. 1946;46:1173–6.
14. Minkowski E. Le temps vécu. Etudes phénoménologiques et psychopathologiques. Paris: Coll. de l'Evolution Psychiatrique; 1933.
15. Minkowski E. La schizophrénie. Psychopathologie des schizoïdes et des schizophrènes. Paris: Payot; 1927.
16. Binswanger L. Drei Formen missglückten Daseins. Verstiegenheit, Verschrobenheit, Manieriertheit. Tübingen: Niemeyer; 1956.
17. Tellenbach H. Geschmack und Atmosphäre. Medien menschlichen Elementarkontaktes. Salzburg: Otto Müller Verlag; 1968.

18. Heimann P. On countertransference. Int J Psychoanal. 1950;31:81–4.
19. Racker H. The meanings and uses of countertransference. Psychoanal Q. 1957;26(3):303–57.
20. Kernberg O. Notes on countertransference. J Am Psychoanal Assoc. 1965;13:38–56.
21. Klein M. Notes on some schizoid mechanisms. Int J Psychoanal. 1946;27:99–110.
22. Spillius E, O'Shaughnessy E, editors. Projective identification: the fat of a concept. New York: Routledge; 2012.
23. Kohut H. Forms and transformations of narcissism. J Am Psychoanal Assoc. 1966;14(2):243–72.
24. Bolognini S. L'empatia psicoanalitica. Torino: Bollati Boringhieri; 2002.
25. Winnicott DW. Hate in the countertransference. Int J Psychoanal. 1949;30:69–74.
26. Grube M. Towards an empirically based validation of intuitive diagnostic: Rümke's 'praecox feeling' across the schizophrenia spectrum: preliminary results. Psychopathology. 2006;39:209–17.
27. Ungvari GS, Xiang Y, Hong Y, Leung HCM, Chiu HFK. Diagnosis of schizophrenia: reliability of an operationalized approach to 'praecox-feeling'. Psychopathology. 2010;43:292–9.
28. Brody EM, Farber BA. The effects of therapist experience and patient diagnosis on countertransference. Psychotherapy. 1996;33:372–80.
29. McIntyre SM, Schwartz RC. Therapists' differential countertransference reactions toward clients with major depression or borderline personality disorder. J Clin Psychol. 1998;54(7):923–31.
30. Holmqvist R. Staff feelings and patient diagnosis. Can J Psychiatr. 2000;45(4):349–56.
31. Schwartz RC, Smith SD, Psychotherapists CB. Countertransference reactions toward clients with antisocial personality disorder and schizophrenia: an empirical test of theory. Am J Psychother. 2007;61(4):375–93.
32. Thylstrup B, Hesse M. Substance abusers' personality disorders and staff members' emotional reactions. BMC Psychiatry. 2008;8:21.
33. Røssberg JI, Karterud S, Pedersen G, Friis S. Psychiatric symptoms and countertransference feelings: an empirical investigation. Psychiatry Res. 2010;178(1):191–5.
34. Betan E, Heim AK, Conklin CZ, Westen D. Countertransference phenomena and personality pathology in clinical practice: an empirical investigation. Am J Psychiatry. 2005;162(5):890–8.
35. Røssberg JI, Karterud S, Pedersen G, Friis S. An empirical study of countertransference reactions toward patients with personality disorders. Compr Psychiatry. 2007;48:225–30.
36. Colli A, Tanzilli A, Di Maggio G, Lingiardi V. Patient personality and therapist response: an empirical investigation. Am J Psychiatry. 2014;171:102–8.
37. Tanzilli A, Colli A, Muzi L, Lingiardi V. Clinician emotional response toward narcissistic patients: a preliminary report. Res Psychother Psychopathol Process Outcome. 2015;18(1):1–9.
38. Lingiardi V, Tanzilli A, Colli A. Does the severity of psychopathological symptoms mediate the relationship between patient personality and therapist response? Psychotherapy. 2015;52(2):228–37.
39. Pallagrosi M, Fonzi L, Picardi A, Biondi M. Assessing clinician's subjective experience during interaction with patients. Psychopathology. 2014;47:111–8.
40. Pallagrosi M, Fonzi L, Picardi A, Biondi M. Association between clinician's subjective experience during patient evaluation and psychiatric diagnosis. Psychopathology. 2016;49:83–94.
41. Gabbard GO. Countertransference issues in psychiatric treatment. Washington DC: American Psychiatric Press; 1999.
42. Picardi A, Pallagrosi M, Fonzi L, Biondi M. Psychopathological dimensions and clinician's subjective experience. Psychiatry Res. 2017;258:407–14.
43. APA. Diagnostic and statistical manual of mental disorders. 4th ed Rev (DSM-IV-TR) ed. Washington: American Psychiatric Publishing; 2000.
44. World Health Organization. The ICD-10 classification of mental and Behavioural disorders. Clinical descriptions and diagnostic guidelines. Geneva: World Health Organization; 1992.
45. Lukoff D, Nuechterlein KH, Ventura J. Manual for expanded brief psychiatric rating scale. Schizophr Bull. 1986;12:594–602.

46. Ventura J, Lukoff D, Nuechterlein KH, Liberman RP, Green M, Shaner A. Appendix 1: brief psychiatric rating scale (BPRS) expanded version (4.0) scales, anchor points and administration manual. Int J Methods Psychiatr Res. 1993;3:227–44.
47. Overall JE, Gorham DR. The brief psychiatric rating scale. Psychol Rep. 1962;10:799–812.
48. Overall JE. The brief psychiatric rating scale in psychopharmacology research. In: Pichot P, Olivier-Martin E, editors. Psychological measurements in psychopharmacology: modern problems in psychopharmacology. Basel: Karger; 1974. p. 67–78.
49. Morosini P, Roncone R, Impallomeni M, Marola V, Casacchia M. Presentazione dell'adattamento italiano della Brief Psychiatric Rating Scale, versione 4.0 ampliata (BPRS 4.0). Riv Riabil Psichiatr Psicosoc. 1995;3:195–8.
50. Roncone R, Ventura J, Impallomeni M, Falloon IR, Morosini PL, Chiaravalle E, Casacchia M. Reliability of an Italian standardized and expanded brief psychiatric rating scale (BPRS 4.0) in raters with high versus low clinical experience. Acta Psychiatr Scand. 1999;100:229–36.
51. Dazzi F, Shafer A, Lauriola M. Meta-analysis of the brief psychiatric rating scale - expanded (BPRS-E) structure and arguments for a new version. J Psychiatr Res. 2016;81:140–51.
52. Tabachnick B, Fidell L. Using multivariate statistics. New York: Harper Collins College Publishers; 1996.
53. Picardi A, Toni A, Caroppo E. Stability of alexithymia and its relationships with the 'big five' factors, temperament, character, and attachment style. Psychother Psychosom. 2005;74:371–8.
54. Picardi A, Martinotti G, Paci M, Simi C, Caroppo E. Reliability of self-reported attachment style in patients with severe psychopathology. J Pers Assess. 2011;93:491–9.
55. Picardi A, Caroppo E, Porcelli P, Di Maria G, Munittola G, Martinotti G. Alexithymia and severe psychopathology: a study on psychiatric inpatients. Psychopathology. 2012;45:159–66.
56. Phillips J, Frances A, Cerullo MA, Chardavoyne J, Decker HS, First MB, Ghaemi N, Greenberg G, Hinderliter AC, Kinghorn WA, LoBello SG, Martin EB, Mishara AL, Paris J, Pierre JM, Pies RW, Pincus HA, Porter D, Pouncey C, Schwartz MA, Szasz T, Wakefield JC, Waterman GS, Whooley O, Zachar P. The six most essential questions in psychiatric diagnosis: a pluralogue. Part 4: general conclusion. Philos Ethics Humanit Med. 2012;7:14.

Phenomenology of Temporality and Dimensional Psychopathology

10

Thomas Fuchs and Mauro Pallagrosi

10.1 Introduction

The dimensional approach to diagnosis, as discussed in this book, can provide an individual profile of a patient's psychopathology. However, it still operates within a third-person framework. According to this approach, a subject (the clinician) observes an object (the patient), under the assumption that the mind is ultimately a product of brain activities. In contrast to the categorical perspective, the dimensional approach cuts across categorical boundaries, offering a more complex and refined view of the patient's psychopathological condition. It may thus lead to a more specific psychopharmacological or psychotherapeutic intervention. For example, a broad general category such as major depression can be differentiated into multiple forms of depression, each displaying relatively distinct symptomatological dimensions, which can then be treated accordingly.

The combination of categorical and dimensional diagnostics can thus convey a very accurate view of the patient. However, since both these diagnostic approaches have their foundations in a third-person perspective, they appear to lack a holistic comprehension of the different symptoms. As described in depth in the previous chapter, an approach that disregards the subjective experience underlying a symptomatological assessment risks not conveying a picture of the person in his or her totality. Moreover, it can lead to an incomplete understanding of the patient's way of *being-in-the-world* [1, 2], through which it might be possible to identify a *trouble générateur* on the basis of the various manifestations of the illness. In other words, "there is a lack of a suitable psychopathological framework that could integrate

T. Fuchs (✉)
Psychiatric Department, University of Heidelberg, Heidelberg, Germany
e-mail: thomas.fuchs@urz.uni-heidelberg.de

M. Pallagrosi
Department of Human Neurosciences, Sapienza University of Rome, Rome, Italy
e-mail: mauro.pallagrosi@uniroma1.it

M. Biondi et al. (eds.), *Dimensional Psychopathology*,
https://doi.org/10.1007/978-3-319-78202-7_10

single symptoms and neuropsychological dysfunction into a coherent whole of altered conscious experience" [3].

In contrast, the identification of a basic disturbance by means of an in-depth description of the patient's subjectivity could indeed be highly significant for diagnosis, for research, as well as for treatment purposes. Some scholars have pointed out that subjectivity and intersubjectivity are basic categories of *being-in-the-world* and thus represent intrinsic aspects of a thorough psychiatric assessment [3–5]. Phenomenology in the Husserlian tradition seems ideally suited as a conceptual framework for a precise description including the integration of single anomalous experiences into more encompassing intentional structures. It "helps to define mental disorders on the basis of their structural features, linking apparently disconnected phenomena together" [6]. Hence, its contribution to psychopathological analysis could essentially enrich the objectifying perspective of categorical and dimensional studies, giving depth and substance to clinical observations based on these two paradigms.

Phenomenology can be defined as a descriptive and analytical science of consciously lived experiences and the objects of these experiences. Classical phenomenologists investigate consciousness by putting aside causal explanations and focusing on the way it shows itself in subjective experience. "Consciousness manifests itself as a 'becoming', a temporal 'streaming' of a unity of intertwined experiences. This streaming is not an amorphous mass of contents, but is organized into a field of consciousness, which exhibits certain structures involving intentionality, temporality, embodiment, self-awareness, and intersubjectivity" [7].

Classical psychopathology has been profoundly influenced by phenomenological analysis, and many authors, such as Jaspers, Minkowski, Binswanger, Tellenbach, and Blankenburg, have adopted the phenomenological method to explore the alteration of consciousness in mental illness. Indeed, their attempts to comprehend the lived experience of affected people have also played a role in better understanding the normal functioning of consciousness, providing philosophical contributions to the general theory of lived experience: "core features of subjectivity, including fundamental aspects of self-experience, can be sharply illuminated through a study of their pathological distortions" [8].

The basic structures investigated in psychopathology have been mainly the same ones as those of phenomenology in general: notions of self, self-awareness, temporality, intersubjectivity, and embodiment. Modern psychiatrists such as Parnas, Stanghellini, Callieri, and others have carried forward this exploration, in an ongoing debate with philosophers both in the phenomenological field and, more recently, in the cognitive sciences [9]. In the previous chapter, some issues about intersubjectivity, and how its disruption may be analysed in order to achieve a breakthrough in the exploration of mental pathology, were discussed. In particular, the potential of an overall clinical view that takes into account both a standardised diagnostic approach and an examination of subjective experience has been considered.

In this chapter, we aim to describe and investigate another core notion through which phenomenology can substantially contribute to psychopathological progress: the notion of temporality. Temporality is indeed one of the most central and complex topics in phenomenological psychopathology. Many authors such as Straus [10], Minkowski [11], von Gebsattel [12], Binswanger [13], Tellenbach [14], Blankenburg [15], and Kimura [16] have focused their research on the altered experience of time in different clinical contexts. Temporality has also been considered an essential element of schizophrenia. Even the well-known Minkowskian concepts of *schizoidy* and *syntony* are grounded on a temporal concept of rhythmic modalities, namely, a subject's synchronisation and desynchronisation with his environment.

Temporality is thus not only an essential way of investigating consciousness and the self but also has relevance for studying subjectivity in people who are affected by mental disorders. Similar to intersubjectivity (see Chap. 9), subjectivity is a basic feature of the self, and it can be regarded as belonging to a more fundamental level of explanation, underlying the different symptom manifestations. For this reason, in this chapter we will first discuss its relevance in the evaluation of psychopathological conditions and then the convenience of complementing the positivistic third-person approach with such a perspective rooted in phenomenology.

10.2 Temporality from a Phenomenological Perspective

Mental illness not only interrupts the continuity of normal life but can also be accompanied by a radical change in subjective temporality, even to the point of a fragmentation of the experience of the self in time.

The phenomenological analysis of lived time, carried out by psychopathologists such as Jaspers [17], Minkowski [11], Binswanger [13], and Tatossian [18], deals with this crucial problem. These authors attempted to develop a systematic framework for the investigation of conscious experience in mental disorders, giving particular attention to the problem of altered lived experience of time. As stated by Tatossian, the impairment of lived time should not be considered as simply one symptom among others. On the contrary, it expresses a fundamentally altered mode of existence which cannot be reduced to brain dysfunctions. Indeed, a lack of attention to this aspect implies the loss of the possibility to articulate the subjectivity of the affected person in a comprehensive way, leading to a flawed psychopathological understanding.

Phenomenological philosophy has explored the concept of time as a basic structure of the human self. According to its perspective, human beings are *time-producing* organisms, and their awareness of being is imbued with the sense of a lived duration of experience. In the phenomenological model, in fact, the flow of consciousness proceeds, on a pre-reflective level, along with a specific sense of *for-me-ness* or *mineness*. In other words, every experience that appears in the stream of consciousness belongs to the experiencing subject in a unique way, even in the pre-linguistic phase of development. This sense of *mineness*, which is considered as the essential feature of the *minimal* or *experiential self* [8], entails the intrinsic capacity of experiencing the *block of duration*, that is, the continuity bridging discrete experiences. Without it, we would have an infinite sequence of single moments, which means that there would not be any *experience* at all; for as Zahavi [19] writes, "every experience is a temporally extended lived presence".

Husserl has examined the temporality of consciousness in depth, speaking of the *width* of the present as an interlacement of three elements: *retention*, *presentation,* and *protention* [20]. Each present moment with its primary impression (*presentation*) still maintains an awareness of the just-passed moments (*retention*), which inexorably fade from the consciousness, although they can be recollected as memories later on. At the same time, the present also contains a sort of expectation of the next possible future (*protention*), an implicit anticipation of what is going to happen in the flow. According to Husserl's classic example, when we hear a melody we experience it in its temporal duration, at each moment being still aware of the notes just heard and also prepared for the tones to come; we do not experience music as simply isolated tones that replace each other abruptly.

If *retention*, *presentation*, and *protention* describe the continuity of temporality, another concept grasps its energetic foundation, still at a pre-reflective or implicit level, namely, *conation* [21]. *Conation* is conceived as the basic energetic momentum of mental life, which can be expressed by such concepts as drive, striving, urge, or affect. It may also be regarded as an affective "energy", which is at the root of our spontaneity, directedness, attention, and tenacious pursuit of a goal and which also sustains what Merleau-Ponty called the *intentional arc* [22]. Through this concept, Merleau-Ponty describes a "unity of synthesis", which bridges sequential moments of consciousness by an intentional, dynamic, and affective directedness and which connects the lived body to an intended goal of action:

> Let us therefore say […] that the life of consciousness—cognitive life, the life of desire, or perceptual life—is subtended by an 'intentional arc', which projects round about us our past, our future, our human setting, our physical, ideological, and moral situation, or, rather, which results in our being situated in all these respects. It is this intentional arc which brings about the unity of the senses, of intelligence, of sensibility and motility.(Merleau-Ponty 1962, 120)

Thus, *conation* may be regarded as the result of the encounter between the potentialities of our body and the corresponding affective qualities of the environment.

The fundamental structure of *implicit time* is thus established by a synthesis of two components, which are only conceptually distinguishable: on the one hand, the *protention*, *presentation*, and *retention* system and, on the other hand, the "energetic" moment or *conation.* At the same time, these components are the conditions for a basic (or minimal) *sense of coherent self*, which is essentially temporal and inherent in the stream of consciousness. According to Merleau-Ponty, "we must understand time as the subject and the subject as time" [22]. In the same respect, Zahavi proposes the notion of *minimal* or *experiential self*, which is not socially constituted but is instead inherent in any experience as such. He defines the minimal self as "the very subjectivity of experience and [...] not something that exists independently of the experiential flow" [8].

Since Zahavi does not speak about "energy", our proposal is to complement the notion of *experiential self* with that of a bodily and affective drive that "energises" the flow of consciousness. Moreover, we propose that implicit time is also related to *the others* with whom we share a basic "contemporality" [21]. Since their very birth, infants experience the presence of others, mainly through interbodily resonance, coordination of utterances, and affect attunement. Thus, implicit time starts with the shared rhythms and dynamics of early interactions or with *primary intersubjectivity* [23]. Even the conative-affective momentum of conscious life is not only an individual, monadic force; it is always embedded in social relationships. Infants move forward into a promising future because they feel contemporal with caring adults who structure the world to be an inviting place.

So far, we have described time experience as it is implicitly lived, i.e. thoroughly intertwined with the concept of the pre-reflective self. However, phenomenological analysis describes the self as a complex, multilayered phenomenon. The pre-reflective self is considered as a basic prerequisite for a more complex sense of self. Similarly, a different kind of intersubjectivity, which is experienced at a reflective level and consciously shared (i.e. the *secondary intersubjectivity* according to Trevarthen [24]), implies a different mode of living time.

Based on a primary sense of self-awareness, which probably starts in prenatal stages, humans develop a more complex form of self that depends on interactions with others in the intersubjective field. As a result of a long developmental process, particularly including the acquisition of language, there unfolds a *narrative self* related to autobiographical temporality. Whereas the *minimal self* is connected with the lived experience of implicit time, the emerging *narrative self* is connected to the explicit dimension of temporality. *Explicit time* superimposes itself on the implicit one when the tacit undercurrent of experience becomes consciously or reflectively lived. For example, when we look at a child obliviously absorbed in his play, we can assume that he does not reflectively experience the passing of time: time is lived at an implicit level, as an unimpeded flow. However, if we ask the child to tell a story about his play, we lead him to interrupt that flow and to share with us a narrative sequence, with a beginning, a circumscribed duration, and a "historical" reference. This is a sequence of explicit time, which indeed seems to be produced primarily through a disturbance or negation of the implicit time of pure becoming. In fact, explicit temporality frequently arises in states of desynchronisation, produced by a retardation or an acceleration of inner time in relation to external or social processes—a desynchronisation that is frequently perceived as unpleasant [21].

Explicit time can be divided into the three dimensions of *present*, *past*, and *future*, which, unlike the *presentation*, *retention*, and *protention* system, are actively reflected and synthesised by the subject. This is why they require an extended personal, or *narrative*, self, which is capable of engaging in a reflective relationship towards itself and is thus in the position, on the one hand, to project itself into the future and, on the other hand, to appropriate its own life story in the form of autobiographical narratives.

To summarise, we can distinguish between:

- **Implicit, pre-reflective, passive temporality**, based on the threefold structure of time consciousness (*retention*, *presentation*, *protention*) and on *conation*. Together, these dimensions of implicit time form the structure of the *intentional* *arc*. Included in this level is a component of basic intersubjectivity or contemporality.

- **Explicit, conscious, or reflected temporality**, actively constructed by the reflective or *narrative self*. This is substantially, but not entirely, intersubjectively constituted. The *narrative self* connects *past*, *present*, and *future* through an active synthesis, according to one's autobiographical or narrative identity.

10.3 Psychopathology of Temporality

Here we present a brief account of temporality in schizophrenia, melancholic depression, and borderline personality disorder, as paradigmatic cases for a psychopathology of temporality. We have chosen to focus on these three disorders in order to give an account of the phenomenological method and to illustrate how major symptoms of these disorders may be regarded as manifesting a disturbance of the synthesis of different levels of time consciousness.

10.3.1 Temporality in Schizophrenia

In schizophrenia, we encounter a weakening and temporal fragmentation of self-experience, which, according to phenomenological concepts, should be considered as a generative disturbance of the illness. Especially symptoms like thought disorder, thought withdrawal, or thought insertion, passivity experiences or the more basic "loss of natural self-evidence" [15] may be regarded as resulting from a fragmentation of the *intentional arc*, which is essential to all our perceiving, thinking, and acting. This disturbance of self-coherence affects intersubjective synchronisation as well, so that schizophrenia always appears as a disturbance of basic intersubjectivity or contemporality.

The continuity of the *intentional arc* disintegrates, arguably due to an impairment of the *protentional* function, thus creating gaps in the flow of consciousness, which in severe cases are experienced as thought blockages or thought withdrawal. Generally, *protention* presents a vaguely determined expectation or openness towards the future. It opens up a "cone of probability" [25]. This cone originates in the present and continuously moves forward. If the *protentional*, and thus preparatory or anticipatory, process now fails to function, events will start coming too rapidly for conscious apperception. The *protentional* function will be "overwhelmed", and perplexity results when patients try to interpret the meaning of what intrudes on them. This model resembles the concept of aberrant assignment of salience to the elements of one's experience, as proposed by Kapur [26].

The subject then loses the ability to be actively directed towards the future and is instead left with focusing on what just turned up in his consciousness or on the sensory feedback of his just-passed movement. This *transcendental delay* may be regarded as the essence of major schizophrenic self-disturbances. Acts or thoughts are no longer embedded in the continuity of basic self-experience but appear as being inserted or—if further externalised—as auditory hallucinations. The temporal disintegration of the *intentional arc* thus results in an externalisation of the fragments [21].

The synthesis of inner time consciousness is also bound up with an implicit self-awareness. If this synthesis is disturbed, the patient not only loses the feeling that particular conscious events belong to herself, but the continuity of her self-experience is affected as well. The continuity of the sense of self depends, in fact, on the constant linking of the primal impression with protention and retention. If this linkage is disturbed, the sense of self can no longer be recovered by a subsequent recording of what has been experienced.

In sum, from a phenomenological point of view, key schizophrenic symptoms such as thought disorder, thought insertion, auditory hallucinations, and passivity experiences may be described as disturbances of transcendental constitution of inner time consciousness or of the microstructure of temporality. There is increasing evidence for a structural homology between phenomenological and cognitive neuroscience views of schizophrenia with regard to the temporal order of mental life [25, 27, 28]. Several authors have pointed out the parallel between Husserl's tripartite concept of time consciousness and Fuster's analysis of the cognitive functions of the prefrontal cortex, where integration across time plays a cardinal role in the temporal organisation of behaviour [29]. This integration is served by working memory, selective attention, and preparatory set. The dorsolateral prefrontal and the anterior cingulate cortex seem to play essential roles in the neural network underlying these functions [27]. Although Husserl would have certainly opposed a neuropsychological explanation of consciousness, the fact that he attributed the intentional structure of time consciousness to passive syntheses, i.e. to functions not performed by the subject, indicates that it is reasonable to look for their possible neurobiological correlates.

In the basic stages of the illness, however, subtler disturbances in self-coherence can be found which do not yet have the character of breaks in the *intentional arc* but rather indicate a weakness of the self-awareness or *ipseity* (basic sense of self) inherent in it [30]. Patients can no longer trust the continuity and identity of their experience, which is undermined by the loss of implicit *mineness* and familiarity. Moreover, the disintegration and alienation of routine units of activity often force patients to produce every single movement intentionally in a way that one could call a Cartesian effect of the mind on the body: the body's implicit knowledge is lost and has to be substituted by "hyper-reflexive" self-observation and self-control. As Sass and Parnas have put it, the patient's mental processes "are no longer permeated with the sense of selfhood but have become more like introspected objects, with increased reified, spatialized and externalized qualities" [30].

The weakening of basic self-coherence affects intersubjective temporality in every phase of the illness. Patients do not develop a certainty of contemporality, the unquestioned assurance of living with others in a shared time of emotional resonance and synchrony. Considering this, autistic withdrawal can also be understood as an attempt to reduce the complexity of the social sphere and to compensate for the lack of ability to synchronise, namely, by avoiding overstimulating and potentially overwhelming interactions.

Finally, schizophrenic delusion can be understood as a failure to take the other's perspective: what is typical of delusions is the reinterpretation of all opposing evidence according to a rigid cognitive schema, at the price of a decoupling from intersubjective exchange. Delusions seem to permit the patient to reintegrate the irritating fragments generated by the basal disintegration of time into a coherent, though distorted framework. In other words, the intrusions, inserted thoughts, passivity phenomena, and other fragments of the broken *intentional arc* are "re-temporalised" at the explicit level, namely, reintegrated into a fixed delusional narrative. Using Heidegger's terminology, one could say that in schizophrenic delusions the "ontological" (existential) threat presented by the imminent loss of the self is replaced by the "ontic" (inner-worldly) threat posed by presumed persecutors. Thus, the frozen reality of delusion arrests the course of biographical and intersubjective time in order to compensate for the fragmentation of the more basic lived time.

To summarise: the fundamental disorder or *trouble générateur* of schizophrenia consists of a weakening and temporal fragmentation of basic self-experience. It appears in premorbid or chronic phases as a lack of a sense of self-coherence that undermines the habitual conduct of life and has to be compensated for through rational reconstruction at the explicit level of time. In acute phases, it manifests itself on the micro-level of time consciousness as an increasing fragmentation of the *intentional arc* and the self-coherence connected to it. This results in the appearance of major self-disturbances, such as thought withdrawal or thought insertion, hallucinations, and delusions of influence. In all phases, this disturbance of self-constitution is accompanied by a profound desynchronisation of intersubjective temporality, which culminates in delusion—a "frozen reality" detached from the ongoing intersubjective constitution of a shared world.

10.3.2 Temporality in Melancholic Depression

In melancholic depression, lived time becomes explicit or even object-like and turns into a constant burden of guilt. Time is reified to the point of becoming an irreversible facticity of the past on the one hand and an inevitable, predetermined future on the other. The psychotic culmination of this experience in delusions of indelible guilt or imminent death also indicates a basal disturbance of constitutive temporality. In contrast to schizophrenia, however, there is no fragmentation of the stream of consciousness, but rather a retardation or inhibition. The schizophrenic incoherence and blockade of thought to the point of thought withdrawal is fundamentally different from this inhibition, since in depression it is not the coherence of the stream of consciousness, but its *conative-affective dynamics* that is affected [21]. Moreover, we propose that in depression the intersubjective dimension of temporality, both at the basic level of the minimal self and at the level of the narrative self, undergoes a *desynchronisation*.

Tellenbach [14] characterised the *typus melancholicus* by the patient's inability to let go of the past (which also means an inability to grieve) and therefore a failure to live his own present. The hyper-conformism that characterises the melancholic personality type could also be interpreted as an attempt to avoid or nullify former ruptures or desynchronisations early in life that were experienced as extremely painful by the subject. It could then be hypothesised that in his early affective interactions with others, the patient had experienced some painful loss of resonance. This would imply that later events of desynchronisation (i.e. falling short of goals, painful losses, or separations) strongly resonate with that experience, appearing as highly distressing breaks in the continuity of time. In such a situation, the *narrative self* will fail to perform the active synthesis of biographical time and to continuously integrate one's past with one's future. This process of active synthesis indeed includes the capacity for closure of the past, as a prerequisite for not falling victim to time and becoming dominated by it.

At this point, a depressive illness may occur, corresponding to a switch from an intersubjective or existential desynchronisation into a biological one (overall organismic stress reaction connected with various disturbances of biorhythms, sleep, appetite, etc.). With the resulting loss of drive and *conation,* the depressive psychopathology further increases the social desynchronisation. Temporality is affected at the level of the conative momentum, but the constitutive (protentional, presentational, retentional) synthesis of inner time consciousness remains intact. What is lacking is the affective tension and energy that carries the intentional arc forward. Since conation implies the affective interaction with others, this dimension is of an intrinsically intersubjective nature as well.

In terms of *explicit time*, in depression the past remains always present as a constant accusation. Future is experienced as a process leading to an irreversible and fatal end that is already known from the past. Complete desynchronisation from intersubjective time is marked by the transition to melancholic delusion. Its climax in nihilistic delusion—the idea that one has already died or the world does not exist any longer—comes close to the schizophrenic's depersonalisation; however, it is ultimately based on the loss of conative-affective dynamics instead of a breakdown of the transcendental synthesis of temporality.

10.3.3 Temporality in Borderline Personality Disorder

In reactive, neurotic, or personality disorders, it is only the biographical level of temporalisation that is affected, whereas the fundamental dimension of implicit time is maintained. However, in severe cases this may well lead to a fragmentation of narrative identity [31].

As stated above, the self is a multifaceted and multilayered concept, and some of its most important features belong to the narrative domain, which is based on a temporally enduring *self-identity*. Such a self-identity "relies upon the ability to maintain memories, personality traits, goals, and values within a coherent narrative structure" [8]. In other words, it entails the capacity to articulate one's own historical continuity, thus unfolding the sense of a growing but stable identity. The French philosopher Paul Ricoeur places the very essence of the human being in the temporal relationship that we have towards ourselves through a narrative identity that implies a process of integration, or at least a quest for coherence of the personal past, present, and future [32]. This process is basically social; it starts from the first relationships in our childhood and continues for the rest of our lives. Furthermore, personal identity is rooted in a complex interaction with others who are not only the implicit auditors and witnesses but in a sense also "co-authors" of our life stories.

In borderline personality disorder (BPD), we find marked disturbances of identity; indeed, a specific form of self-fragmentation is exhibited. Patients with BPD lack the strength to establish and maintain a coherent self-concept. They tend to switch from one present to the next, being totally identified with their momentary state of affect. This leads to a temporal splitting of the self, with a tendency to neglect or exclude past and future dimensions of relationships such as constancy, commitment, responsibility, and identity.

Affect dysregulation and impulsivity—highly represented clinical features in this disorder—express the patient's inability to contain and regulate overwhelming moods and affects. The patients undergo intense and abrupt mood changes, including anxiety, dysphoria, anger, shame, depression, or short-lived enthusiasm and euphoria. They are unable to draw on the experiences of the past in order to determine their own future through reflective decisions.

Thus, BPD individuals show a characteristic structure of temporality: they identify themselves with a short-lived, rather flat and empty present. They often describe lasting feelings of emptiness and boredom since their transitory present has no depth. It lacks the fulfilment that grows from the integration of past experience and anticipated future into the present. Bin Kimura speaks of a kind of absolute "now", which he calls the *intra festum* type of temporality [16]. In fact, for borderline individuals, the present moment loses its relation to the past or the future, lacking the continuity of coherent narratives and instead acquiring the features of inflated spontaneity, ecstasy, and oblivion. Others have defined borderline temporality as a cyclical movement without any historical progression [33].

This typical structure of temporality is deeply intertwined with the incoherence of autobiographical experiences and the fragmentation of identity seen in BPD patients. This fragmentation is increased by their tendency to dissociate as a result of traumatic experiences and adverse early environments. Whilst working as a defence mechanism against trauma-related distressing emotions, in the long term, dissociations can undermine the coherence of the life narrative. At the same time, regarding intersubjective relations, the patients do not succeed in integrating series of interactions to form a coherent concept of the other. BPD patients lack *object constancy* in the sense of being able to retain a positive image of important others in spite of temporary separation or rejection. Again, the result is a fragmentation of the *narrative self*: a shifting view of oneself and others, with sharp discontinuities, rapidly changing roles and relationships, and an underlying feeling of inner emptiness.

Since the patients' lack of a stable sense of self may be derived from deficits in early attunement and resulting attachment disorders, BPD may also be regarded as a disorder of intersubjective temporality. If experiences of stable, trusting relationships are missing, the child will not be able to establish the inner schemes of being with others that are necessary to form coherent narratives of oneself.

Both paradigmatic illnesses studied here—**schizophrenia** and **melancholic depression—**primarily **affect the basal level of lived time**. In schizophrenia, there is a weakening and fragmentation of temporal self-coherence rooted in *ipseity*, whereas we find a phasic inhibition of *conation* and *affectivity* in melancholic depression. Thus, in both of these disorders, the explicit dimension of time experience is not sufficient to capture the crucial temporal disturbance.

On the other hand, in **borderline personality disorder** and other neurotic spectrum disorders, it is mainly the level of **explicit temporality** and **narrative identity** that is affected, implying disturbances of the intersubjective dimension of time as well.

Conclusions

The dimensional approach to patient evaluation, as discussed throughout the book, provides information about the most prominent target symptoms and the personal symptom profile of each patient. This approach can play a substantial role in targeting the specific needs of the patient, especially in terms of a psychopharmacologically tailored treatment. Indeed, a dimensional approach can be more sensitive to the different profiles of symptoms, allowing more individualised treatment, as well as subtler diagnostic differentiation among the overly broad categories of the DSM.

In addition, dimensionality, in terms of description of the psychopathologic experience, seems to support a less "labelling attitude" on the part of the clinician, reducing the risk of the "narrative traps" represented by categorical diagnosis. In fact, these can trap the clinician's mind in preformed narrative contents, which can discourage the effort to really get to know the patient as a unique individual.

In Chap. 9, the dimensional model has been related to a very different epistemological perspective, i.e. a phenomenological one. Exploring in particular the intersubjective experience during the clinical encounter, the authors have proposed an integration of different epistemologies and pointed out how this may enable a more comprehensive understanding of mental illness. This may also help to better address the patient's needs and further a stronger therapeutic alliance. As we have stated in our introduction, the phenomenological method is different from the objectifying or third-person approach to psychic suffering that is characteristic of both the categorical and dimensional perspective. In contrast, it is aimed at a holistic reconstruction of the experience of the mentally ill person in terms of his or her fundamental characteristics of *being-in-the-world*.

Temporal experience, which has been the object of our present discussion, may represent a similar "cornerstone" in a multifaceted overview of the psychiatric assessment. Similar to alterations of intersubjective experience, to which temporality is thoroughly related, the disruptions of lived time are not simply psychopathological symptoms. Rather, they represent a fundamental break in one of the most basic human functions, since temporal experience also structures the experience of self. Hence, we can assume that distortions of lived time are always involved in the core of psychic disturbances and give rise to different symptomatic manifestations. From this perspective, it follows that each psychopathological symptom or dysfunction acquires its meaning in relation to the broader picture of an altered existence, and temporality is one of the main dimensions giving sense to that picture.

Further research on the relationships between the "subject-object" epistemology, particularly the dimensional one, and the phenomenological "subject-subject" epistemology seems very promising. Since every person affected by mental illness presents many needs, an intervention model that allows the clinician to design a systematically tailored treatment should be sought, without losing the sense of the patient's existential wholeness. Recently, a number of researchers have attempted to transfer some phenomenological concepts to standardised and operational instruments [34–36], including in one case a specific section dedicated to temporal experience [36]. This may be the first step towards an integrated psychopathological and psychotherapeutic approach.

Moreover, an empirical study of the distortions of lived time may improve the analytical understanding of such experiences and of how they evolve over the course of a pathological condition and its treatment. As we have seen, from a phenomenological point of view, significant desynchronisations and failures of attunement at different levels of temporality might produce a rupture in the process of temporal synthesis, both at the pre-reflective and reflective level.

We thus propose an integration of different approaches to diagnosis and treatment, as a necessary step towards more effective pharmacological and relational interventions. Ultimately, they should both help to re-establish interpersonal relatedness for and with the patient. If temporality represents a central dimension of subjectivity, its distortions are the implicit targets of interventions aimed to restore at least moments of synchronisation and mutual understanding in the therapeutic relationship.

Despite our research efforts, time could well represent a dimension of reality that we cannot change in terms of a *restitutio ad integrum*, since we are time-producing beings and our life story is stored irreversibly in our living body. Still, as clinicians, we should try to alleviate the suffering of patients by restoring, as much as possible, their ability to unfold their own time, both in terms of subjectivity and intersubjectivity. In this sense, it would be beneficial to investigate the therapeutic effects of establishing rhythmic patterns of interaction, not only in psychotherapeutic settings but in every clinical encounter, even during the initial diagnostic process [37, 38]. In fact, every clinical encounter entails this opportunity, and any kind of intervention, be it psychotherapeutic or psychopharmacological, should promote the lived relation with another human being, primarily in the form of a sharing of resonance and rhythmic time. This implies a recommendation of analysing the patient's mode of temporalisation and, last but not least, the consideration of the enormous value, in terms of therapeutic potential, of our personal time shared with them.

References

1. Heidegger M. Sein und Zeit. Halle: Niemeyer Verlag; 1927.
2. Binswanger L. Zur phänomenologischen anthropologie. Bern: Francke; 1947.
3. Fuchs T. Subjectivity and intersubjectivity in psychiatric diagnosis. Psychopathology. 2010;43:268–74.

4. Stanghellini G. The grammar of the psychiatric interview. A plea for the second-person mode of understanding. Psychopathology. 2007;40(2):69–74.
5. Nordgaard J, Sass LA, Parnas J. The psychiatric interview: validity, structure, and subjectivity. Eur Arch Psychiatry Clin Neurosci. 2013;263(4):353–64.
6. Fuchs T. Phenomenology and psychopathology. In: Schmicking D, Gallagher S, editors. Handbook of phenomenology and cognitive science. Dordrecht: Springer; 2010. p. 546–73.
7. Parnas J, Sass LA, Zahavi D. Rediscovering psychopathology: the epistemology and phenomenology of the psychiatric object. Schizophr Bull. 2013;39(2):270–7.
8. Zahavi D. Subjectivity and selfhood: investigating the first-person perspective. Cambridge, MA: MIT Press; 2008.
9. Gallagher S, Zahavi D. The phenomenological mind: an introduction to philosophy of mind and cognitive science. New York: Routledge; 2008.
10. Straus E. Das Zeiterleben in der depression und in der psychopathischen Verstimmung. Mschr Psychiat Neurol. 1928;68:640–56.
11. Minkowski E. Le temps vécu: études phénoménologiques et psychopathologiques. Paris: D'Artrey; 1933.
12. von Gebsattel E. Zeitbezogenes Zwangsdenken in der Melancholie. In: von Gebsattel E, editor. Prolegomena einer medizinischen antropologie. Berlin: Springer; 1954. p. 1–18.
13. Binswanger L. Melancholie und Manie. Pfullingen: Neske; 1960.
14. Tellenbach H. Melancholie. Zur Problemgeschichte, Typologie, Pathogenese und Klinik. Berlin-Göttingen-Heidelberg: Springer; 1961.
15. Blankenburg W. Der Verlust der natürlichen Selbstverständlichkeit. Berlin: Springer; 1971.
16. Kimura B. Ecrits de psychopathologie phénoménologique. Paris: Presses Universitaires de France; 1992.
17. Jaspers K. Allgemeine Psychopathologie. Berlin: Springer; 1913.
18. Tatossian A. La Phénoménologie des psychoses. Paris: Masson; 1979.
19. Zahavi D. Self and other. Exploring subjectivity, empathy and shame. Oxford: Oxford University Press; 2014.
20. Husserl E. Vorlefungen zur Phänomenologie des inneren Zeitbewußtfeins. Niemeyer Verlag: Halle; 1928.
21. Fuchs T. Temporality and psychopathology. Phenomenol Cogn Sci. 2013;12(1):75–104.
22. Merleau-Ponty M. Phénoménologie de la perception. Paris: Gallimard; 1945. English edition: Merleau-Ponty M. Phenomenology of perception (trans: Smith C). London: Routledge & Kegan Paul; 1962
23. Trevarthen C. Communication and cooperation in early infancy. A description of primary intersubjectivity. In: Bullowa M, editor. Before speech: the beginning of human communication. London: Cambridge University Press; 1979. p. 321–47.
24. Trevarthen C, Hubley P. Secondary intersubjectivity: confidence, confiding and acts of meaning in the first year. In: Lock A, editor. Action, gesture and symbol: the emergence of language. London: Academic Press; 1978. p. 183–229.
25. Fuchs T. The temporal structure of intentionality and its disturbance in schizophrenia. Psychopathology. 2007;40:229–35.
26. Kapur S. Psychosis as a state of aberrant salience: a framework linking biology, phenomenology, and pharmacology in schizophrenia. Am J Psychiatry. 2003;160:13–23.
27. Kaiser S, Weisbrod M. Intentionality as a link between the neuropsychology and the symptoms of schizophrenia. Psychopathology. 2007;40:221–8.
28. Vogeley K, Kupke C. Disturbances of time consciousness from a phenomenological and a neuroscientific perspective. Schizophr Bull. 2007;33:157–65.
29. Fuster JM. Cortex and mind—unifying cognition. Oxford: Oxford University Press; 2003.
30. Sass LA, Parnas J. Schizophrenia, consciousness and the self. Schizophr Bull. 2003;29(3):427–44.
31. Fuchs T. Fragmented selves. Temporality and identity in Borderline personality disorder. Psychopathology. 2007;40:379–87.
32. Ricoeur P. Soi-même comme un autre. Paris: Seuil; 1990.

33. Muscatello CF, Scudellari P. Anger and narcissism: between the void of being and the hunger for having. Psychopathology. 2000;33:227–32.
34. Parnas J, Moller P, Kircher T, Thalbitzer J, Jansson L, Handest P, Zahavi D. EASE: examination of anomalous self-experience. Psychopathology. 2005;38(5):236–58.
35. Pallagrosi M, Fonzi L, Picardi A, Biondi M. Assessing clinician's subjective experience during interaction with patients. Psychopathology. 2014;47:111–8.
36. Sass LA, Pienkos E, Skodlar B, Stanghellini G, Fuchs T, Parnas J, Jones N. EAWE: examination of anomalous world experience. Psychopathology. 2017;50(1):10–54.
37. Ramseyer F, Tschacher W. Synchrony in dyadic psychotherapy sessions. In: Marks-Tarlow T, Vrobel S, Rössler OE, editors. Simultaneity: temporal structures and observer perspectives. Singapore: World Scientific; 2008. p. 329–47.
38. Ramseyer F, Tschacher W. Nonverbal synchrony of head-and body-movement in psychotherapy: different signals have different associations with outcome. Front Psychol. 2014;5:979.

MIX
Papier aus verantwortungsvollen Quellen
Paper from responsible sources
FSC® C105338

If you have any concerns about our products,
you can contact us on
ProductSafety@springernature.com

In case Publisher is established outside the EU,
the EU authorized representative is:
Springer Nature Customer Service Center GmbH
Europaplatz 3, 69115 Heidelberg, Germany

Printed by Libri Plureos GmbH
in Hamburg, Germany